Nature, Nurture, and the
Transition to Early Adolescence

NATURE, NURTURE, AND THE TRANSITION TO EARLY ADOLESCENCE

Edited by
Stephen A. Petrill
Robert Plomin
John C. DeFries
John K. Hewitt

OXFORD
UNIVERSITY PRESS
2003

OXFORD
UNIVERSITY PRESS

Oxford New York
Auckland Bangkok Buenos Aires Cape Town Chennai
Dar es Salaam Delhi Hong Kong Istanbul Karachi Kolkata
Kuala Lumpur Madrid Melbourne Mexico City Mumbai Nairobi
São Paulo Shanghai Taipei Tokyo Toronto

Copyright © 2003 by Oxford University Press, Inc.

Published by Oxford University Press, Inc.
198 Madison Avenue, New York, New York 10016

www.oup.com

Oxford is a registered trademark of Oxford University Press

Library of Congress Cataloging-in-Publication Data
Nature, nurture, and the transition to early adolescence / edited by
Stephen A. Petrill . . . [et al.].
    p. cm.
Includes bibliographical references and index.
  ISBN 0-19-515747-8
    1. Nature and nurture—Longitudinal studies.   2. Individual differences in adolescence—
Longitudinal studies.   3. Adopted children—Colorado—Psychology—Longitudinal studies.
4. Colorado Adoption Project.   I. Petrill, Stephen A., 1968–
  BF341 .N387   2003
  155.4′5—dc21   2002003903

9 8 7 6 5 4 3 2 1

Printed in the United States of America
on acid-free paper

Contents

Contributors

Maricela Alarcón, Ph.D.
Department of Neurology, University of California at Los Angeles

Laura A. Baker, Ph.D.
Department of Psychology, SGM 501, University of Southern California

Madeline Becker
Department of Psychology, Boston University

E. G. Bishop, Ph.D.
Department of Psychology, University of Houston

Coleen Carlson, Ph.D.
Department of Psychology, University of Houston

Stacey S. Cherny, Ph.D.
Wellcome Trust Centre for Human Genetics, University of Oxford

Jeanie Clifford
Department of Psychology, University of California, San Diego

Robin P. Corley, Ph.D.
Institute for Behavioral Genetics, University of Colorado

Kirby Deater-Deckard, Ph.D.
Department of Psychology, University of Oregon

John C. DeFries, Ph.D.
Institute for Behavioral Genetics, University of Colorado

Jeffrey R. Gagne
Department of Psychology, Boston University

Nicholas Giardino, Ph.D.
Department of Rehabilitation Medicine, University of Washington

John K. Hewitt, Ph.D.
Institute for Behavioral Genetics, University of Colorado

Rebecca Hobson
Department of Psychology, University of Houston

Alessandra C. Iervolino
Social, Genetic, and Developmental Psychiatry Research Centre, London, U.K.

Beth Manke, Ph.D.
Department of Human Development, California State University at Long Beach

Shirley McGuire, Ph.D.
Department of Psychology, University of San Francisco

Jenae M. Neiderhiser, Ph.D.
Center for Family Research, Department of Psychiatry and Behavioral Science, George Washington University

Stephen A. Petrill, Ph.D.
Department of Biobehavioral Health, Center for Developmental and Health Genetics, Pennsylvania State University

Robert Plomin, Ph.D.
Social, Genetic, and Developmental Psychiatry Research Centre, London, U.K.

Richard Rende, Ph.D.
Department of Psychiatry and Human Behavior, Centers for Behavioral and Preventive Medicine, Brown University School of Medicine

Sally-Ann Rhea
Institute for Behavioral Genetics, University of Colorado

Kimberly J. Saudino, Ph.D.
Department of Psychology, Boston University

Stephanie Schmitz, Ph.D.
Institute for Behavioral Genetics, University of Colorado

Erica L. Spotts, Ph.D.
Department of Medical Epidemiology, Karolinska Institutet, Stockholm, Sweden

Sally J. Wadsworth, Ph.D.
Institute for Behavioral Genetics, University of Colorado

Michelle Ward, M.A.
Department of Psychology, SGM 501, University of Southern California

Nature, Nurture, and the
Transition to Early Adolescence

STEPHEN A. PETRILL
ROBERT PLOMIN
JOHN C. DEFRIES
JOHN K. HEWITT

I

Nature, Nurture, and Adolescent Development

An Introduction

Man is not meant to remain a child. He leaves childhood behind him at the time appointed by nature; and this critical moment, short enough in itself, has far reaching consequences.
—Rousseau, 1957, p. 172

Adolescence as a universal phenomenon among the young of all social classes is a product of modern civilization.
—Cole, 1936, p. 3

After a long period of relative inactivity, research in adolescent development has begun to flourish (Hoffman, 1996). In 1996, an influential special issue of *Developmental Psychology* documented not only the activity around the study of adolescent development but also the gaps in research and methodology, and the need for further research during this period of development, particularly with respect to biopsychosocial interactional processes—the nexus of biology, psychological development, and social interactions (e.g., Hoffman, 1996; Lerner, 1996; Zahn-Waxler, 1996). Adolescence is a point in human development where these biopsychosocial interactional processes are likely to be particularly intense. Adolescence is triggered by puberty, a genetically mediated event, but concluded when the individual crosses a culturally determined threshold of social, emotional, intellectual, educational, and economic competence. Some of the major developmental challenges of adolescents are developing healthy peer, parental, and intimate relationships in the context of emergent physical, cognitive, and social-emotional competencies.

Most investigations of adolescent behavioral development have been normative, focusing on average effects of adolescence in general and puberty in particular (see Brooks-Gunn, Lerner, & Petersen, 1991).

3

Despite the fact that many of the most important developmental and social issues surrounding adolescence involve individual differences, much less attention has been given to the differences found within adolescents. Why do some adolescents develop behavioral problems while others do not? Why do some adolescents show cognitive and academic delays? Why are some families with adolescents rife with discord while other families are more harmonious?

A Focus on Genetically Informative Samples

At the heart of many of these issues is the extent to which adolescent cognitive, social, and behavioral development is the product of genetic and/or environmental influences. Although the relative balance of nature vs. nurture in adolescence has historically been the subject of intense debate (e.g., Bandura & Walters, 1959; Gesell & Ilg, 1943; Mead, 1928–1973), it is now largely accepted that both genes and the environment are important to adolescent development.

Behavioral genetic theory focuses on the causes of individual differences in development rather than normative processes. It partitions causes of variation into distinct categories (genetic influences, environmental influences shared by family members, and experiences unique to the individual), while recognizing that these factors interact and correlate (Plomin, DeFries, McClearn, & McGuffin, 2000). For each of these components we can ask important, but straightforward, developmental questions that go beyond merely estimating its magnitude. For example, are genetic and environmental influences that are important at one age still important at another, or do entirely new influences arise? How do these influences combine and interact to determine outcomes? How do the relationships between variables change over time, and what biological or environmental influences drive these changes? Although there is increasing recognition of the importance of genetic influences on cognitive and behavioral development, studies analyzing genetic and environmental sources of developmental individuality using quantitative genetic designs are still relatively rare, especially longitudinal studies that span infancy through adolescence, which are critical for the study of age-to-age change and continuity (e.g., Hoffman, 1996; Plomin, 1986; Zahn-Waxler, 1996).

Although family and twin designs yield important information, it is generally recognized that the adoption design provides the most convincing evidence upon which to base estimates of genetic and environmental influences. Even the most vociferous critics of the field recognize the power of the adoption design. For example, Leon Kamin states: "There is no way, in ordinary families, of separating the effects of genes from those of environment. The great virtue of studies of adoptive families is that, in theory at least, they allow us to separate genetic from environmental transmission. The adoptive parent provides his or her child with environment but not with genes. Thus the IQ correlation between adopted child and adoptive parent is of considerable theoretical interest—particularly when it is compared to other relevant correlations" (Kamin, 1981, p. 114).

In nonadoptive families, observed relationships between parents and offspring (or between environmental measures and offspring development) can be mediated both genetically and environmentally. The adoption design cleaves these two major classes of developmental influence by studying the resemblance between parents and children who share only family environment (adoptive parents and adoptees, or two adoptive siblings), as well as the resemblance between parents and children who share only heredity (biological parents and their adopted-away children, or two biologically related siblings). In addition to estimating genetic and environmental components of variance, the adoption design facilitates identification of specific environmental influences unconfounded by heredity, analyses of the role of heredity in ostensibly environmental relationships, and assessment of genotype-environment interactions and correlations. These analyses are made even more valuable by the addition of a longitudinal dimension of repeated assessments of the children's development and their home environments to permit analyses of the etiology of change and continuity in development (Plomin, 1986).

A Focus on Individual Differences in Normal Development

The focus of much previous research on adolescence has been on problem behaviors, such as violence, substance abuse, eating disorders, or depression (Hoffman, 1996). Further, the few available genetic studies have focused on psychopathology (Eaves et al., 1997; Hewitt et al., 1997). In contrast, our focus is on genetic and environmental influences on individual differences across normal dimensional variation. Although variability across the normal range may not be as dramatic as a focus on psychopathology, there are several reasons why the normal range of variability has profound implications for understanding child development and mental health.

First, as has been noted by Offord (1995) and many others, it remains an open question whether a boundary between normal and abnormal behavior in childhood can be determined in a way that meaningfully distinguishes between phenomenologically or etiologically distinct categories. The arbitrary quality of the thresholds implied by diagnostic definitions is underscored by periodic changes in the numbers of symptoms, or in the severity, duration, impairing consequences, and pervasiveness of symptoms required for psychiatric diagnoses (Hewitt et al., 1997). Second, "genetic studies have suggested that phenotypically distinct disorders, such as generalized anxiety disorder and major depressive disorder, may be highly genetically correlated or represent the manifestation of a single pathway of genetic vulnerability (Kendler, Neale, Kessler, Heath, & Eaves, 1992). Moreover, genetic vulnerability might be better thought of as a continuously distributed dimension, perhaps polygenic [or oligogenic], rather than in terms of distinct classes. In the case of anxiety and depression, [for example,] an index of the genetic liability might be provided by a dimensional assessment of the personality trait of neuroticism, a trait that can best be studied developmentally in community samples (Jardine, Martin, & Henderson, 1984; Eaves, Eysenck, & Martin, 1989)" (Hewitt et al., 1997, p. 945). Third, we would like to study development in individuals who do not yet (and may never) present with a clinical disorder, and to understand the developmental mediation of risks (both genetic and environmental). Fourth, because most environmental and genetic influences operate within the normal range, it is this range of influence that comprises the raw material with which educational or child-rearing interventions must work. Fifth, the extremes of the dimensions of normal variability can best be understood in the context of normal behavioral variability and its associated environmental and genetic influences. This critical issue of the etiological relationship between the normal and the abnormal has been the focus of numerous methodological advances and empirical studies (DeFries & Fulker, 1985, 1988; Deater-Deckard, Reiss, Hetherington, & Plomin, 1997; Fulker & Cherny, 1995; Plomin, 1991; Petrill et al., 1997, 1998).

Our approach is to study genetically informative families (e.g., adoptive and biological siblings) using dimensional phenotypic assessments including cognitive abilities, educational achievement, personality, interests, attitudes, adjustment, and problem behaviors. A dimensional approach to behavioral variation during the transition to adolescence, studied in genetically informative community samples, allows us to assess the impact of genetic and environmental factors on cognitive and socioemotional functioning, independent of specific clinical outcomes.

A Focus on Development

Although several behavioral genetic studies have used adolescent subjects, there have been relatively few behavioral genetic studies of adolescent *development*—that is, studies that consider phenomena most relevant to adolescence and the transition from childhood through adolescence. In an investigation of genetic and environmental influences on general cognitive ability, the classic longitudinal adoption study of Skodak and Skeels (1949) suggested some increase in genetic influence on IQ from middle childhood to adolescence as seen in increasing correlations between biological mothers and their adopted-away children. Similarly, one of the most important contemporary longitudinal twin studies in behavioral genetics, the Louisville Twin Study (Wilson, 1983), also shows an increase in genetic influence on IQ from middle childhood to 15 years of age. However, these studies assessed children only once during adolescence and their focus was general intelligence. Although personality and problem behaviors in adolescence have received some attention in behavioral genetic studies (e.g., Billing, Hershberger, Iacono, & McGue, 1996; Eaves et al., 1997; Ge et al., 1996; Hewitt et al., 1997; Macaskill, Hopper, White, & Hill, 1994; McGue, Sharma, & Benson, 1996; Pike, McGuire, Hetherington, Reiss, & Plomin, 1996; Rende, Plomin, Reiss, & Hetherington, 1993), few were longitudinal studies of the developmental process. Further, relatively little is known about such important traits as specific cognitive abilities, academic performance, and interests and attitudes as they interact with the social and biological milestones of adolescence.

The Colorado Adoption Project

Beginning in 1975, the Colorado Adoption Project (CAP) was initiated with the goal of examining a sample of adopted and nonadopted children and their families from 1 to 16 years. To date, the CAP has examined the children, adopted parents, biological parents, and home environments of 245 adoptive families and 245 matched nonadoptive families since the children were infants. Home visits were conducted when the children (and their younger siblings) were 1, 2, 3, and 4 years of age. Phone interviews were conducted with parents when the children were ages 5, 6, and 8. The children were studied initially in the laboratory following first grade, then again following sixth grade and at age 16. Telephone interviews were conducted with parents and children in the summers following third through fifth grades, as well as after seventh through ninth grades; telephone cognitive assessments were conducted as well during the third-, fourth-, and eighth-grade sessions. Parent questionnaires were administered at all of these ages and teacher questionnaires were included during the school years.

Nature, Nurture, and Adolescent Development

This is the fourth book in a series describing the CAP results. The first three books presented results from infancy (Plomin & DeFries, 1985), early childhood (Plomin, DeFries, & Fulker, 1988), and middle child-hood (DeFries, Plomin, & Fulker, 1994). In this book we examine the CAP results from ages 9, 10, 11, and 12 years.

This volume presents new findings on several key aspects of adolescent development. First, we describe research that examines cognitive development in early adolescence on the following topics: developmental analyses of general cognitive ability (chapter 2); parent-offspring analyses of specific cognitive abilities (chapter 3); stability and change in reading performance (chapter 4); and memory ability in middle childhood and early adolescence (chapter 5).

Next we examine adjustment and behavioral problems in adolescence in the following topics: Somatic complaints from early childhood to early adolescence (chapter 6), attention problems (chapter 7), and adolescent adjustment (chapter 8).

Third, we examine the development of mood and temperament in adolescence. Several chapters examine: change in depressive symptoms (chapter 9); loneliness (chapter 10); temperament (chapter 11); the relationship between temperament and behavioral problems (chapter 12); and humor use, temperament, and well-being (chapter 13).

Fourth, we present several studies that examine the relationship between measured environments and adolescent outcomes. These chapters examine the relationship between prenatal smoking and behavioral problems in adolescence (chapter 14), the relationship between negative family environments and externalizing behavioral problems (chapter 15), maternal differential treatment (chapter 16), life events (chapter 17), and the developmental trajectory of gene-environment processes (chapter 18).

Finally, we present a summary chapter integrating these findings and outlining future research directions (chapter 19).

As was true in the third CAP book, we are pleased that it was written in collaboration with so many colleagues, staff, former students, and current students who have been involved with the CAP during the past quarter-century. This strategy has not only led to a broader perspective but also reflects the strength of a research team where each of its members has contributed to the substantive, methodological, and conceptual advances that have kept this project so vibrant for so many years. This strategy has greatly broadened the perspective of this book and added to the expertise brought to bear on adolescence. However, this is not just another edited volume: we are all part of a team committed to the CAP. None of these analyses could be undertaken were it not for the industrious and conscientious efforts of the many testers who have worked on the CAP since its inception. We thank the many research assistants who worked on data collection at younger ages along with those who collected information from the young adolescents. Of note are those who conducted 100 or more early adolescent tests: Bo Bishop,

Kim Corley, Lara Cunning, Betty-Jane Luzietti, Sara McIntosh, Stephanie Runck, Stephanie Bogott, and John Steinbach. Annie Johnson's efforts were particularly valuable, as she was the primary tester for the pilot work and the CAP testing, as well as the trainer for other testers and the data manager for the project. We also wish to thank Elana Pyle, who was instrumental in moving this book to publication. Finally, we are especially grateful to the staff members of the Lutheran Family Services and the Denver Catholic Community Services, whose cooperation made the CAP possible, as well as to the families who have continued to provide their time and goodwill to this project.

We are grateful to the following agencies that have supported CAP over the past quarter-century. CAP was launched in 1976 though funds from the University of Colorado's Biomedical Research Support Grant and a small grant from the National Institute of Mental Health (MH-28076). The William T. Grant Foundation supported the project from 1976 to 1979, as well as the testing of CAP children at age 7. The Spencer Foundation provided support from 1982 to 1984 for the purpose of testing younger adopted and nonadopted siblings of the original CAP children and again from 1985 to 1989 for the extension of CAP testing into early adolescence. The National Institute of Child Health and Human Development has continuously supported CAP since 1977 (HD-10333 and HD-18426). Since 1988, the National Institute of Mental Health (MH-43899) has supported CAP research in early adolescence. Since 1978, the National Science Foundation has enabled us to examine mother-child and sibling interactions (BNS-7826204, BNS-8200310, BNS-8505692, BNS-8643938, BNS-8806589, and BNS-9180744). Finally, we profited greatly from our participation in the Early Childhood Transitions Research Network of the John D. and Catherine T. MacArthur Foundation.

We dedicate this book to the memory of David W. Fulker. Modern quantitative genetic analysis began with his paper with John Jinks in 1970 in *Psychological Bulletin*. They showed how behavioral genetic data can be tested for its fit to a model of genetic and environmental transmission that makes assumptions explicit. This so-called model-fitting approach swept through the field and is now the only way data are analyzed and also played a major role in the acceptance of behavioral genetic research in the behavioral sciences. David continued to be at the forefront of development of these quantitative genetic techniques. He brought model-fitting to the complex CAP design in the early 1980s and, when he came to the Institute for Behavioral Genetics in 1983, he assumed responsibility for the adolescence phase of the project. David was also at the forefront of the merger of quantitative genetics and molecular genetics in the development of techniques to identify quantitative trait loci (QTLs), realizing early on that quantitative genetic model-fitting techniques could incorporate specific DNA markers to find QTLs. One of David's most novel contributions here was in recognizing the power of selecting from the extremes of a distribution and, especially, studying the extremes in the context of normal variation. This symmetry in bringing the two worlds of genetics, quantitative genetics and molecular genetics, together was greatly pleasing to David. He was a brilliant scholar, wonderful colleague, caring mentor, and much-missed friend who contributed so much to the success of the field of behavioral genetics as well as to CAP.

REFERENCES

Bandura, A., & Walters, R. H. (1959). *Adolescent aggression*. New York: Ronald Press.

Billing, J. P., Hershberger, S. L., Iacono, W. G., & McGue, M. (1996). Life events and personality in late adolescence: Genetic and environmental relations. *Behavior Genetics, 26*, 543–554.

Brooks-Gunn, J., Lerner, R., & Petersen, A. (Eds.). (1991). *The encyclopedia of adolescence*. New York: Garland.

Cole, L. (1936). *Psychology of adolescence*. New York: Farrar & Rinehart.

Deater-Deckard, K., Reiss, D., Hetherington, E. M., & Plomin, R. (1997). Dimensions and disorders of adolescent adjustment: A quantitative genetic analysis of unselected samples and selected extremes. *Journal of Child Psychology and Psychiatry, 38*, 515–525.

DeFries, J. C., & Fulker, D. W. (1985). Multiple regression analysis of twin data. *Behavior Genetics, 15*, 467–473.

DeFries, J. C., & Fulker, D. W. (1988). Multiple regression analysis of twin data: Etiology of deviant scores versus individual differences. *Acta Geneticae Medicae et Gemellologiae: Twin Research, 37*, 205–216.

DeFries, J. C., Plomin, R., & Fulker, D. W. (1994). *Nature and nurture during middle childhood*. Oxford: Blackwell.

Eaves, L. J., Eysenck, H. J., Martin, N. G. (1989). *Genes, culture and personality: An empirical approach*. San Diego: Academic Press (Harcourt Brace Jovanovich).

Eaves, L. J., Silberg, J. L., Meyer, J. M., Maes, H. H., Simonoff, E., Pickles, A. Rutter, E., Neale, M. C., Reynolds, C., Erickson, M., Heath, A. C., Loeber, R., Truett, K., & Hewitt, J. K. (1997). Genetics and developmental psychopathology: 2. The main effects of genes and environment on behavioral problems in the Virginia Twin Study of Adolescent Behavioral Development. *Journal of Child Psychology and Psychiatry, 38*, 965–980.

Fulker, D. W., & Cherny, S. S. (1995). Genetic and environmental influences on cognition during childhood. *Population Research and Policy Review, 14*, 283–300.

Ge, X., Conger, R. D., Cadoret, R. J., Neiderhiser, J. M., Yates, W., Troughton, E., & Stewart, M. A. (1996). The developmental interface between nature and nurture: A mutual influence model of child antisocial behavior and parent behaviors. *Developmental Psychology, 32*, 574–589.

Gesell A., & Ilg, F. L. (1943). *Infant and child in the culture of today.* New York: Harper and Row.

Hewitt, J. K., Eaves, L. J., Silberg, J. L., Rutter, M., Simonoff, E., Meyer, J. M., Maes, H. H., Pickles, A., Neale, M. C., Loeber, R., Erickson, M., Kendler, K. S., Heath, A. C., Truett, K., & Reynolds, C. (1997). Genetics and developmental psychopathology: 1. Phenotypic assessment in the Virginia Twin Study of Adolescent Behavioral Development. *Journal of Child Psychology and Psychiatry, 38,* 943–963.

Hoffman, L. W. (1996). Progress and problems in the study of adolescence. *Developmental Psychology, 32,* 777–780.

Jardine, R., Martin N. G., & Henderson, A. S. (1984). Genetic variation between neuroticism and the symptoms of anxiety and depression. *Genetic Epidemiology, 1,* 89–107.

Kamin, L. (1981). Studies of adopted children. In H. J. Eysenck & L. Kamin (Eds.), *The intelligence controversy.* New York: John Wiley.

Kendler, K. S., Neale, M. C., Kessler, R. C., Heath, A. C., & Eaves, L. J. (1992). Major depression and generalized anxiety disorder: Same genes (partly) different environments? *Archives of General Psychiatry, 49,* 716–722.

Lerner, R. M. (1996). Relative plasticity, integration, temporality, and diversity in human development: A developmental contextual perspective about theory, process, and method. *Developmental Psychology, 32,* 781–786.

Mead, M. (1928/1973). *Coming of age in Samoa: A psychological study of primitive youth.* New York: American Museum of Natural History.

Macaskill, G. T., Hopper, J. L., White, V., & Hill, D. (1994). Genetic and environmental variation in Eysenck Personality Questionnaire scales measured on Australian adolescent twins. *Behavior Genetics, 24,* 481–491.

McGue, M., Sharma, A., & Benson, P. (1996). The effect of common rearing on adolescent adjustment: Evidence from a U. S. adoption cohort. *Developmental Psychology, 32,* 604–613.

Offord, D. R. (1995). Child psychiatric epidemiology: Current status and future prospects. *Canadian Journal of Psychiatry, 40,* 284–288.

Petrill, S. A., Saudino, K. J., Cherny, S. S., Emde, R. N., Hewitt, J. K., Fulker, D. W., & Plomin, R. (1997). Exploring the genetic etiology of low general cognitive ability from 14 to 36 months. *Developmental Psychology, 33,* 544–548.

Petrill, S. A., Saudino, K. J., Cherny, S. S., Emde, R. N., Hewitt, J. K., Fulker, D. W., & Plomin, R. (1998). Exploring the genetic etiology of high cognitive ability in 14 to 36 month-old twins. *Child Development, 69,* 68–74.

Pike, A., McGuire, S., Hetherington, E. M., Reiss, D., & Plomin, R. (1996). Family environment and adolescent depressive symptoms and antisocial behavior: A multivariate genetic analysis. *Developmental Psychology, 32,* 590–603.

Plomin, R. (1986). *Development, genetics, and psychology.* Hillsdale, NJ: Lawrence Erlbaum Associates.

Plomin, R. (1991). Genetic risk and psychosocial disorders: Links between the normal and abnormal. In M. Rutter & P. Casaer (Eds.), *Biological risk factors for psychosocial disorders* (pp. 101–138). Cambridge, UK: Cambridge University Press.

Plomin, R., & DeFries, J. C. (1985). *Origins of individual differences in infancy: The Colorado Adoption Project.* Orlando, FL: Academic Press.

Plomin, R., DeFries, J. C., & Fulker, D. W. (1988). *Nature and nurture in infancy and early childhood.* New York: Cambridge University Press.

Plomin, R., DeFries, J. C., McClearn, G. E., & McGuffin, P. (2000). *Behavioral genetics* (4th ed.). New York: Worth Publishers.

Rende, R. D., Plomin, R., Reiss, D., & Hetherington, E. M. (1993). Genetic and environmental influences on depressive symptomatology in adolescence: Individual differences and extreme scores. *Journal of Child Psychology & Psychiatry & Allied Disciplines, 34,* 1387–1398.

Rousseau, J. J. (1957). *Emile.* New York: E. P. Dutton.

Skodak, M., & Skeels, H. M. (1949). A final follow-up of one hundred adopted children. *Journal of Genetic Psychology, 75,* 85–125.

Wilson, R. S. (1983). The Louisville Twin Study: Developmental synchronies in behavior. *Child Development, 54,* 298–316.

Zahn-Waxler, C. (1996). Environment, biology, and culture: Implications for Adolescent Development. *Developmental Psychology, 32,* 571–573.

E. G. BISHOP
STACEY S. CHERNY
JOHN K. HEWITT

2

Developmental Analysis of IQ

Introduction

The psychometric approach to the measurement of cognitive development has occupied researchers for the entirety of the 20th century, beginning with the work of Binet and Simon and their "holistic" view of intelligence (1905), Spearman's (1927) common underlying general factor or "g," and on through to Cattell's (1971) crystallized and fluid intelligence model. General cognitive ability may be the area in which we know the most from a behavioral genetic perspective. Bouchard and McGue (1981) noted over 140 studies of this domain, which yielded the largely consistent result that genetic differences account for approximately 50% of the observed variability in general cognitive ability. This 50% estimate has also been reported more recently by Plomin, DeFries, McClearn, & Rutter (1997).

Longitudinal adoption study data can be used to address two aspects of the developmental process. The first aspect concerns the sources of observed variation in individual differences in general cognitive ability at each age of assessment. By comparison of correlations from adoptive and biological siblings, as well as identical or monozygotic (MZ) and fraternal or dizygotic (DZ) twins, behavioral genetic methodology allows us to partition the observed variation into variation due to differences between individuals in genetic makeup and environmental differences between individuals. These environmental differences can be further subdivided into those shared by members of the family and those nonshared environmental influences that are unique to individuals. We can then examine whether the relative magnitudes of heritable and environmental contributions change across the developmental period in question.

The second and more interesting aspect of the developmental process involves continuity and change. Researchers have hypothesized major changes in the structure of mental functioning across development (e.g., McCall, 1979). A pioneering study by Honzik, MacFarlane, and Allen (1948) found that the age of the child at the time of first testing had an impact on the prediction of IQ in later childhood. For example, the correlation between ages 2 and 5 was only .32, but this correlation increased to .70 between ages 5 and 8 and .85 between ages 9 and 12. Other early studies also found little or no correlation between IQ measured in infancy and early and middle childhood (e.g., Anderson, 1939; Bayley, 1943). McCall, Applebaum, and Hogarty (1973) found changes in IQ measured in the same child from 2.5 to 17 years of age ranging as much as 40 points. If such changes were occurring, we might expect some discontinuity in cognitive ability scores over this period, with genetic variation being one salient cause. The extent to which genetic and environmental sources of individual differences contribute to continuity and change in general cognitive ability can be estimated and explicitly tested.

The purpose of this chapter is to describe some aspects of the development of general cognitive ability from infancy through the transition to early adolescence. This will be done through analysis of data from the Colorado Adoption Project (CAP; DeFries, Plomin, & Fulker, 1994; Plomin & DeFries, 1985; Plomin, DeFries, & Fulker, 1988). In this chapter, we focus on the development of individual differences in general cognitive ability from ages 1 to 12. This updates previous reports of CAP general ability data (Cardon, Fulker, DeFries, & Plomin, 1992; Cherny & Cardon, 1994; Cherny, Fulker, & Hewitt, 1997; Fulker, Cherny, & Cardon, 1993). The extent that phenotypic differences are correlated over time implies continuity in development; the extent that they are not correlated implies change. The relationship between genetic and environmental influences across time indicates the degree to which these processes of continuity and change are driven by genetic factors and by the environment.

The structural equation modeling approach brings powerful methods to bear on the issue of behavioral development. Predictions for family resemblance and phenotypic variation that follow from the structural models can be fit, using maximum likelihood procedures, to the observations (in the form of either summary statistics or directly to the raw data). The procedure yields: (1) estimates of the magnitude of the contributions to the phenotypic variance of genetic and environmental influences; (2) a likelihood ratio chi-square statistical test of the goodness of fit of the model to the data; and, importantly, (3) tests of the statistical significance of individual parameters of the model (e.g., genetic influence on the phenotype) achieved by sequential fitting of hierarchically nested models and use of the difference chi-square test. The model-fitting and estimation procedures can be implemented using the program Mx (Neale, Boker, Xie, & Maes, 1999), a program specifically developed for analyzing genetically informative data.

This approach to analyzing behavior genetic data is quite flexible and able to provide insights into the causes of interrelationships between different aspects of behavior, as well as behavioral development. For example, researchers using these methods have found genetic variance for specific cognitive abilities independent of the genetic factors involved in general cognitive ability (Cardon, 1994; Cardon & Fulker, 1994; Fulker & Cardon, 1993). Similarly, the multivariate modeling approach has been used to explore the origins of covariation between cognitive ability and academic achievement. In early childhood, scholastic achievement was found to be largely genetically mediated, as was the covariation between cognitive ability and academic achievement; in contrast, differences between these domains are environmental in origin (Cardon, DiLalla, Plomin, DeFries, & Fulker, 1990; Wadsworth, 1994; Wadsworth, DeFries, Fulker, & Plomin, 1995a,b).

Developmental continuity can be modeled by allowing two processes to be formalized in the structural equations. First, genetic and environmental effects expressed in the child's phenotype at one age may have direct consequences for the phenotype expressed at later ages; this is sometimes referred to as the simplex component of the model. Second, genes or environments may be consistent in the nature of their influence during development; this is the common-factor component of the model.

Alongside developmental continuity, gene expression and the influence of the environment can change with age. Such changes are incorporated into the developmental model by allowing independent genetic and environmental innovations, or specific sources of variance, at each age. An important deduction from these models is that different kinds of developmental processes lead to different correlations between relatives as a function of their age; this permits the empirical resolution of mechanisms of development.

It should be noted that the analysis reported is an environmental analysis as much as it is a genetic one. For example, quantitative genetic analyses can estimate the magnitude of environmental variance common to family members or specific to individuals (e.g., Dunn & McGuire, 1994; Plomin & Daniels, 1987). The use of genetically informative designs, in conjunction with environmental assessment, is also a powerful tool to identify specific environmental effects free of genetic bias (Braungart, 1994; Plomin, 1995; Plomin & Bergeman, 1991), to isolate environmental influences that affect some individuals but not others (genotype-environment interaction; Hershberger, 1994; Plomin & Hershberger, 1991), and to assess the extent to which children create their own environments (genotype-environment correlation; Hershberger, 1994; Loehlin & DeFries, 1987). Results of such analyses support the idea that human behavioral genetic methodologies serve equally to enhance our understanding of environmental and genetic influences on behavioral development in children.

A Behavioral Genetic Model for Individual Differences

The behavioral genetic model presently employed recognizes three sources of individual differences in general intelligence. In this model, for a pair of siblings measured on a particular phenotype, P, G represents additive genetic differences among individuals, C represents common environmental influences shared by children reared together in the same home, and E represents nonshared environmental influences unique to the individual. These three variables are latent variables—that is, they are not directly observable. However, use of genetically informative data allows us to estimate the magnitude of the contribution of these sources of individual differences. The genetic correlation (r_G) between genotypes for pairs of individuals in the present samples is zero in the case of adoptive siblings and half for nonadoptive siblings. The correlation (r_C) between the shared environment of sibling 1 with that of sibling 2 is, by definition, unity. The correlation (r_E) between nonshared environmental influences is, by definition, zero.

The impacts of these three sources of variation—genetic, shared environmental, and nonshared environmental—are denoted as h, c, and e, respectively, and the variance explained by each is the square of these quantities, h^2, c^2, and e^2. The quantity h^2 is referred to as the *narrow-sense heritability*. In the absence of genetic dominance or epistasis (gene × gene interaction), this parameter describes the total variation due to genetic differences between individuals. Should sources of nonadditive genetic variation be important, the combined study of twins and siblings will permit their evaluation. In addition, information from the parents is capable of resolving the effects of assortative mating and genotype-environment correlation, if these should prove important.

A Behavioral Genetic Model for Development

Given the assessment of these basic sources of variation at each age point, we then need a developmental model, in addition to the basic genetic and environmental model, to evaluate the relationships among the genetic and environmental variables over time. For the present analysis, the model that was fitted to these data was one first proposed by Eaves, Long, and Heath (1986) and represents a combination of a single general factor present at all ages and a simplex model of age-to-age transmission effects. The general factor implies a static process where influences are global across all ages. The simplex implies a more dynamic process in which new variation arises at each age and persists to the next age. The two processes are the static common factor and dynamic simplex. In the behavioral genetic literature, a parameterization called the *Cholesky decomposition* is also frequently used to describe developmental data (e.g., Cherny et al., 1994; Cherny et al., 1997; Fulker et al., 1993). While examination of the Cholesky decomposition can often answer most of the questions that the present developmental model can address, the simplex-factor model formalizes particular hypotheses about the underlying developmental processes and therefore is better suited to testing those hypotheses.

A simplex model implies a matrix of correlations (between measures taken across time) where measures taken closer together are more highly correlated than measures taken further apart in time. The matrix would then have higher correlations nearer the diagonal and decreasing correlations the further one goes from the diagonal. This is typically observed with longitudinal data and is the case for our studies of cognitive ability. When developmental models of this kind have been applied to longitudinal cognitive data at the phenotypic level (Humphreys & Davey, 1988), a simplex was found to provide a better account of the data than a common factor from ages 1 through 10.

The full longitudinal behavioral-genetic model involves a threefold expansion of the developmental model just described, to allow for genetic, shared, and nonshared environmental levels of variation and covariation. The expected covariance matrices for nonadoptive and adoptive siblings, implied by the full model, can be derived using three parameter matrices adapted from the LISREL model (Jöreskog & Sörbom, 1989)—β, Γ, and Ψ—at each of the genetic, shared, and non-shared environmental levels, or nine parameter matrices in total. The β matrices contain the age-to-age transmission parameters, the Γ matrices contain the common factor loadings, and the Ψ matrices contain the time-specific variances or the new variation at each age. The parameters in Ψ are constrained to be positive, as must be the case for variances. The expectations for the genetic component of covariance, G, would be given by

$$G = (I - \beta_G)^{-1}(\Gamma_G \Gamma'_G + \Psi_G)(I - \beta'_G)^{-1}.$$

The expectations for the shared and nonshared environmental components of covariance, C and E, respectively, are obtained in an analogous manner. In summary, the model has three types of parameters to explain continuity and change: (1) the loadings of the common set of genes or environmental influences on the measures (P_i) at all ages, symbolized by γ_i; (2) new genetic or environmental influences or innovations appearing at each age (ψ_i); and (3) age-to-age transmissions of genetic and environmental influences (β_i). Both the common-factor variance and the transmission of genetic (or environmental) influences from age to age are what developmentalists generally refer to as *continuity*. The new variation at each age reflects developmental change. Changes in the structure of cognitive functioning (e.g., McCall, 1979) might be expected to manifest themselves as the appearance of such new genetic variation during development.

Fitting Models to Data from the Colorado Adoption Project

The tests used were the Bayley Mental Development Index (Bayley, 1969) at ages 1 and 2, the Stanford-Binet IQ test (Terman & Merrill, 1973) at ages 3 and 4, a composite including the Wechsler Intelligence Scale for Children-Revised (WISC-R; Wechsler, 1974) at age 7, the first principal component from a telephone-administered cognitive test battery at ages 9 and 10, and the WISC-R at age 12. The telephone battery was designed to assess verbal, spatial, perceptual speed, and memory abilities. The tests follow those described by Kent and Plomin (1987) and have similar reliabilities to in-person versions of the same tests, which were those used in the Hawaii Family Study of Cognition (DeFries et al., 1974). The present battery also has similar factor structure to an in-person version (Cardon, Corley, DeFries, Plomin, & Fulker, 1992).

To test the adequacy of these models and to estimate the importance, or magnitude, of each type of influence, we compare the observed patterns of test scores, both among sib pairs and across ages, with those expected from the model. A first step is to compute the matrix of variances and covariances among family members and across ages that we expect on the basis of the model. The nonadoptive and adoptive expected covariance matrices take a special form whereby they are partitioned into four equal quadrants. The top left and bottom right quadrants contain the within-pair variances and covariances and therefore contain the phenotypic variances and covariances. The other two quadrants contain the cross-sibling variances and covariances. These are expected to differ across the four groups of varying genetic and environmental similarity. The expected covariance matrices are estimated as:

$$\Sigma = \begin{Bmatrix} G+C+E & rG+C \\ rG+C & G+C+E \end{Bmatrix}$$

where r is the genetic correlation and equal to $\frac{1}{2}$ for nonadoptive siblings and 0 for adoptive siblings. As can be seen, the observed variances and covariances are composed of the sum the additive genetic covariance matrix (G), the shared environmental covariance matrix (C), and the nonshared environmental covariance matrix (E), which are each in turn a function of the model parameters, as described above.

Due to the incomplete nature of data from developmental studies, where individuals may be missing an assessment at one or more ages, we fit the model directly to the raw data rather than to observed covariance matrices. The data were first standardized within each age, across all individuals participating in Institute for Behavioral Genetics (IBG) family and twin studies as a single group. This standardization procedure effectively eliminates age differences in variances, which most likely are merely a result of using different tests at different ages, while preserving adoptive, nonadoptive, sib 1, and sib 2 variance differences. Fitting to the standardized raw data (rather than summary variance-covariance statistics), we maximize the log likelihood function:

$$LL = \sum_{i=1}^{N}\left[\frac{k}{2}\log(2\pi) - \frac{1}{2}\ln|\Sigma_i| - \frac{1}{2}(\mathbf{x}_i - \boldsymbol{\mu}_i)'\Sigma_i^{-1}(\mathbf{x}_i - \boldsymbol{\mu}_i)\right]$$

where k is the number of observed data points in the sibship, x_i is the vector of scores for sib pair i, Σ_i is the expected covariance matrix appropriate to the type of sibling pair (e.g., complete data or a specific pattern of missing data), N is the total number of pairs, and μ_i is the vector of mean scores for a particular type of sibling pair. This maximum likelihood estimation procedure for raw data analysis is readily implemented in Mx (Neale et al., 1999). Testing the relative fit of nested models (e.g., model A and model B) makes use of the fact that -2 ($LL_B - LL_A$) is distributed as a χ^2 statistic with degrees of freedom equal to the difference in the number of free parameters in the two models.

Because of the large number of possible models, we have followed Cherny et al. (1997) in considering each source of genetic and environmental influence in the order of nonshared environment, shared environment, and then genetic influences. For each of these sources, we first considered the alternative models for continuity. We fit models that dropped only the common factor, only the transmission effects, or both of these together. We retained the model for continuity that best accounted for the data using the Akaike Information Criterion (AIC = $\chi^2 - 2$ * degrees of freedom), taking into account both goodness of fit and parsimony. We then tested whether age-specific effects could be dropped without a significant worsening of fit. Because of the large total

TABLE 2.1. Longitudinal Correlations for Tests of General Mental Ability[a]

	Age in Years							
	1	2	3	4	7	9	10	12
1	1.00	.37	.23	.22	.22	.07	.05	.06
	(691)	(631)	(589)	(586)	(583)	(573)	(586)	(575)
2		1.00	.51	.46	.37	.28	.28	.34
		(647)	(593)	(584)	(567)	(554)	(568)	(557)
3			1.00	.60	.39	.32	.28	.36
			(609)	(582)	(550)	(538)	(550)	(544)
4				1.00	.49	.34	.36	.41
				(604)	(547)	(533)	(545)	(535)
7					1.00	.61	.60	.58
					(613)	(578)	(588)	(566)
9						1.00	.81	.62
						(607)	(592)	(575)
10							1.00	.66
							(621)	(590)
12								1.00
								(647)

[a]Sample size in parentheses

number of observations, we have also followed Cherny et al. (1997) in using a conservative 0.01 alpha level as a guide to the significance of parameter sets. Finally, we must note that although the common factor and the transmission effects predict different patterns of correlations, when there are high correlations between occasions the two patterns become almost indistinguishable. In the limiting case of perfect correlations between occasions, the common factor and transmission models are equivalent. As a consequence, common factor loadings and transmission parameters are not independent. Retaining both in a model can lead to negative dependencies between factor loadings and transmission parameters. This can result in negative parameter estimates that, while providing an optimal "mathematical" fit of a particular data set, are not robust (i.e., would likely flip-flop with minor changes in the data) and make no theoretical sense. To avoid this, we have emphasized parsimony over statistical significance in arriving at the best model.

The Results of Behavioral Genetic Modeling

Table 2.1 gives the phenotypic correlations for general mental ability based on the CAP samples through age 12 years. Recall that general mental ability was assessed using the Bayley Mental Development Index (Bayley, 1969) at ages 1 and 2 years, the Stanford-Binet (Terman, & Merrill, 1973) at ages 3 and 4, the WISC-R (Wechsler, 1974) at age 12,

TABLE 2.2. Longitudinal Sibling Correlations for Measures
of General Cognitive Ability[a]

	Control Sibs	n	Adopted Sibs	n
Year 1	**.38**	107	.07	87
Year 2	**.36**	99	.04	88
Year 3	**.37**	95	**.26**	84
Year 4	**.28**	98	.06	87
Year 7	**.47**	99	.04	88
Year 9	**.40**	104	**.24**	83
Year 10	**.35**	106	**.25**	85
Year 12	**.29**	98	.12	88

[a]Significant correlations shown in **bold** ($p < .05$)

and the first principal component for tests of specific cognitive ability at ages 7, 9, and 10 years. As can be seen, these correlations show a simplex pattern, with measurements taken closer together in time being more highly correlated, and also increasing stability at later ages.

In table 2.2, Pearson correlations at each age are given for control sibling (biological siblings living together) and adoptive sibling (non-biological siblings living together) pairs. Because this is a longitudinal study, the cross correlations between occasions provide additional data about genetic and environmental influences on development that no cross-sectional study can yield.

To test which developmental processes may be operating for general cognitive ability from ages 1 through 12, and to arrive at the most parsimonious model that could explain these data, we performed the series of model comparisons described above, beginning with tests of the non-shared environmental processes. These tests appear beginning in table 2.3. The first test was whether the common factor could be dropped from the model without a decrement in fit. We found that, indeed, a nonshared environmental common factor was unnecessary in explaining these data (model 2). Next, the transmission parameters were tested and also found unnecessary (model 3). Then, the common factor and transmission parameters were tested as a set and were found to be unnecessary (model 4). Occasion-specific nonshared environmental influences could not be dropped (model 5).

Tests of the developmental processes present at the level of the shared sibling environment appear in table 2.4. When we keep age-to-age transmission in our model, the common factor is not significant (model 6). When we keep the common factor in the model, age-to-age transmission is not significant (model 7). We can, in fact, drop both of these sources of continuity (model 8). Furthermore, age-specific shared environmental variance can be dropped without significant worsening

TABLE 2.3. Tests of Nonshared Environment Development Patterns

Model	Description	−2LL[a]	NPAR[b]	x^2	df^c	p	AIC[c]
1	Full Model	7,259.889	69				
2	Model 1, drop common factor	7,267.442	61	7.553	8	.478	−8.667
3	Model 1, drop transmission	7,268.030	62	8.141	7	.320	−5.859
4	Model 1, drop both common and transmission	7,276.984	54	17.095	15	.313	−47.095
5	Model 4, drop specifics	*	48	*	8	<.001	*

[a]Log-likelihood function; [b]Number of free parameters; [c]Akaike's Information Criterion; *Cannot be estimated

TABLE 2.4. Tests of Shared-Environment Development Patterns

Model	Description	−2LL[a]	NPAR[b]	x^2	df^c	p	AIC[c]
	Model 4	7,276.984	54				
6	Model 4, drop common factor	7,282.685	46	5.701	8	.681	−10.299
7	Model 4, drop transmission	7,282.781	47	5.797	7	.564	−8.203
8	Model 4, drop both common and transmission	7,289.587	39	12.603	15	.633	−17.397
9	Model 4, drop specifics	7,290.051	31	13.067	23	.951	−32.933

[a]Log-likelihood function; [b]Number of free parameters; [c]Akaike's Information Criterion

of the fit (model 9). Thus, there is insufficient evidence in these data to conclude that there are significant shared environmental influences, even though the positive adoptive sibling correlations suggest that there are some shared environmental influences.

Table 2.5 contains the tests of the genetic components in the developmental model. Although the common factor was statistically significant (model 9), it was relatively unimportant in contrast to the simplex transmission parameters that clearly represent the largest contribution to developmental continuity. Similarly, age-specific genetic influences were substantial at earlier ages and highly significant. A model that drops all the common factor loadings, except those on ages 2 and 12, which must be equated, fits the data best. The final reduced model is, therefore, model 13, shown in figure 2.1, with parameter estimates shown in table 2.6. In contrast to the environment, genetic influences arise at each of the earlier ages and persist to later ages.

TABLE 2.5. Tests of Genetic Development Patterns

Model	Description	$-2LL^a$	$NPAR^b$	x^2	df^c	p	AIC^c
	Model 9	7,290.051	31				
10	Model 9, drop common factor	7,330.905	23	40.854	8	.000	24.854
11	Model 9, drop transmission	7,617.900	24	327.849	7	.000	313.849
12	Model 9, drop both common and trans	8,631.973	16	1,341.946	15	.000	1,311.946
13	Model 9, drop common factor except at ages 2 & 12	7,304.171	25	14.120	7	.049	0.12
14	Model 13, drop specifics	*	17	*			

^{a}Log-likelihood function; ^{b}Number of free parameters; ^{c}Akaike's Information Criterion; *Cannot be estimated

TABLE 2.6. Standardized Model Parameters for the Reduced Model of Cognitive Development

Path	Year 1	Year 2	Year 3	Year 4	Year 7	Year 9	Year 10	Year 12
Ψ_G	.75	0	.26	.01	.52	.16	0	0
β_G		.65	.75	.99	.58	.94	1.02	.64
Γ_G		.73						.29
Ψ_E	.25	.31	.35	.36	.26	.19	.15	.45

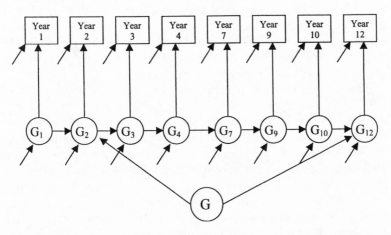

FIGURE 2.1. Reduced model of cognitive development. Residual variances on the G factors are parameters in Ψ_g. Residual variances loading directly on the phenotypes are parameters in Ψ_E. All other estimates are standardized path coefficients.

TABLE 2.7. Heritability Estimates (and percentage of genetic and environmental variance newly arising at each age)

	Y1	Y2	Y3	Y4	Y7	Y9	Y10	Y12
h^2	.75	.69 (77%)	.65 (40%)	.64 (2%)	.74 (70%)	.81 (20%)	.85 (0%)	.55 (0%)
e^2	.25	.31 (100%)	.35 (100%)	.36 (100%)	.26 (100%)	.19 (100%)	.15 (100%)	.45 (100%)

This final developmental model gives an account of the data not significantly worse than the initial full model ($\chi^2 = 44.282$ for 52 df, $p = .768$). Table 2.7 gives the estimates of heritability (h^2) and nonshared environmental variation (e^2) under this model. Also shown in this table is the percentage of each source of variation that is newly arising at each age of assessment.

Conclusions

The final model shows distinct types of contribution to the developmental process from genetic and environmental sources, essentially replicating the results reported by Cherny and Cardon (1994) through age 9 years. First, the environmental influences shared by virtue of being reared in the same household (e.g., SES) appear to be of relatively little importance throughout this period. The positive adoptive sibling correlations suggest that shared environments do influence IQ in this age period, but there is insufficient power in these data alone to be certain about this. Second, it appears that genetic variation drives a developmental process involving some new variation at each of the younger ages with reliable transmission from one age to the next, especially after early infancy. Finally, the nonshared environmental influences are transitory in early childhood (neither transmission nor common factor parameters are significant).

This analysis confirms that the most important source of individual differences in general cognitive ability is the genetic inheritance of the individual child. Nonshared environmental variation is substantial, although it must be kept in mind that this includes measurement error. Environmental influences shared by the family are relatively small and not significant in our developmental analysis of these data alone. Perhaps the most interesting aspect of these results is the pattern of genetic influences during development. During infancy and early childhood, genes contribute in novel ways at each age of assessment, strongly suggesting that the biological underpinnings of general cognitive ability are not fixed at birth but continue to change along with the still developing brain. Later, by age 10 or 11 years, we witness the end of this developmental process and the beginning of the stabilization (or complete continuity) of the heritable influences on general cognitive ability.

REFERENCES

Anderson, L. D. (1939). The predictive value of infant tests in relation to intelligence at 5 years. *Child Development, 10,* 202–212.

Bayley, N. (1943). Mental growth during the first three years. In R. G. Barker, J. S. Lounin, & H. F. Wright (Eds.), *Child behavior and development.* New York: McGraw-Hill.

Bayley, N. (1969). *Manual for the Bayley scales of infant development.* New York: Psychological Corporation.

Benson, J. B., Cherny, S. S., Haith, M. M., & Fulker, D. W. (1993). Rapid assessment of infant predictors of adult IQ: Midtwin-midparent analyses. *Developmental Psychology, 29,* 434–447.

Binet, A.., & Simon, T. (1905) Méthodes nouvell pour le diagnostic du niveau Intellectual des anormaux. *L'Année Psychologique, 11,* 245–336.

Bouchard, Jr., T. J., & McGue, M. (1981). Familial studies of intelligence: A review. *Science, 212,* 1055–1059.

Braungart, J. M. (1994). Genetic influences on "environmental" measures. In J. C. DeFries, R. Plomin, & D. W. Fulker (Eds.), *Nature and nurture during middle childhood* (pp. 233–248). Oxford: Blackwell.

Cardon, L. R. (1994). Specific cognitive abilities. In J. C. DeFries, R. Plomin, & D. W. Fulker (Eds.), *Nature and nurture during middle childhood* (pp. 57–76). Oxford: Blackwell.

Cardon, L. R., Corley, R. P., DeFries, J. C., Plomin, R., & Fulker, D. W. (1992). Factorial validation of a telephone test battery of specific cognitive abilities. *Personality and Individual Differences, 13,* 1047–1050.

Cardon, L. R., DiLalla, L. F., Plomin, R., DeFries, J. C., & Fulker, D. W. (1990). Genetic correlations between reading performance and IQ in the Colorado Adoption Project. *Intelligence, 14,* 245–257.

Cardon, L. R., & Fulker, D. W. (1994). A model of developmental change in hierarchical phenotypes with application to specific cognitive abilities. *Behavior Genetics, 24,* 1–16.

Cardon, L. R., Fulker, D. W., DeFries, J. C., & Plomin, R. (1992). Continuity and change in general cognitive ability from 1 to 7 years of age. *Developmental Psychology, 28,* 64–73.

Cattell, R. B. (1971). *Abilities: Their structure, growth, and action.* Boston: Houghton Mifflin.

Cherny, S. S., & Cardon, L. R. (1994). General cognitive ability. In J. C. DeFries, R. Plomin, & D. W. Fulker (Eds.), *Nature and nurture during middle childhood* (pp. 45–56). Oxford: Blackwell.

Cherny, S. S., Fulker, D. W., Emde, R. N., Robinson, J., Corley, R. P., Reznick, S. J., & DeFries, J. C. (1994). A developmental-genetic analysis of continuity and change in the Bayley Mental Development Index from 14 to 24 months of age: The MacArthur Longitudinal Twin Study. *Psychological Science, 5,* 354–360.

Cherny, S. S., Fulker, D. W., & Hewitt, J. K. (1997). Cognitive development from infancy to middle childhood. In R. J. Sternberg & E. L. Grigorenko (Eds.), *Intelligence: Heredity and environment* (pp. 463–482). New York: Cambridge University Press.

DeFries, J. C., Plomin, R., & Fulker, D. W. (1994). *Nature and nurture during middle childhood.* Oxford: Blackwell.

DeFries, J. C., Vandenberg, S. G., McClearn, G. E., Kuse, A. R., & Wilson, J. R. (1974). Near identity of cognitive structure in two ethnic groups. *Science, 183*, 338–339.

DiLalla, L. F., Thompson, L. A., Plomin, R., Phillips, K., Fagan III, J. F., Haith, M. M., Cyphers, L. H., & Fulker, D. W. (1990). Infant predictors of preschool and adult IQ: A study of infant twins and their parents. *Developmental Psychology, 26*, 759–769.

Dunn, J., & McGuire, S. (1994). Young children's nonshared experiences: A summary of studies in Cambridge and Colorado. In E. M. Hetherington, D. Reiss, & R. Plomin (Eds.), *Separate social worlds of siblings: Importance of nonshared environment on development* (pp. 111–128). Hillsdale, NJ: Lawrence Erlbaum Associates.

Eaves, L. J., Long, J., & Heath, A. C. (1986). A theory of developmental change in quantitative phenotypes applied to cognitive development. *Behavior Genetics, 16*, 143–162.

Emde, R. N., Plomin, R., Robinson, J. A., Reznick, J. S., Campos, J., Corley, R., DeFries, J. C., Fulker, D. W., Kagan, J., & Zahn-Waxler, C. (1992). Temperament, emotion, and cognition at 14 months: The MacArthur Longitudinal Twin Study. *Child Development, 63*, 1437–1455.

Fulker, D. W., & Cardon, L. R. (1993). What can twins tell us about the structure and correlates of cognitive abilities? In T. J. Bouchard, Jr. (Ed.), *Twins as a tool of behavioral genetics.* Chichester, England UK: John Wiley.

Fulker, D. W., Cherny, S. S., & Cardon, L. R. (1993). Continuity and change in cognitive development. In R. Plomin & G. E. McClearn (Eds.), *Nature, nurture, and psychology* (pp. 77–97). Washington, DC: American Psychological Association.

Hershberger, S. L. (1994). Genotype-environment interaction and correlation. In J. C. DeFries, R. Plomin, & D. W. Fulker (Eds.), *Nature and nurture during middle childhood* (pp. 281–294). Oxford: Blackwell.

Honzik, M. P., MacFarlane, J. W., & Allen, L. (1948). The stability of mental test performance between two and eighteen years. *Journal of Experimental Education, 17*, 309–324.

Humphreys, L. G., & Davey, T. C. (1988). Continuity in intellectual growth from 12 months to 9 years. *Intelligence, 12*, 183–197.

Jöreskog, K. G., & Sörbom, D. (1989). *LISREL 7: A guide to the program and applications* (2nd ed.). Chicago: SPSS, Inc.

Kent, J., & Plomin, R. (1987). Testing specific cognitive abilities by telephone and mail. *Intelligence, 11*, 391–400.

Loehlin, J. C., & DeFries, J. C. (1987). Genotype-environment correlation and IQ. *Behavior Genetics, 17*, 263–277.

McCall, R. B. (1979). The development of intellectual functioning in infancy and the prediction of later IQ. In J. D. Osofsky (Ed.), *The handbook of infant development* (pp. 707–741). New York: John Wiley.

McCall, R. B., Applebaum, M. I., & Hogarty, P. S. (1973). Developmental changes in mental performance. *Monographs of the Society for Research in Child Development, 38* (Serial No. 150).

Neale, M. C., Boker, S. M, Xie, G., & Maes, H. H. (1999). *Mx: Statistical modeling* (5th ed.) Richmond: Department of Psychiatry, Medical College of Virginia.

Plomin, R. (1995). Genetics and children's experiences in the family. *Journal of Child Psychology and Psychiatry, 36,* 33–68.

Plomin, R., & Bergeman, C. S. (1991). The nature of nurture: Genetic influence on "environmental" measures. *Behavioral and Brain Sciences, 14,* 414–424.

Plomin, R., Campos, J., Corley, R., Emde, R. N., Fulker, D. W., Kagan, J., Reznick, J. S., Robinson, J., Zahn-Waxler, C., & DeFries, J. C. (1990). Individual differences during the second year of life: The MacArthur Longitudinal Twin Study. In J. Columbo & J. Fagan (Eds.), *Individual differences in infancy: Reliability, stability, & predictability* (pp. 431–455). Hillsdale, NJ: Lawrence Erlbaum Associates.

Plomin, R., & Daniels, D. (1987). Why are children in the same family so different from each other? *Behavioral and Brain Sciences, 14,* 414–424.

Plomin, R., & DeFries, J. C. (1985). *Origins of individual differences in infancy: The Colorado Adoption Project.* Orlando, FL: Academic Press.

Plomin, R., DeFries, J. C., & Fulker, D. W. (1988). *Nature and nurture in infancy and early childhood* (3rd ed.). Cambridge: Cambridge University Press.

Plomin, R., DeFries, J. C., McClearn, G. E., & Rutter, M. (1997). *Behavioral genetics* (3rd ed.). New York: W. H. Freeman and Company.

Plomin, R., Emde, R. N., Braungart, J. M., Campos, J., Corley, R., Fulker, D. W., Kagan, J., Reznick, J. S., Robinson, J., Zahn-Waxler, C., & DeFries, J. C. (1993). Genetic change and continuity from fourteen to twenty months: The MacArthur Longitudinal Twin Study. *Child Development, 64,* 1354–1376.

Plomin, R., & Hershberger, S. (1991). Genotype-environment interaction. In T. D. Wachs & R. Plomin (Eds.), *Conceptualization and measurement of organism-environment interaction* (pp. 29–43). Washington, DC: American Psychological Association.

Santrock, J. W. (1998). *Child development* (8th ed.). Boston, MA: McGraw-Hill.

Scarr, S. (1993). Biological and cultural diversity: The legacy of Darwin for development. *Child Development, 64,* 1333–1353.

Spearman, C. E. (1927). *The abilities of man.* New York: Macmillan.

Terman, L. H., & Merrill, M. A. (1973). *Stanford-Binet intelligence scale: 1972 norms edition.* Boston: Houghton Mifflin.

Wadsworth, S. J. (1994). School achievement. In J. C. DeFries, R. Plomin, & D. W. Fulker (Eds.), *Nature and nurture during middle childhood* (pp. 86–101). Oxford: Blackwell.

Wadsworth, S. J., DeFries, J. C., Fulker, D. W., & Plomin, R. (1995a). Cognitive ability and academic achievement in the Colorado Adoption Project: A multivariate analysis of parent-offspring and sibling data. *Behavior Genetics, 25,* 1–15.

Wadsworth, S. J., DeFries, J. C., Fulker, D. W., & Plomin, R. (1995b). Covariation among measures of cognitive ability and academic achievement in the Colorado Adoption Project: Sibling analysis. *Personality and Individual Differences, 18,* 63–73.

Wechsler, D. (1974). *Manual for the Wechsler Intelligence Scale of Children-Revised.* New York: Psychological Corporation.

MARICELA ALARCÓN
ROBERT PLOMIN
ROBIN P. CORLEY
JOHN C. DEFRIES

3

Multivariate Parent-Offspring Analyses of Specific Cognitive Abilities

Introduction

Although the etiology of individual differences in general cognitive ability has been investigated for over a century (Plomin, DeFries, McClearn, & Rutter, 1997), specific cognitive abilities (SCA) have been studied less extensively. Nevertheless, during the last few decades family, twin, and adoption designs have been used to begin to assess the genetic and environmental etiologies of SCA.

The largest family study of SCA was the Hawaii Family Study of Cognition, which included test data from 1816 intact nuclear families (DeFries et al., 1979). The families were mostly Americans of European (AEA) or of Japanese (AJA) ancestry who were administered 15 tests of specific cognitive abilities. Regressions of offspring on midparent, single-parent/single-child correlations, and sibling correlations provided evidence for the familial transmission of SCA. For example, the regressions of midchild on midparent for verbal, spatial, perceptual speed, and memory principal component scores were .54, .60, .41, and .31, respectively, for the AEA families and .48, .42, .34, and .18, respectively, for the AJA families. Thus, midparent-offspring resemblance was higher for verbal and spatial abilities than for perceptual speed and memory factors in both ethnic groups. In contrast, little evidence for assortative mating was found in either group: spouse correlations for the four principal component factors ranged from .03 to .22 for AEA and from 0.02 to .25 for AJA.

Measures of parent-offspring resemblance can only be considered upper-bound estimates of heritability. Thus, family studies can provide conclusive evidence for the familiality of a trait, but not for its genetic etiology. In contrast, results obtained from twin studies can provide estimates of heritability. The twin design facilitates the partitioning of phenotypic variances and covariances into genetic, shared environmental and nonshared environmental components. Since the similarity of MZ twin pairs is due to both additive genetic (h^2) and shared environmental (c^2) influences, any phenotypic correlation between the members of the pairs estimates $h^2 + c^2$. However, because DZ twins share half of their segregating genes and all of the shared environmental influences, their corresponding phenotypic correlation estimates $\frac{1}{2}h^2 + c^2$. Therefore, twice the difference between the MZ and DZ phenotypic correlations estimates heritability (h^2). An estimate of the proportion of the total variance due to shared environmental effects (c^2) is obtained by subtracting the heritability estimate from the MZ correlation. In addition, by subtracting the MZ correlation from 1, an estimate of the proportion of the variance due to nonshared environmental effects (e^2) is obtained.

Results obtained from studies of twins provide evidence for the heritable nature of SCA. Consistent with the results of the HFSC, results obtained from early twin studies suggest that heritability estimates for verbal and spatial abilities tend to be somewhat higher than those for perceptual speed and especially memory (DeFries, Vandenberg, & McClearn, 1976; Plomin, 1988). Nonverbal memory continues to show lower heritability in more recent studies (e.g., McGue & Bouchard, 1989; Pedersen, Plomin, Nesselroade, & McClearn, 1992; Thompson, Detterman, & Plomin, 1991); however, this estimate may vary as a function of the type of test used (Thapar, Petrill, & Thompson, 1994).

Results from two twins-reared-apart studies confirm the heritable nature of SCA (McGue & Bouchard, 1989; Pedersen et al., 1992). McGue and Bouchard (1989) reported heritabilities of .57, .71, .53, and .42 for verbal, spatial, perceptual speed, and memory abilities, respectively, in 74 pairs of twins participating in the Minnesota Study of Twins Reared Apart (MSTRA). The corresponding average estimates obtained by Pedersen et al. (1992) from 146 twin pairs reared-apart and 156 twins reared-together from the Swedish Adoption/Twin Study of Aging (SATSA) were .58, .46, .58, and .38, respectively. These h^2 estimates are somewhat higher than those obtained in previous twin studies, possibly due to the fact that both twins-reared-apart studies included adult participants: the mean age of the MSTRA twins was 39.9 years and the SATSA twins had a mean age of 65.6 years at the time of testing.

Evidence for higher heritabilities of SCA has recently been reported in elderly twins. In a sample of Swedish twins who were at least 80 years of age, heritability estimates for verbal, spatial, perceptual speed, and memory abilities were .55, .32, .62, and .52, respectively (McClearn et al., 1997). This evidence for higher heritabilities in adult and elderly samples supports the hypothesis that genetic influences on individual differences in SCA may increase as a function of age (Plomin, 1986).

In addition to providing evidence for genetic influences on SCA, twin studies show evidence for shared-environmental influences (Nichols, 1978). However, estimates of c^2 (i.e., the proportion of the total variation that is due to shared environmental influences) obtained from twin studies may be inflated because identical and fraternal twins may experience more similar environments than nontwin siblings (Plomin, DeFries et al., 1997). In support of this hypothesis are data from SATSA (Pedersen et al., 1992), a study of twins reared apart, where the heritability estimates across measures were highly similar to the MZ reared-apart correlations and shared family environment accounted for less than 15% of the variance in two verbal and two spatial tests. In addition, c^2 estimates from an analysis of data from the Western Reserve Twin project were negligible, ranging from .01 (for speed) to .08 (for verbal) (Thompson et al., 1991).

Although results obtained from twin and family studies suggest that SCA are differentially heritable, adoption studies can provide even more compelling evidence for the etiology of individual differences for these abilities and for their covariation. For example, when adopted children and their biological and adoptive parents are tested at the same ages, adoption studies yield direct estimates of genetic and environmental contributions to individual differences, genetic and environmental correlations among different abilities, and tests of familial environmental transmission parameters and of genotype-environment correlations. Adoption studies are especially valuable for assessing shared-environmental influences because they provide a direct estimate of c^2 based on the resemblance of genetically unrelated family members. In contrast, the twin-design estimates shared environment indirectly, as residual twin resemblance not explained by genetic parameters.

Previous analyses of CAP data have included multivariate parent-offspring and sibling analyses of individual differences in verbal, spatial, perceptual speed, and visual memory abilities at various stages of development. For example, Rice, Carey, Fulker, & DeFries (1989) examined the etiology of individual differences in SCA by fitting a multivariate conditional path model to parent-offspring data from the CAP when the offspring were 4 years of age. Assuming complete isomorphism (i.e., that the genetic and environmental etiologies of individual differences are the same) between the adult and child measures, resulting h^2 estimates for verbal, spatial, perceptual speed, and visual memory were .12,

.31, .21, and .06, respectively. However, as noted previously by Rice, Fulker, & DeFries (1986), these putative h^2 estimates actually equal $h_c r_{gc-a} h_a$, where h_c and h_a are the square roots of heritabilities in children and adults and r_{gca} is the genetic correlation between the two measures—that is, an index of genetic stability between childhood and adulthood (DeFries, Plomin, & LaBuda, 1987). Thus, the low h^2 estimates obtained from CAP parent-offspring analyses when the children were 4 years of age may have been due to the low genetic stability of SCA between early childhood and adulthood.

The relationship or covariation among the specific cognitive abilities can be examined by a multivariate genetic analysis. Multivariate genetic analysis partitions the phenotypic covariation among traits into genetic and environmental sources of covariation (Plomin, 1986). More specifically, the phenotypic correlation (r_P) between traits X and trait Y can be expressed as follows: $r_P = h_X h_Y r_G + e_X e_Y r_E$, where h_X and h_Y are the square roots of the heritabilities of traits X and Y, e_X and e_Y are the square roots of the environmentalities, r_G is the genetic correlation between X and Y, and r_E is the environmental correlation. Two key concepts in the application of multivariate methods are bivariate heritability and genetic correlation. Bivariate h^2 may be defined as the proportion of the phenotypic correlation that is due to heritable influences—that is, $h_X h_Y r_G / r_P$ (Plomin & DeFries, 1979), whereas the genetic correlation is an index of the extent to which two traits are influenced by the same genes. Thus, the bivariate h^2 for traits X and Y can be low even though the overlap of genes that determine these traits may be considerable (Plomin & DeFries, 1979).

An important question in cognitive developmental research is the extent to which cognitive processes are modular (specific and independent) or molar (general and dependent). Genetic research on the normal range of genetic and environmental sources of individual differences in cognitive processes can bring a powerful empirical approach to the issue of modularity because it can investigate the extent to which genetic effects on one cognitive process covary with genetic effects on other cognitive processes. From a genetic perspective, modularity implies genetically distinct abilities; genetic effects on one cognitive ability should be independent of genetic effects on other cognitive abilities. The issue of genetic overlap among traits can be addressed by multivariate genetic analysis, which focuses on the covariance among traits rather than the variance of each trait considered separately (Plomin, DeFries et al., 1997).

Although multivariate analyses of subtests for general cognitive ability have been well documented (Casto, DeFries, & Fulker, 1995; Tambs, Sundet, & Magnus, 1986), few multivariate genetic analyses of SCA data have been previously reported. Rice et al. (1989) presented evidence for significant genetic covariation among SCA in parent-offspring conditional path analysis of data from the CAP when the offspring were 4 years of age. Moreover, the genetic correlations among the four SCA were not significantly different from each other, or from unity, suggesting that a single cognitive ability factor may underlie the covariation among the verbal, spatial, perceptual speed, and memory abilities.

Several investigators have attempted to examine the genetic influences unique to SCA independent of the effects of general cognitive ability. For example, Pedersen, Plomin, and McClearn (1994) partialed out the effects of general cognitive ability, or g, using a bivariate Cholesky decomposition model and tested whether the remaining genetic variation was significant in data from SATSA. Their results show that 12 of 13 cognitive tests are influenced by specific genetic effects independent of those due to general cognitive ability. Cardon, Fulker, DeFries, & Plomin (1992) fitted a sophisticated multivariate hierarchal model with Schmid-Leiman transformations to SCA data from the CAP siblings at 7 years of age, and results support the hypothesis of genetic influences on specific cognitive abilities (verbal, spatial, and memory) independent of those due to general cognitive ability. Subsequently, the same analysis was conducted on a larger sample of SCA data from CAP siblings at 3, 4, 7, and 9 years of age (Cardon, 1994). Evidence for the presence of a general cognitive ability factor was consistent across ages. Genetic influences specific to each ability were more variable: specific effects on verbal and spatial abilities were present at all ages, whereas perceptual speed was influenced by unique effects during the early years (3 and 4 years of age) and, lastly, memory was affected by specific genetic influences at 3, 4, and 7 years of age.

These results (Cardon et al., 1992; Cardon, 1994) have been confirmed by an analysis of SCA data from 283 elementary school twin pairs in the Western Reserve Twin Project (Luo, Petrill, & Thompson, 1994). SCA were influenced by both a common set of genes and sets of genes unique to each ability.

The findings of twin and adoption research suggest that genetic overlap among specific cognitive abilities is considerable, although some genetic effects are unique to each ability. Thus, results obtained from the few multivariate genetic analyses conducted to date suggest cognitive abilities are molar (i.e., correlated) rather than modular (i.e., independent). The primary purpose of the present report is to present the results of multivariate genetic analyses of CAP parent-offspring data at 7 and 12 years of age, and to compare them to those obtained by Rice et al. (1989) when the CAP children were only four years of age. Based on the previous results of SCA studies, we expected to find evidence for genetic overlap among verbal, spatial, perceptual speed, and, to a lesser extent, memory abilities.

Materials and Methods

Participants

The Lutheran Social Services of Colorado and Catholic Community Services, two large adoption agencies in Denver, facilitated the recruitment of the CAP biological parents and provided contact with adoptive parents who were asked to participate in the study after the completion of the adoption process. The nonadoptive or control families were recruited through Denver area hospitals and were matched to the adoptive families based on the proband's sex, number of children in the family, father's age, occupation, and years of education. More detailed descriptions of the CAP have been presented elsewhere (DeFries, Plomin, & Fulker, 1994; Plomin, DeFries, & Fulker, 1988; Plomin & DeFries, 1985). The current CAP sample includes 178 adopted children tested at 7 years of age and their adoptive (178 mothers and fathers) and biological parents (178 mothers and 33 fathers). The nonadoptive (control) sample tested at age 7 includes 198 sets of parents and nonadoptive children. The sample tested at 12 years of age includes 175 adopted individuals tested at 12 years of age and their adoptive (175 mothers and fathers) and biological parents (175 mothers and 34 fathers). The corresponding nonadoptive sample includes 209 sets of parents and nonadoptive 12-year-old children.

Test Battery

All adults were administered an extensive test battery of specific cognitive abilities (DeFries, Plomin, Vandenberg, & Kuse, 1981) similar to the battery used in the Hawaii Family Study of Cognition. Multiple regression analyses were used to correct each test score for age and sex. As described by Rice et al. (1986), the residual scores were standardized and summed to compute four specific cognitive ability scales. The verbal scale was obtained from the Educational Testing Service (ETS) Things Categories, a Vocabulary measure (Primary Mental Abilities Vocabulary and ETS Vocabulary), and ETS Word Beginnings and Endings. The spatial visualization scale was computed using an adapted version of the Minnesota Paper Form Board, ETS Card Rotations, ETS Hidden Patterns, ETS Identical Pictures and Raven's Progressive Matrices. The ETS Subtraction and Multiplication test and the Colorado Perceptual Speed constituted the perceptual speed scale. And the Picture Memory (Immediate and Delayed Recognition) and Names and Faces (Immediate and Delayed Recognition) were summed to yield the visual memory scale.

Each test score from the batteries administered to children at 7 and 12 years of age was also age- and sex-corrected simultaneously in the adopted and control groups and the residuals were then standardized and summed to compute four specific cognitive ability scales. For the measures obtained at age 7, the WISC-R Vocabulary and Verbal Fluency subtests were summed to compute the verbal scale. The spatial visualization scale was a combination of the Primary Mental Abilities rotation test and the WISC-R Block Design test. The Colorado Perceptual Speed (CPS) test and the ETS Identical Pictures test were used to obtain the perceptual speed scale. And the memory scale was the sum of the Picture Memory (Immediate and Delayed Recognition) and Names and Faces (Immediate and Delayed Recognition).

For the data obtained at age 12, the WISC-R Similarities and the WISC-R Vocabulary subtest were used to obtain the verbal scale. The spatial visualization scale was computed by summing the ETS Hidden Patterns and the ETS Card Rotations. The ETS Finding A's and the Colorado Perceptual Speed tests were used to compute the perceptual speed scale. And, as for the parent- and age-7 data, the sum of the Picture Memory (Immediate and Delayed Recognition) and Names and Faces (Immediate and Delayed Recognition) constituted the visual memory scale. The verbal, spatial, perceptual speed, and memory score means do not differ significantly between the adopted and control children at ages 7 or 12.

Analyses

Three variance-covariance matrices were computed for each analysis using SPSS-X (1988): one for complete adoptive families including both adoptive parents, both biological parents and the adopted child; a second for adoptive families without the biological father; and a third for nonadoptive or control families. The covariance matrix for the complete adoptive families includes four specific cognitive ability scores for each of the adoptive fathers, adoptive mothers, biological fathers, biological mothers, and adopted children and, thus, its dimensions are 20 × 20. The corresponding covariance matrix for the adoptive families where the biological father had missing data is 16 × 16; and the covariance matrix for the control families, which includes four measures for the control father, control mother, and their children, is 12 × 12.

Model

A parent-offspring multivariate conditional path model (Rice et al., 1989) was fit to the observed covariance matrices calculated from data obtained at ages 7 and 12 using a maximum-likelihood function in Mx (Neale, 1999). Figures 3.1 and 3.2 depict the CAP path diagrams,

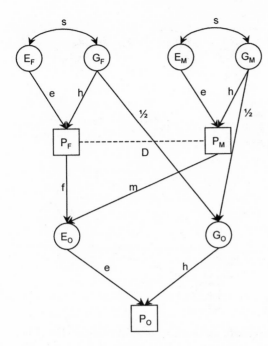

FIGURE 3.1. Path diagram for genetic and environmental transmission in nonadoptive families using delta paths.

assuming isomorphism of the child and adult measures. For nonadoptive families (figure 3.1), the column vectors of observed scores are represented by P_F, P_M, and P_O for the control fathers, mothers, and offspring, respectively. The corresponding vectors of additive genetic values and environmental deviations are G_F, G_M, and G_O, and E_F, E_M, and E_O. For adoptive families (figure 3.2), the vectors of observed scores for the biological fathers and mothers are symbolized by P_{BF} and P_{BM}, respectively. The corresponding vectors of additive genetic values and environmental deviations are G_{BF} and G_{BM}, and E_{BF} and E_{BM} for biological fathers and mothers, respectively. Figure 3.2 also depicts the column vectors of observed scores (P_{AF}, P_{AM}, P_{AO}), additive genetic values (G_{AF}, G_{AM}, G_{AO}), and environmental deviations (E_{AF}, E_{AM}, E_{AO}) for the adoptive fathers, adoptive mothers, and adopted offspring, respectively.

Diagonal matrices of the square roots of heritability and environmentality are represented by the paths h and e, respectively, and 1/2 is a diagonal matrix in both figures 3.1 and 3.2. In this analysis, cultural transmission is modeled as the influence of the parents' phenotype on the environment of the offspring, and maternal (m) and paternal (f) transmission parameters are allowed to differ. Parental transmission of both genetic and environmental influences will yield genotype-environment correlations (represented by the full nonsymmetric s

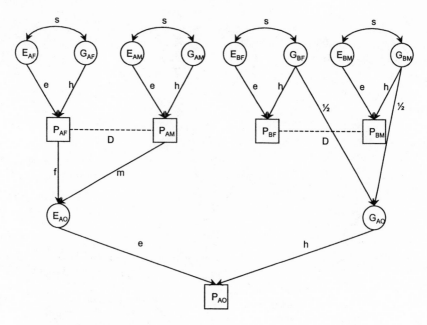

FIGURE 3.2. Path diagram for genetic and environmental transmission in adoptive families using delta paths.

matrix); thus, s is a function of m and f and leads to the following constraint:

$$s = (1/2)(R_G h' + se')[(I + D'R_p)m' + (I + DR_p)f']$$

where I is an identity matrix, R_G is a genetic correlation matrix, R_P is a phenotypic correlation matrix, and D is a matrix representing assortment.

Assortative mating at the phenotypic level is modeled using delta paths (Phillips, 1989; Van Eerdewegh, 1982). The parameters for assortative mating (D) were calculated prior to the analysis and then fixed in the parent-offspring analyses to reduce the number of parameters to be estimated in the model. Adopted and nonadopted individuals may have different within-person phenotypic correlations (R_P). For nonadopted individuals, R_P is a function of s and is constrained as follows:

$$R_p = hR_G h' + hse' + es'h' + eR_E e',$$

where R_E is an environmental correlation matrix. For nonadopted individuals, the R_P constraint fixes the diagonal elements of the phenotypic correlation matrices to unity and estimates e as residuals. The phenotypic correlations for adopted children may be less than the estimate

TABLE 3.1. Expected Familial Covariances for a Multivariate Conditional Path Model with a Full Adoption Design

Variable	Expected Covariance
Nonadoptive Families	
P_M, P_F'	$V_P^{1/2}[R_P\ DR_P']\,V_P^{1/2}$
P_O, P_M'	$V_O^{1/2}[\frac{1}{2}h((R_Gh' + se)(I + D'R_P')) + e(mR_P + fR_PD'R_P')]V_P^{1/2}$
P_O, P_F'	$V_O^{1/2}[\frac{1}{2}h((R_Gh' + se)(I + DR_P')) + e(fR_P + mR_PDR_P')]V_P^{1/2}$
$P_M, P_M' = P_F, P_F'$	$V_P^{1/2}[R_P]\,V_P^{1/2}$
P_O, P_O'	$V_O^{1/2}[R_P]\,V_O^{1/2}$
Adoptive Families	
$P_{BM}, P_{BF}' = P_{AM}, P_{AF}'$	$V_P^{1/2}[R_P\ DR_P']\,V_P^{1/2}$
P_{AO}, P_{BM}'	$V_O^{1/2}[\frac{1}{2}h((R_Gh' + se)(I + D'R_P'))]V_P^{1/2}$
P_{AO}, P_{BF}'	$V_O^{1/2}[\frac{1}{2}h((R_Gh' + se)(I + DR_P'))]V_P^{1/2}$
P_{AO}, P_{AM}'	$V_O^{1/2}[e(mR_P + fR_PD'R_P')]V_P^{1/2}$
P_{AO}, P_{AF}'	$V_O^{1/2}[e(mR_P + fR_PDR_P')]V_P^{1/2}$
$P_{BM}, P_{BM}' = P_{BF}, P_{BF}' = P_{AM}, P_{AM}' = P_{AF}, P_{AF}'$	$V_P^{1/2}[R_P]\,V_P^{1/2}$
P_{AO}, P_{AO}'	$V_O^{1/2}[hR_Gh' + eR_Fe']V_O^{1/2}$

for nonadopted individuals because the genotype-environment correlations (s) do not contribute to the R_P of adopted individuals.

A previous analysis of specific cognitive abilities found the influence of selective placement to be nonsignificant (Rice et al., 1986), thus, it is not modeled in age 7 or age 12 analysis.

Table 3.1 shows the expected familial covariances previously obtained by Rice et al. (1989) using path tracing rules for the multivariate case (Vogler, 1985) and for delta paths (Van Eerdewegh, 1982), where $V_P^{1/2}$ and $V_O^{1/2}$ represent diagonal matrices of parent and offspring phenotypic standard deviations which were estimated and fixed prior to the analyses.

Results

Parents

Prior to fitting the full parent-offspring model, the spousal phenotypic correlations (M) and assortment parameters (D) were calculated (Alarcón, Plomin, Fulker, Corley, & DeFries, 1998) and are presented in table 3.2. These correlations and assortment parameters did not differ significantly among the adoptive, biological, and control group parents ($\chi^2 = 43.79$, $df = 32$, $p = .080$). As expected based on the results of the SCA data at 4 years of age reported by Rice et al. (1986, 1989), the within- and between-trait spousal correlations are low, with the

TABLE 3.2. Spousal Correlations (m) and Assortative Mating (D) Parameters

	Perceptual			
	Verbal (V)	Spatial (S)	Speed (P)	Memory (M)
m^a				
V	.332	.192	.180	.040
S	.077	.139	.084	.031
P	.139	.143	.155	.133
M	.040	.088	.011	.145
D^a				
V	.383	−.017	.025	−.102
S	−.133	.125	−.031	−.056
P	−.022	−.017	.107	.132
M	−.049	.069	−.081	.145

[a]Rows, mothers; columns, fathers. $N = 418$ spouses

exception of verbal ability. The spousal correlation for verbal ability is .33 and its corresponding assortment parameter is .38 in the full model.

Age 7

The genetic and environmental parameter estimates for the full model fitted to the age 7 data are presented in tables 3.3 and 3.4. As shown in table 3.4, the full model did not fit the data (model 1, $\chi^2 = 467.88$, $df = 382$, $p = .002$); however, the χ^2 goodness-of-fit index is a function of sample size (Bollen, 1989; Hayduk, 1987) and the large sample size included in this analysis ($N = 1339$) may be the cause of the poor fit of the full model to the data. The pattern of correlations in the adopted families with biological fathers is irregular and may have also contributed to the lack of fit of the full model to the data. As shown in table 3.5, the correlations in the adopted families with biological fathers ($N = 33$) are somewhat larger than those in adopted families without biological fathers ($N = 145$). Although these correlation matrices are not significantly different ($\chi^2 = 153.18$, $df = 136$, $p = .149$), the exclusion of the 33 adopted families with biological fathers from the analysis resulted in a good fit of the full model to the data ($\chi^2 = 186.40$, $df = 172$, $p = .214$). Thus, the irregular pattern of correlations observed in the group of adopted families with biological fathers may be the cause of the poor fit of the full model to the data including all three groups. Nevertheless, since there is a small number of adopted families with biological fathers ($N = 33$), this group is given less weight in the analysis and, not surprisingly, the parameter estimates obtained from the full model excluding these families were similar to those reported in this chapter when the model was fitted to the complete sample.

TABLE 3.3. Parameter Estimates from the Full Model Fitted
to SCA Data Obtained at Age 7

	Perceptual			
	Verbal (V)	Spatial (S)	Speed (P)	Memory (M)
h				
V	.429			
S		.590		
P			.486	
M				.441
e				
V	.870			
S		.807		
P			.874	
M				.898
m				
V	−.009	−.073	.079	−.020
S	−.054	−.006	−.002	.087
P	−.086	.009	.061	−.060
M	−.103	.088	.047	.080
f				
V	.095	.076	.099	−.107
S	−.048	.191	−.011	.095
P	.089	−.048	−.002	−.002
M	.007	.116	.058	.008
s				
V	.032	.051	−.011	.062
S	.074	.032	.006	.074
P	.058	.053	.001	.089
M	−.009	.017	−.012	.003

TABLE 3.4. Summary of Model Comparisons for Age 7 Data

Model	χ^2	vs.	χ^2_{diff}	df	p	AIC
1. Full	467.88			382	.002	−296.12
2. $\mathbf{m} = \mathbf{f} = \mathbf{s} = 0$	511.23	1	43.35	32	.087	−316.77
3. $\mathbf{m} = \mathbf{f} = \mathbf{s} = 0$ $\mathbf{h}$ equal	517.88	2	6.65	3	.084	−316.12
4. $\mathbf{m} = \mathbf{f} = \mathbf{s} = 0$ $\mathbf{h}$ equal, $\mathbf{R}_G$ equal	525.87	3	7.99	5	.157	−318.13
5. $\mathbf{m} = \mathbf{f} = \mathbf{s} = 0$ $\mathbf{h}$ equal, $\mathbf{R}_G = 0$	545.64	4	19.77	1	<.001	−300.36

TABLE 3.5. Phenotypic Correlations and Variances (on the diagonal) for Adoptive (AF, AM, BF, BM, AO) and Nonadoptive (F, M, O) Families When the Offspring Were 7 Years of Age (V, Verbal; S, Spatial; P, Perceptual Speed; M, Memory)

	AF_V	AF_S	AF_P	AF_M	AM_V	AM_S	AM_P	AM_M	BF_V	BF_S	BF_P	BF_M	BM_V	BM_S	BM_P	BM_M	AO_V	AO_S	AO_P	AO_M
AF_V	4.96	.51	.40	.39	.43	.04	.11	.01	.07	-.01	-.01	.18	.36	.08	.08	.30	.05	-.02	-.15	.04
AF_S		7.89	.46	.27	.25	.15	.17	.29	-.06	.10	.24	.29	.12	-.05	.13	-.04	.31	.29	-.05	.15
AF_P			3.06	.54	.22	.17	.25	.07	-.27	.03	-.09	-.08	.19	.08	.09	.28	.03	.11	.09	.08
AF_M				2.47	.25	.02	.13	-.02	.04	-.04	-.05	-.11	.08	.06	-.15	.44	-.02	.24	-.46	.08
AM_V					4.20	.40	.31	.20	-.06	.03	.15	-.17	.04	-.17	-.04	.16	.20	-.24	-.22	.12
AM_S						12.41	.43	.07	-.43	-.07	.24	-.30	-.07	-.07	-.04	.10	-.10	-.28	.03	-.03
AM_P							1.80	-.03	-.02	.11	-.17	-.10	-.28	-.06	.03	.08	-.08	-.18	.09	.20
AM_M								1.16	.06	.04	.29	.08	-.02	-.17	.14	-.10	-.01	.15	.03	.35
BF_V									4.96	.45	.43	.44	.02	-.07	.07	.18	.14	-.11	-.17	.10
BF_S										7.05	.28	.40	.15	.05	.37	.34	.08	-.08	.36	.32
BF_P											2.85	.23	.23	-.01	.11	.09	.19	-.11	-.07	.05
BF_M												2.67	-.01	-.11	.21	.14	.31	.06	.19	.24
BM_V													2.82	.12	.37	.32	.10	.11	.11	.18
BM_S														1.49	.31	.10	-.25	.02	-.04	-.24
BM_P															2.03	.01	-.06	-.01	.29	.20
BM_M																1.83	.06	.02	.11	.21
AO_V																	2.45	-.05	-.08	.35
AO_S																		2.10	-.02	.20
AO_P																			1.67	.25
AO_M																				4.34

	F_V	F_S	F_P	F_M	M_V	M_S	M_P	M_M	O_V	O_S	O_P	O_M
F_V	5.05											
F_S	.46	12.03										
F_P	.48	.47	3.00									
F_M	.30	.15	.25	2.53								
M_V	.34	.16	.22	.08	4.27							
M_S	.08	.09	.08	.06	.46	11.34						
M_P	.17	.01	.16	.18	.48	.39	2.98					
M_M	.07	-.08	-.03	.22	.19	.21	.05	2.13				
O_V	.28	.17	.21	-.01	.21	.07	.11	.04	2.36			
O_S	.14	.29	.15	.11	.21	.25	.13	.06	.11	2.67		
O_P	.10	.09	.08	.07	.05	.13	.16	.03	.24	.36	2.89	
O_M	.22	.12	.15	.16	.10	.12	.13	.15	.14	.28	.24	2.92

TABLE 3.6. Univariate and Bivariate Heritability Estimates Obtained from Fitting the Full Model to SCA Data Measured at Ages 4, 7, and 12

	Verbal (V)	Spatial (S)	Perceptual Speed (P)	Memory (M)
Age 4[a]				
V	.116	.222	.129	.224
S		.313	.495	.318
P			.211	.654
M				.062
Age 7				
V	.242	.619	.269	.665
S		.349	.615	−.213
P			.236	.195
M				.194
Age 12[b]				
V	.261	.180	.352	.611
S		.347	.569	.586
P			.384	.685
M				.534

[a]Estimates calculated from Rice et al., 1989. [b]Estimates reported in Alarcón et al., 1998

The heritability estimates obtained from the fit of the full model to the data for verbal, spatial, perceptual speed, and memory abilities measured when the offspring were 7 years old are .24, .35, .24, and .19, respectively (table 3.6). As expected, these h^2 estimates at 7 years of age are somewhat higher than those reported at age 4 (Rice et al., 1986, 1989). The maternal (m) and paternal (f) transmission parameter estimates are small, as shown in table 3.3, suggesting that familial environmental influences do not have a large impact on the offsprings' specific cognitive abilities. Since the genotype-environment correlations are a function of the parental transmission parameters and these are low, estimates of s are also negligible. As shown in table 3.7, the phenotypic correlations (R_P) are moderate and range from .18 to .44; verbal, spatial, and perceptual speed are more highly correlated with each other than with memory. The genetic correlations (R_G) range from −.15 to .91 and are higher than the corresponding environmental correlations (R_E) which range from .08 to .36.

Estimates of bivariate heritability, the proportion of the phenotypic correlation due to heritable influences, range from −.21 (for spatial ability and memory) to .67 (for verbal ability with memory), as shown in table 3.6. The average bivariate h^2 is .36, suggesting that the phenotypic correlations among the four specific cognitive ability measures are due substantially to genetic effects.

TABLE 3.7. Phenotypic, Genetic, and Environmental Correlations Obtained from the Full Model Fitted to SCA Data Obtained at Ages 4, 7, and 12

	Verbal (V)	Spatial (S)	Perceptual Speed (P)	Memory (M)
Age 4[a]				
R_P				
V	1.00	.48	.48	.29
S		1.00	.45	.24
P			1.00	.18
M				1.00
R_G				
V	1.00	.49	.34	.62
S		1.00	.76	.51
P			1.00	.85
M				1.00
R_E				
V	1.00	.45	.43	.20
S		1.00	.27	.19
P			1.00	.06
M				1.00
Age 7				
R_P				
V	1.00	.40	.41	.22
S		1.00	.44	.22
P			1.00	.18
M				1.00
R_G				
V	1.00	.79	.49	.63
S		1.00	.91	−.15
P			1.00	.13
M				1.00
R_E				
V	1.00	.18	.37	.08
S		1.00	.23	.30
P			1.00	.15
M				1.00
Age 12[b]				
R_P				
V	1.00	.45	.42	.21
S		1.00	.50	.22
P			1.00	.21
M				1.00
R_G				
V	1.00	.27	.46	.34
S		1.00	.78	.30
P			1.00	.31
M				1.00
R_E				
V	1.00	.50	.35	.11
S		1.00	.37	.17
P			1.00	.12
M				1.00

[a]Estimates reported in Rice et al., 1989. [b]Estimates reported in Alarcón et al., 1998

The goodness-of-fit comparisons for the nested models are presented in table 3.4. Consistent with the results obtained when the offspring were 4 years old (Rice et al., 1986, 1989), familial environmental transmission and the genotype-environment correlations were not significantly different from zero (model 2; $\chi^2_{diff} = 43.35$, $df = 32$, $p = .087$). Thus, individual differences in specific cognitive abilities during both early and middle childhood are not substantially influenced by familial environmental transmission. The heritability estimates for specific cognitive abilities obtained in the full model range from .19 to .35, and these could be equated without a significant reduction of fit of the model to the data (model 3; $\chi^2_{diff} = 6.65$, $df = 3$, $p = .084$).

To test for differential genetic covariances among the four measures at 7 years of age, the six genetic correlations (R_G) were constrained to be equal (model 4). Since the reduced model did not worsen the fit of the model to the data ($\chi^2_{diff} = 7.99$, $df = 4$, $p = .157$), the results at age 7 agree with those obtained during early childhood—that is, specific cognitive abilities appear to be influenced by a pervasive genetic factor whose contribution to each ability does not differ substantially. In order to test the null hypothesis that the traits were genetically independent, the genetic correlations were fixed to zero in model 5. The results indicate a significant genetic covariation among the four specific cognitive abilities ($\chi^2_{diff} = 19.77$, $df = 1$, $p < .001$).

When the most parsimonious model (model 4) was fitted to the data, the resulting heritability estimate for each of the specific cognitive abilities is .27 and the genetic correlation between them is .57.

Age 12

As shown in table 3.6, the heritability estimates obtained from the fit of the full model to the data for verbal, spatial, perceptual speed, and memory are .26, .35, .38, and .53, respectively. These h^2 estimates at 12 years of age are substantially higher than those reported at age 4 (Rice et al., 1986, 1989) and somewhat higher than those estimated at age 7, suggesting that genetic influences may increase in late childhood. However, this apparent change in heritability could also be due to an increasing genetic stability of cognitive abilities between childhood and early adolescence (DeFries et al., 1987). If the genetic influences on these measures in adulthood are more highly correlated with those at age 12 than at age 4, putative heritability estimates obtained from parent-offspring data would appear to increase across age. Another possible explanation for the change in heritability is an increase in reliability of the measures as a function of age; however, the Cronbach alpha reliabilities of the four specific cognitive abilities are very similar on average at 4 and 12 years of age, namely, .67 and .70, respectively. In either case, these results support the hypothesis that genes that affect adult cognitive abilities are manifested increasingly between childhood and early adolescence (Plomin et al., 1997). In contrast, and consistent with the previous age 4 and age 7 results, the maternal (m) and paternal (f) transmission parameter estimates are small, indicating that familial environmental influences do not significantly affect specific cognitive abilities in children ($\chi^2_{diff} = 36.27$, $df = 32$, $p > .250$). As shown in table 3.7, the phenotypic correlations (R_P) range from .21 to .50 and, again, verbal, spatial, and perceptual speed are more highly correlated with each other than with memory. The genetic correlations (R_G) range from .27 to .78 and are consistently higher than the corresponding environmental correlations (R_E), which range from .11 to .50.

Bivariate heritabilities, presented in table 3.6, range from .18 (for verbal with spatial abilities) to .69 (for perceptual speed with memory). The average bivariate h^2 when the offspring were 12 years old (.50) is substantially higher than the corresponding average estimates when the children were 4 and 7 years of age, and indicate that about 50% of the phenotypic covariation among the four specific cognitive ability measures is due to genetic effects.

To test for differential genetic covariances among the four measures at 12 years of age, the six genetic correlations (R_G) were constrained to be equal. The results at age 12 agree with those obtained during early childhood: the genetic correlations are not significantly different from each other ($\chi^2_{diff} = 7.72$, $df = 4$, $p > .100$) but are significantly different from zero ($\chi^2_{diff} = 21.91$, $df = 1$, $p < .005$), suggesting that a general genetic factor may substantially influence specific cognitive abilities. The most parsimonious model was fitted to the age 12 data resulting in a heritability estimate for each of the specific cognitive abilities of .33 and a genetic correlation of .48.

Discussion

In this chapter we report the results of the first multivariate genetic analysis of CAP specific cognitive abilities data when the children were 7 years of age. In addition, these results were compared to those obtained from a prior analysis when the CAP children were 4 years of age and to those from a recent analysis of data collected when the children were 12 years of age (Alarcón et al., 1998).

The etiologies of individual differences for each of the four measures—verbal, spatial, perceptual speed, and memory—and their co-variation were assessed by fitting a parent-offspring multivariate conditional path model to CAP specific cognitive abilities data. Consistent with the results obtained from multivariate genetic analyses of the data when the offspring were 4 years old (Rice et al., 1989), familial environmental transmission did not significantly affect individual differences in specific cognitive abilities when the offspring were 7 or 12 years of age. Therefore, maternal and paternal cultural transmission effects do not account for substantial variability in specific cognitive abilities during either childhood or early adolescence.

The results obtained from the present study are consistent with the hypothesis that the heritabilities of specific cognitive abilities increase with age (Plomin, 1986). Assuming isomorphism between the child and adult measures, heritability estimates of specific cognitive abilities range from .06–.31, .19–.35, and .26–.53 when the offspring were 4 (Rice et al., 1986, 1989), 7, and 12 years of age, respectively.

The parameter estimates, more specifically the genetic correlations, obtained from the fit of the full model to the age 7 and age 12 data suggest that verbal, spatial, and perceptual speed abilities may have a common genetic factor while memory appeared to be influenced by a unique set of genes provide support for the hypothesis of a common genetic factor of general cognitive ability. Consistent with results that were obtained when the offspring were four years of age (Rice et al., 1989), the covariation between specific cognitive abilities during childhood and early adolescence may be influenced by a single general cognitive ability factor. Moreover, the bivariate h^2 estimates increased from age 7 to age 12 and suggest that by early adolescence on average about half of the moderate phenotypic correlations among specific cognitive abilities are due to genetic influences. These results at various stages of development are consistent with the results of other multivariate genetic analyses that support the hypothesis of cognitive molarity in which processes are general and overlapping instead of modular or independent.

If the covariation (molarity) of specific cognitive abilities is due to genetic influences, this implies that environmental factors may be responsible for their independence (modularity). However, our results suggest that these environmental factors may not be caused by cultural transmission from parent to offspring. Thus, environmental factors related to parents' cognitive abilities may have no long-term effect on the development of their children's cognitive abilities. However, other aspects of parenting might affect cognitive development, such as attitudes about education and achievement, parental warmth and support, schools or peers. If so, it would be of considerable interest to assess the extent to which such environmental correlates account for variation and covariation in specificity of cognitive abilities.

REFERENCES

Alarcón, M., Plomin, R., Fulker, D. W., Corley, R., & DeFries, J. C. (1998). Multivariate path analysis of specific cognitive abilities data at 12 years of age in the Colorado Adoption Project. *Behavior Genetics, 28*, 255–264.

Bollen, K. A. (1989). *Structural equations with latent variables*. New York: John Wiley.

Cardon, L. R. (1994). Specific cognitive abilities. In J. C. DeFries, R. Plomin, & D. W. Fulker (Eds.), *Nature and nurture during middle childhood* (pp. 57–76). Oxford: Blackwell.

Cardon, L. R., Fulker, D. W., DeFries, J. C., & Plomin, R. (1992). Multivariate genetic analysis of specific cognitive abilities in the Colorado Adoption Project at age 7. *Intelligence, 16*, 383–400.

Casto, S. D., DeFries, J. C., & Fulker, D. W. (1995). Multivariate genetic analysis of Wechsler intelligence scale for children-revised (WISC-R) factors. *Behavior Genetics, 25*, 25–32.

DeFries, J. C., Johnson, R. C., Kuse, A. R., McClearn, G. E., Polovina, J., Vandenberg, S. G., & Wilson, J. R. (1979). Familial resemblance for specific cognitive abilities. *Behavior Genetics, 1*, 23–48.

DeFries, J. C., Plomin, R., & Fulker, D. W. (1994). *Nature and nurture during middle childhood*. Oxford, UK: Blackwell.

DeFries, J. C., Plomin, R., & LaBuda, M. C. (1987). Genetic stability of cognitive development from childhood to adulthood. *Developmental Psychology, 23*, 4–12.

DeFries, J. C., Plomin, R., Vandenberg, S. G., & Kuse, A. R. (1981). Parent-offspring resemblance for cognitive abilities in the Colorado Adoption Project: Biological, adoptive, and control parents and one-year-old children. *Intelligence, 5*, 245–277.

DeFries, J. C., Vandenberg, S. G., & McClearn, G. E. (1976). Genetics of specific cognitive abilities. *Annual Review of Genetics, 10*, 179–207.

Hayduk, L. A. (1987). *Structural equation modeling with LISREL*. Baltimore: The Johns Hopkins University Press.

Luo, D., Petrill, S. A., & Thompson, L. A. (1994). An exploration of genetic g: Hierarchical factor analysis of cognitive data from the Western Reserve Twin Project. *Intelligence, 18*, 335–347.

McClearn, G. E., Johansson, B., Berg, S., Pedersen, N. L., Ahern, F., Petrill, S. A., & Plomin, R. (1997). Substantial genetic influence on cognitive abilities in twins 80 or more years old. *Science, 276*, 1560–1563.

McGue, M., & Bouchard, T. J. (1989). Genetic and environmental determinants of information processing and special mental abilities: A twin analysis. In R. J. Sternberg (Ed.), *Advances in the psychology of human intelligence* (pp. 7–45). Hillsdale, NJ: Lawrence Erlbaum Associates.

Neale, M. C. (1999). *Mx: Statistical modeling* (5th ed.). Richmond: Department of Psychiatry, Medical College of Virginia.

Nichols, R. C. (1978). Twin studies of ability, personality, and interests. *Homo, 29*, 158–173.

Pedersen, N. L., Plomin, R., & McClearn, G. E. (1994). Is there G beyond g? (Is there genetic influence on specific cognitive abilities independent of genetic influence on general cognitive ability?). *Intelligence, 18*, 133–143.

Pedersen, N. L., Plomin, R., Nesselroade, J. R., & McClearn, G. E. (1992). A quantitative genetic analysis of cognitive abilities during the second half of the life span. *Psychological Science, 3*, 346–353.

Phillips, K. (1989). Delta path models for modeling the effects of multiple selective associations in adoption designs. *Behavior Genetics, 3*, 609–620.

Plomin, R. (1986). *Genes, development, and psychology.* Hillsdale, NJ: Lawrence Erlbaum Associates.

Plomin, R. (1988). The nature and nurture of cognitive abilities. In R. J. Sternberg (Ed.), *Advances in the psychology of human intelligence* (Vol. 4, pp.1–33). Hillsdale, NJ: Lawrence Erlbaum Associates.

Plomin, R., & DeFries, J. C. (1979). Multivariate behavioral genetic analysis of twin data on scholastic abilities. *Behavior Genetics, 9,* 505–517.

Plomin, R., & DeFries, J. C. (1985). *Origins of individual differences in infancy: The Colorado Adoption Project.* Orlando, FL: Academic Press.

Plomin, R., DeFries, J. C., & Fulker, D. W. (1988). *Nature and nurture in infancy and early childhood.* Cambridge: Cambridge University Press.

Plomin, R., DeFries, J. C., McClearn, G. E., & Rutter, M. (1997). *Behavioral genetics* (3rd ed.). New York: W. H. Freedman.

Plomin, R., Fulker, D. W., Corley, R., & DeFries, J. C. (1997). Nature, nurture and cognitive development from 1 to 16 years: A parent-offspring adoption study. *Psychological Science, 8,* 442–447.

Rice, T., Carey, G., Fulker, D. W., & DeFries, J. C. (1989). Multivariate path analysis of specific cognitive abilities in the Colorado Adoption Project: Conditional path model of assortative mating. *Behavior Genetics, 19,* 195–207.

Rice, T., Fulker, D. W., & DeFries, J. C. (1986). Multivariate path analysis of specific cognitive abilities in the Colorado Adoption Project. *Behavior Genetics, 16,* 107–125.

Sewell, T. E., & Severson, R. A. (1974). Learning ability and intelligence as cognitive predictors of achievement in first-grade black children. *Journal of Educational Psychology, 66,* 948–955.

SPSS (1988). *SPSS-X user's guide* (3rd ed.). Chicago, IL: SPSS Inc.

Tambs, K., Sundet, J. M., & Magnus, P. (1986). Genetic and environmental contribution to the covariation between the Wechsler adult intelligence scale (WAIS) subtests: A study of twins. *Behavior Genetics, 16,* 475–491.

Thapar, A., Petrill, S. A., & Thompson, L. A. (1994). The heritability of memory in the Western Reserve Twin Project. *Behavior Genetics, 24,* 155–160.

Thompson, L. A., Detterman, D. K., & Plomin, R. (1991). Associations between cognitive abilities and scholastic achievement: Genetic overlap but environmental differences. *Psychological Science, 2,* 158–165.

Van Eerdewegh, P. (1982). *Statistical selection in multivariate systems with applications in quantitative genetics.* Unpublished doctoral dissertation, Washington University, St. Louis, Missouri.

Vogler, G. P. (1985). Multivariate path analysis of familial resemblance. *Genetic Epidemiology, 2,* 35–53.

SALLY J. WADSWORTH
JOHN C. DEFRIES

4

Etiology of the Stability of Reading Performance from 7 to 12 Years of Age and Its Possible Mediation by IQ

Introduction

Results obtained from longitudinal studies (e.g., Byrne, Freebody, & Gates, 1992; LaBuda & DeFries, 1989; Lyytinen, 1997; Maughan, Hagell, Rutter, & Yule, 1994) suggest that early measures of reading ability may be strong predictors of later reading achievement; however, the etiology of the stability of individual differences in reading performance has only recently begun to be addressed (Wadsworth, Fulker, & DeFries, 1999).

In an analysis of longitudinal data from the Colorado Family Reading Study, DeFries and Baker (1983) assessed the stability of reading performance of 51 normally-achieving readers and 51 reading-disabled probands who were 9.5 and 14.9 years of age, on average, at initial and follow-up testing, respectively. Stability correlation coefficients for a composite measure of reading performance (reading recognition, reading comprehension, and spelling) were .58 for the sample of normally-achieving readers and .30 for the reading-disabled probands. Subsequently, Shaywitz, Escobar, Shaywitz, Fletcher, and Makuch (1992) analyzed data from 414 children who were administered tests of general cognitive ability during grades 1, 3, and 5, and tests of reading and mathematics achievement during grades 1 through 6. The correlation between ability-achievement discrepancy scores at grades 1 and 3 was .53, whereas that between scores at grades 3 and 5 was .67. More recently, as a part of the Dunedin Multidisciplinary Health and Development Study, Williams and McGee (1996) reported stability correlations for word recognition in 349 boys and 319 girls tested at ages 7, 9, and 15. Stability coefficients between ages 7 and 9 were .86 for girls and .88 for boys. Corresponding correlations were .76 and .72 between ages 9 and 15, and .67 for both girls and boys between ages 7 and 15. Finally, Wagner et al. (1997) investigated the stability of phonological processing abilities and word-level reading in 244 children tested yearly from kindergarten through fourth grade. Individual differences in both phonological processing (phonological awareness, phonological memory, and serial naming) and word-level reading (word identification and word analysis) were remarkably stable, with stability coefficients for the one-year interval ranging from .69 for word-level reading to 1.00 for phonological memory, and for the four-year interval ranging from .27 to .77 for word-level reading and phonological memory, respectively. Thus, results of these longitudinal studies suggest that individual differences on tests of reading and reading component processes are stable over time.

Although the etiology of stability in reading performance has only recently been explored (Wadsworth et al., 1999), a number of studies have examined genetic and environmental influences on individual differences in reading performance at one or more ages. Results of both twin and adoption studies have provided consistent evidence for genetic influence on reading performance of individuals from age 7 to adulthood, with heritability estimates ranging from .18 to .53 depending on the age of the subjects, type of sample, and measure used (e.g., Alarcón & DeFries, 1995; Cardon, DiLalla, Plomin, DeFries, & Fulker, 1990; Stevenson, Graham, Fredman, & McLoughlin, 1987; Wadsworth, DeFries, Fulker, & Plomin, 1995).

The evidence that reading performance is both relatively stable and heritable suggests that genetic influences on reading-related measures may be continuous throughout development. That is, genetic influences that are important for reading performance at one age may also be important at later ages. As a first step toward assessing the etiology of stability in reading performance, DeFries and Baker (1983) found that over 40% of the stability correlation for normally-achieving readers was related to parental influences. However, because this was a family study, rather than a twin or adoption study, they could not distinguish between genetic and shared-family environmental contributions to stability.

In a more recent study, genetic and environmental influences on stability of individual differences in a measure of reading recognition between 7 and 12 years of age were assessed using data from the Colorado Adoption Project (CAP; Wadsworth et al., 1999). Analysis of data from 203 related and unrelated sibling pairs at age 7 and 148 pairs at age 12 indicated that about 70% of the stability in reading performance between the two ages could be attributed to common genetic influences.

In addition to common genetic influences on reading ability at different ages, other factors may mediate the stability of individual differences in reading performance over time. For example, measures of reading performance are correlated with measures of general cognitive ability both phenotypically and genetically (e.g., Cardon et al., 1990; Thompson, Detterman, & Plomin, 1991; Wadsworth et al., 1995). Further, results of studies focusing on the etiology of stability of general and specific cognitive abilities provide evidence for substantial and significant genetic stability from age to age (Cardon, Fulker, DeFries, & Plomin, 1992; Cardon & Fulker, 1994; Cherny, Fulker, & Hewitt, 1997; Fulker, Cherny, & Cardon, 1993). It is reasonable to hypothesize, therefore, that the stability of individual differences in measures of reading performance may be due, at least in part, to the stability of individual differences in cognitive ability.

The aim of the current study, therefore, was to assess the etiology of stability of individual differences in reading performance from age 7 to age 12, and its possible mediation by IQ, in a sample of related and unrelated sibling pairs tested in the CAP. Because genetic factors have been found to contribute importantly to stability of individual differences in reading measures between 7 and 12 years of age, and because measures of reading performance are correlated with IQ, both phenotypically and genetically, we hypothesized that the stability of reading performance may be mediated both phenotypically and genetically, at least in part, by IQ.

Methods

Subjects and Measures

The present study included IQ and reading performance data from 92 pairs of adopted children and their unrelated siblings (20 same-sex and 72 opposite-sex pairs), and from 103 pairs of nonadopted control children and their related siblings (66 same-sex and 37 opposite-sex pairs), tested in the CAP during the summer following first grade (average age of 7.4 years), as well as data from 92 pairs of unrelated siblings (22 same-sex and 70 opposite-sex pairs) and 98 pairs of related siblings (60 same-sex and 38 opposite-sex pairs) tested during the summer following sixth grade (average age of 12.5 years). These represent the numbers of pairs in which there is either IQ or reading performance data for both members of a sibling pair. Full Scale IQ was measured at age 7 using the Wechsler Intelligence Scale for Children-Revised (Wechsler, 1974), and reading performance was measured at ages 7 and 12 using the

TABLE 4.1. Sample Sizes by Age, Sex, and Relationship

	Year 7		Year 12	
Relationship	Male	Female	Male	Female
Adopted probands	108	91	109	89
Unrelated siblings	45	52	38	56
Control probands	116	99	121	101
Related siblings	60	45	54	44

Reading Recognition subtest of the Peabody Individual Achievement Test (PIAT; Dunn & Markwardt, 1970). Sample sizes by age, sex, and relationship (adoptive proband, unrelated sibling, control proband, related sibling) are given in table 4.1. Because data are not available at both time points for all subjects, and because data from all probands and siblings are included in the stability analyses, the numbers of all individuals for whom either IQ or reading performance data are available at age 7 and for whom reading performance data are available at age 12 are provided.

Analyses

Because of the variability in patterns of missing data in the CAP, a maximum likelihood pedigree approach using the MX statistical modeling package (Neale, 1999) was employed in order to make use of all available data, thereby increasing both power to detect effects and precision of parameter estimates. The pedigree approach involves the calculation of a log-likelihood for each family, and the summation of these across all pedigrees. For model comparisons, twice the difference between the log-likelihoods for the two models is distributed asymptotically as a chi-square, with degrees of freedom equal to the difference in the number of free parameters estimated in fitting each model.

For the purposes of the current study, the relationships among Full Scale IQ at age 7 (Y7FSIQ) and reading performance at ages 7 (Y7READ) and 12 (Y12READ) were explored using a Cholesky decomposition. By fitting this model to the data, the extent to which the stability of reading performance from age 7 to age 12 is due to influences shared with IQ can be examined (figure 4.1). For example, the observed correlation (r_P) between Y7READ and Y12READ is due to influences that are either correlated with or independent of IQ as follows: ($\lambda_{21} \times \lambda_{31}$) + ($\lambda_{22} \times \lambda_{32}$). Thus, the proportion of the observed stability correlation due to influences shared with Y7FSIQ is ($\lambda_{21} \times \lambda_{31}$)/$r_P$. The Cholesky can be partitioned further to include estimates of genetic, shared environmental, and nonshared environmental contributions to

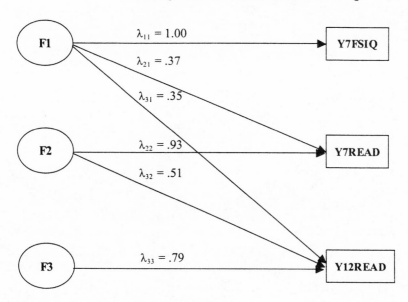

FIGURE 4.1. Results of phenotypic Cholesky.

the variance in each of the measures, as well as to the correlations among the measures at the two ages (figure 4.2).

The adoptive sibling design is based on the comparison of the correlations of adoptive (unrelated) siblings to those of nonadoptive (related) siblings (e.g., Plomin, DeFries & Fulker, 1988). Whereas adoptive siblings are genetically unrelated and share only family environmental influences, related siblings share half of their segregating genes, on average. Therefore, in the absence of genetic nonadditivity, the phenotypic correlation between related siblings is a function of one-half the heritability of the trait, plus shared environmental influences, which are assumed to be no more highly correlated for related siblings than for unrelated siblings. In contrast, the phenotypic correlation between genetically unrelated adoptive siblings results only from shared environmental influences (in the absence of selective placement). Therefore, by analyzing these correlations, the contributions of genetic, shared environmental, and nonshared environmental influences can be estimated.

Figure 4.2 depicts the genetic/environmental model, illustrating the genetic (A1, A2, A3), shared environmental (C1, C2, C3), and nonshared environmental (E1, E2, E3) factor structures underlying Y7FSIQ, Y7READ, and Y12READ for one sibling only. Using this model, the etiology of the correlation among measures of IQ at age 7 and reading performance at ages 7 and 12 can be assessed—that is, the

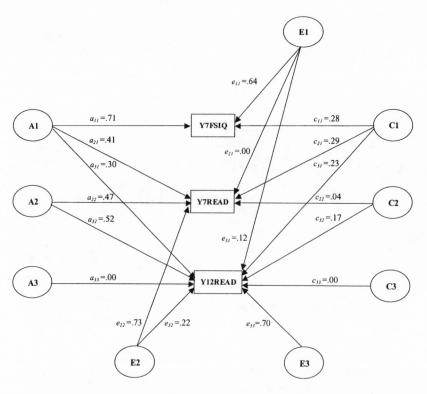

FIGURE 4.2. Results of genetic/environmental Cholesky: Full model.

extent to which the stability of individual differences in reading performance at the two ages is due to common genetic and environmental influences independent of IQ, and the extent to which it is due to
genetic and environmental influences shared with IQ. In addition,
estimates of heritability, shared and nonshared environmentality, and
genetic and environmental correlations among the measures are computed with relative ease.

Proportions of variance in reading performance at each age due to
genetic and environmental influences are calculated from the standardized path coefficients as the sum of the squared paths from common
and specific factors to each measure. For example, from figure 4.2, the
heritability (h^2) of Y7FSIQ is simply the square of the path from A_1 to
Y7FSIQ (i.e., a_{11}^2), whereas the heritability of Y7READ is the square of
the path from A_1 to Y7READ, plus the square of the path from A_2 to
Y7READ (i.e., $a_{21}^2 + a_{22}^2$). Estimates of shared (c^2) and nonshared (e^2)
environmental influences are obtained in a corresponding manner.

Estimates of the genetic, shared environmental, and nonshared environmental correlations between the measures (r_A, r_C, and r_E, respec-

TABLE 4.2. Within-Person and Sibling Correlations Among Measures of IQ at Age 7 and Reading Achievement at Ages 7 and 12[a]

	Proband			Sibling		
	Y7FSIQ	Y7READ	Y12READ	Y7FSIQ	Y7READ	Y12READ
Proband						
Y7FSIQ	**1.00**	.40	.35	.34	.18	.18
Y7READ	.38	**1.00**	.57	.27	.27	.27
Y12READ	.37	.65	**1.00**	.05	.20	.26
Sibling						
Y7FSIQ	.05	.08	−.01	**1.00**	.30	.20
Y7READ	.13	.08	.02	.40	**1.00**	.58
Y12READ	.15	.12	.09	.44	.67	**1.00**

[a]With adoptive correlations below the diagonal and control correlations above

tively) are also obtained from the standardized path coefficients. For example, $(a_{21} \times a_{31}) + (a_{22} \times a_{32}) = h_{Y7READ} \times r_A \times h_{Y12READ}$, the phenotypically standardized genetic correlation (i.e., that part of the observed correlation due to common genetic influences), where h is the square root of the heritability of the indicated measure and r_A is the genetic correlation (i.e., the extent to which common genes are influencing reading performance at the two time points). Thus, the genetic correlation may be estimated from the ratio of the phenotypically standardized genetic correlation to the product of the square roots of the two heritabilities (i.e., $r_A = h_{Y7READ} \times r_A \times h_{Y12READ}/h_{Y7READ} \times h_{Y12READ}$). Further, the ratio of the phenotypically standardized genetic correlation to the phenotypic correlation estimates bivariate heritability, the proportion of the observed covariation due to shared genetic influences—that is, $[(a_{21} \times a_{31}) + (a_{22} \times a_{32})]/r_P$.

Results

Within-person and sibling correlations are presented in table 4.2. Within-person correlations are given separately for adoptive and control probands and siblings. Full-Scale IQ at age 7 correlates moderately with reading performance at both 7 and 12 years of age ($r = .20$ to .44, depending on individual type), whereas the stability correlation for reading performance between the two ages ranges from .57 to .67.

Results of fitting the phenotypic Cholesky to the data are presented in figure 4.1. From the parameter estimates, it can be seen that 22% of the stability between reading performance at the two ages is due to influences shared with IQ, that is $(.37 \times .35)/.60$. Although the parameters representing that portion of the stability correlation ($\lambda_{21}, \lambda_{31}$) could not be omitted from the model without significant reduction in

TABLE 4.3. Genetic and Environmental
Contributions to the Variance in Each Measure

Measure	h^2	c^2	e^2
Y7FSIQ	.51	.08	.41
Y7READ	.39	.09	.53
Y12READ	.36	.08	.55

model fit ($\Delta\chi^2$ = 106.13, df = 2), most of the stability between reading performance at ages 7 and 12 in this study is independent of IQ.

As can be seen from table 4.2, scores on all of the measures, as well as across measures, are more highly correlated for related siblings than for unrelated sibling pairs, suggesting genetic influence on individual differences in each of the measures, as well as on the observed covariation among the measures. Results of fitting the genetic/environmental Cholesky to data from sibling pairs are presented in figure 4.2. The parameter estimates obtained from this analysis confirm substantial contribution of genetic influences to the variance in Y7FSIQ and reading performance at both ages, as well as to the covariance among the measures. Interestingly, there is no indication of any new genetic influence at age 12, as evidenced by the path coefficient of .00 from A_3 to Y12READ (a_{33}). Therefore, the genetic influences manifested at age 12 appear to be the same as those at age 7 in this sample.

In contrast, shared environmental influences are relatively small. However, similar to the genetic results, the shared environmental influences on reading performance at age 12 appear to be the same as those present at age 7. In fact, shared environmental influences among all the measures appear to be explained by a single factor. Further, the only influences contributing to the variance in reading performance at age 12, independent of those at age 7, are environmental influences not shared between siblings (i.e., e_{33}).

Estimates of genetic and environmental contributions to the variance in each of the measures are presented in table 4.3. Genetic influences contribute importantly to individual differences in each measure, with heritability estimates of about 50% for IQ and nearly 40% for reading. However, nonshared environmental influences are also important for all the measures, ranging from .41 to .55. In contrast, shared environmental contributions to individual differences in the measures are negligible.

Estimates of the genetic and environmental correlations are presented in table 4.4. Genetic correlations between Y7FSIQ and the reading measures are substantial. As expected, the genetic correlation between Y7FSIQ and Y7READ is somewhat larger than that between Y7FSIQ and Y12READ (.66 and .49, respectively). In contrast, the

TABLE 4.4. Genetic and Environmental
Correlations Among Measures of IQ at age 7
and Reading Achievement at Ages 7 and 12[a]

	Genetic		
	Y7FSIQ	Y7READ	Y12READ
Y7FSIQ	**1.00**	.78	.60
Y7READ	.66	**1.00**	.61
Y12READ	.49	.98	**1.00**

	Shared Environmental		
	Y7FSIQ	Y7READ	Y12READ
Y7FSIQ	**1.00**	.23	.18
Y7READ	.95	**1.00**	.12
Y12READ	.81	.89	**1.00**

	Nonshared Environmental		
	Y7FSIQ	Y7READ	Y12READ
Y7FSIQ	**1.00**	.00	.22
Y7READ	.00	**1.00**	.27
Y12READ	.16	.30	**1.00**

[a]With bivariate heritabilities above the diagonal

genetic correlation between Y7READ and Y12READ is near unity, suggesting that the same genetic influences are involved in reading performance at the two ages. As can be seen from the estimates of bivariate heritability, shared genetic influences account for 61% of the phenotypic stability of reading between the two ages, with nonshared environmental influences accounting for over 25%. Although nonshared environmental influences contribute modestly to stability, these influences are primarily responsible for change, as indicated in figure 4.2. It is also interesting to note that 78% of the observed correlation between Y7FSIQ and Y7READ, and 60% of that between Y7FSIQ and Y12READ, are due to shared genetic influences.

As may be seen from figure 4.2, the genetic stability between reading performance at ages 7 and 12 can also be partitioned into two parts, that portion which is correlated with IQ ($.41 \times .30$) and that which is independent of IQ ($.47 \times .52$). Thus, although influences shared with Y7FSIQ accounted for only 22% of the phenotypic stability of individual differences in reading performance from age 7 to age 12, they account for one-third of the genetic stability—that is, $(.41 \times .30)/[(.41 \times .30) + (.47 \times .52)] = .33$.

TABLE 4.5. Model Comparisons for the Genetic/Environmental Cholesky[a]

Model	LL	NPAR	χ^2	df	p
1. Full	−2408.62	30			
2. No A or C specifics (drop A_3 & C_3)	−2408.62	28	0.00	2	≥.99
3. No A or C specifics (drop A_3 & C_3) No E common factor (drop e_{21}, e_{31}, e_{32})	−2409.37	25	1.50	5	≥.90
4. No C (drop C_1, C_2, C_3) No A specific (drop A_3) No E common factor (drop e_{21}, e_{31}, e_{32})	−2410.49	20	3.74	10	≥.95
5. No A (drop A_1, A_2, A_3) No C specific (drop C_3) No E common factor (drop e_{21}, e_{31}, e_{32})	−2488.53	20	159.82	10	.00
6. 1 Genetic Common Factor (drop A_2 & A_3) plus E specifics (drop C_1, C_2, C_3, e_{21}, e_{31}, e_{32})	−2413.77	18	10.30	12	≥.50

[a]Chi-square and degrees of freedom for all submodels based on twice the difference between the log-likelihood for the submodel and that for the full model (model 1)

Note: LL = Log-likelihood, NPAR = number of parameters estimated

Results obtained from model comparisons are presented in table 4.5. Because estimates of both genetic and shared environmental specifics were zero (figure 4.2), they were excluded from the model with no deterioration of model fit (model 2). The nonshared environmental common factor was then dropped from the model (model 3), and all shared environmental influences could also be omitted without loss of fit (model 4). However, omission of genetic influences resulted in a highly significant deterioration of fit (model 5), but genetic influences could be restricted to a single factor (model 6). Therefore, a model that includes one genetic common factor, no shared environmental influences and nonshared environmental specifics was the most parsimonious model to fit the data.

Discussion

The purpose of the present study was to use the methods of longitudinal behavioral genetic analysis to assess the etiology of the observed stability of reading performance of children tested at ages 7 and 12, and the extent to which stability is related to IQ at age 7. Results obtained from an analysis of data from related and unrelated sibling pairs confirm earlier findings of substantial longitudinal stability. Further, that stability is mediated, at least in part, by IQ. However, the largest portion of the phenotypic stability of reading performance is independent of general cognitive ability.

Genetic analyses confirmed genetic influence on individual differences in IQ at age 7 (h^2 = .51) and in reading performance at each age (h^2 = .39 at age 7 and .36 at age 12). Environmental influences were primarily those not shared by siblings.

As hypothesized, the observed stability between reading performance at ages 7 and 12 in this sample (.60) was found to be due largely to genetic influences (60%). Further, as suggested by a genetic correlation near unity between Y7READ and Y12READ, the same genetic influences manifested at age 7 appear to be operating at age 12, with no new heritable variation occurring at age 12. Similarly, although shared environmental influences contributed little to either the variance or covariance of the measures, those shared environmental influences present at age 12 appear to be the same as those manifested at age 7. In contrast, nonshared environment did not contribute significantly to the stability of reading performance between the two ages, but provided substantial and significant new variation at age 12. Thus, genetic influences were largely responsible for the observed stability of reading performance in this study, whereas nonshared environmental influences were primarily responsible for change.

Are these genetic influences on the stability of individual differences in reading performance mediated by IQ? Results of these analyses suggest that approximately one-third of the genetic influences shared between reading performance at ages 7 and 12 were due to genetic influences also shared with IQ at age 7. In contrast, the nonshared environmental contribution to stability was due to influences independent of IQ.

Although this and other studies of reading performance suggest significant age-to-age stability, the correlation of .60 between reading performance at age 7 and that at age 12 accounts for less than 40% of the variance at age 12. Further, results of the current study suggest that the independent variation in reading performance at age 12 (as well as over half the total variance) is due to nonshared environmental influences. Such influences may be due in part to individual experiences that make children growing up in the same family different (e.g., instructional methods, teachers, peers, etc.), emphasizing the possible importance of such factors for the development of reading ability from middle childhood to adolescence. Thus, our finding of genetic influences on variation in reading performance, as well as on the stability of reading performance, does not diminish the importance of environmental influences.

In conclusion, results of the current study provide evidence for substantial and significant genetic stability of reading performance between 7 and 12 years of age, with modest influence of IQ at age 7 on this stability. However, the number of same-sex adoptive sibling pairs is relatively small, and we examined the stability of individual differences in reading performance at only two time points. Because the CAP is ongoing, sibling pairs are still being tested at age 12 as well as at later ages. Thus, as more data become available, future CAP analyses will have greater power to test hypotheses about stability and change from middle childhood to early adolescence, and will extend this investigation into late adolescence.

ACKNOWLEDGMENTS

The Colorado Adoption Project (CAP) is supported by grants HD-10333 and HD-18426 from the National Institute of Child Health and Human Development (NICHD) and grant MH-43899 from the National Institute of Mental Health (NIMH). The continued cooperation of the many families participating in the CAP is gratefully acknowledged. The authors also wish to thank Robert Plomin for his helpful comments and suggestions during the preparation of this manuscript.

REFERENCES

Alarcón, M., & DeFries, J. C. (1995). Quantitative trait locus for reading disability: An alternative test (abstract). *Behavior Genetics, 25,* 253.

Byrne, B., Freebody, P., & Gates, A. (1992). Longitudinal data on the relations of word-reading strategies to comprehension, reading time, and phonemic awareness. *Reading Research Quarterly, 27,* 141–151.

Cardon, L. R., DiLalla, L. F., Plomin, R., DeFries, J. C., & Fulker, D. W. (1990). Genetic correlations between reading performance and IQ in the Colorado Adoption Project. *Intelligence, 14,* 245–257.

Cardon, L. R., & Fulker, D. W. (1994). A model of developmental change in hierarchical phenotypes with application to specific cognitive abilities. *Behavior Genetics, 24,* 1–16.

Cardon, L. R., Fulker, D. W., DeFries, J. C., & Plomin, R. (1992). Continuity and change in general cognitive ability from 1 to 7 years of age. *Developmental Psychology, 28,* 64–73.

Cherny, S. S., Fulker, D. W., & Hewitt, J. K. (1997). Cognitive development from infancy to middle childhood. In R. J. Sternberg & E. L. Grigorenko (Eds.), *Intelligence, heredity, and environment.* Cambridge: Cambridge University Press.

DeFries, J. C., & Baker, L. A. (1983). Parental contributions to longitudinal stability of cognitive measures in the Colorado Family Reading Study. *Child Development, 54,* 388–395.

Dunn, L. M., & Markwardt, F. C. (1970). *Examiner's manual: Peabody Individual Achievement Test.* Circle Pines, MN: American Guidance Service.

Fulker, D. W., Cherny, S. S., & Cardon, L. R. (1993). Continuity and change in cognitive development. In R. Plomin & G. E. McClearn (Eds.), *Nature, nurture, and psychology* (pp. 77–97). Washington, DC: American Psychological Association.

LaBuda, M. C. & DeFries, J. C. (1989). Differential prognosis of reading-disabled children as a function of gender, socioeconomic status, IQ, and severity: A longitudinal study. *Reading and Writing: An Interdisciplinary Journal, 1*, 25–36.

Lyytinen, H. (1997). In search of the precursors of dyslexia: A prospective study of children at risk for reading problems. In C. Hulme & M. Snowling (Eds.), *Dyslexia: Biology, cognition and intervention*. London: Whurr Publishers, Ltd.

Maughan, B., Hagell, A., Rutter, M., & Yule, W. (1994). Poor readers in secondary school. *Reading and Writing: An Interdisciplinary Journal, 6*, 125–150.

Neale, M. C. (1999). *Mx: Statistical modeling* (5th ed.). Box 126 MCV, Richmond, VA 23298: Department of Psychiatry.

Plomin, R., DeFries, J. C., & Fulker, D. W. (1988). *Nature and nurture in infancy and early childhood*. Cambridge: Cambridge University Press.

Shaywitz, S. E., Escobar, M. D., Shaywitz, B. A., Fletcher, J. M., & Makuch, R. (1992). Evidence that dyslexia may represent the lower tail of a normal distribution of reading ability. *New England Journal of Medicine, 326*, 144–150.

Stevenson, J., Graham, P., Fredman, G., & McLoughlin, V. (1987). A twin study of genetic influences on reading and spelling ability and disability. *Journal of Child Psychology and Psychiatry, 28*, 229–247.

Thompson, L. A., Detterman, D. K., & Plomin, R. (1991). Association between cognitive abilities and scholastic achievement: Genetic overlap, but environmental differences. *Psychological Science, 2*, 158–165.

Wadsworth, S. J., DeFries, J. C., Fulker, D. W., & Plomin, R. (1995). Cognitive ability and academic achievement in the Colorado Adoption Project: A multivariate analysis of parent-offspring and sibling data. *Behavior Genetics, 25*, 1–15.

Wadsworth, S. J., Fulker, D. W., & DeFries, J. C. (1999). Stability of genetic and environmental influences on reading performance at 7 and 12 years of age in the Colorado Adoption Project. *International Journal of Behavioral Development, 23*, 319–332.

Wagner, R. K., Torgesen, J. K., Rashotte, C. A., Hecht, S. A., Barker, T. A., Burgess, S. R., Donahue, J., & Garon, T. (1997). Changing relations between phonological processing abilities and word-level reading as children develop from beginning to skilled readers: A 5-year longitudinal study. *Developmental Psychology, 33*, 468–479.

Wechsler, D. (1974). *Examiner's manual: Wechsler Intelligence Scale for Children-Revised*. New York: The Psychological Corporation.

Williams, S., & McGee, R. (1996). Reading in childhood and mental health in early adulthood. In J. H. Beitchman, N. J. Cohen, M. M. Konstantareas, & R. Tannock (Eds.), *Language, learning, and behavior disorders: Developmental, biological, and clinical perspectives*. Cambridge: Cambridge University Press.

STEPHEN A. PETRILL
SALLY-ANN RHEA

5

Memory Ability during Middle Childhood and Early Adolescence in the Colorado Adoption Project

Introduction

Although human memory has typically been examined using experiment-based mean-differences approaches, many behavioral genetic studies have also begun to examine the etiology of individual differences in memory performance. One approach has been to examine memory compared to other specific cognitive abilities such as verbal ability, spatial ability, and perceptual speed (see DeFries, Plomin, Vandenberg, & Kuse, 1981). In this context, the major questions are whether memory is more or less heritable than other cognitive abilities and whether the genetic variance in memory is related to the genetic variance in other cognitive tasks. Researchers have suggested that the heritability of memory is significant, but lower than other cognitive abilities. Additionally, the genetic variance associated with memory is largely correlated with other nonmemory based measures of cognitive ability (see Plomin, DeFries, McClearn, & Rutter, 1997, for a review).

Recently, behavioral genetic researchers have extended these findings to examine memory ability in adult and elderly populations. Finkel, Pedersen, and McGue (1995) examined digit span, Thurstone's memory, and figure memory in a sample of Minnesota and Swedish adult twins. Finkel et al. (1995) suggested that the heritability of memory ranges from .53 in text recall to .62 in digit span, and that the heritability of memory in older populations was similar to the heritability found in adult populations. Other studies of elderly populations suggest that genetic influences relating to memory are highly correlated with the genetic variance of other cognitive abilities (Finkel et al., 1995; Pedersen, Plomin, & McClearn, 1994; Petrill et al., 1998).

Some researchers have criticized these findings, suggesting that the heritability of memory is dependent upon the type of memory measure employed. In the phenotypic literature, memory ability was once conceptualized as a unitary construct. However, this view began to change in the 1950s as the information-processing approach began to uncover a distinction between short-term and long-term memory. Since then, cognitive psychologists and neuropsychologists have uncovered several dissociations, such as recall vs. recognition memory, implicit vs. explicit memory, declarative vs. procedural memory, and episodic vs. semantic memory (see Squire, 1992, for a review). Given these important distinctions in memory, Thapar, Petrill, and Thompson (1994) examined whether the heritability of memory varies as a function of the type of memory. Thapar et al. (1994) noted that the heritability of memory appeared to be higher in previous behavioral genetic studies that employed measures of associative or meaningful memory (e.g., Bouchard, Segal, & Lykken, 1990; Finkel, McGue, & Fox, 1991; Partanen, Brunn, & Markkanen, 1996; Schoenfeldt, 1968), as opposed to the majority of behavioral genetic studies that employ measures of picture memory. Thapar et al. then replicated these findings in a sample of 6- to 13-year-old twins that simultaneously measured digit span, picture memory, and associative/ meaningful memory.

The purpose of this chapter is to explore memory ability in the Colorado Adoption Project (CAP) at ages 9, 10, 12, and 14. Although memory has been examined longitudinally in conjunction with other cognitive abilities (Cardon, 1994), this chapter will examine memory ability in CAP more systematically using isomorphic tests measured during the transition from middle childhood to adolescence. Given that heritability of memory may vary by the type of memory measure employed (e.g., Thapar et al., 1994), we will first attempt to examine different aspects of memory through both phenotypic and univariate genetic analyses before exploring their longitudinal relationships. It is hypothesized that like other cognitive abilities such as verbal ability, spatial ability, and perceptual speed (see Petrill, 1997), genes will be largely responsible for the similarity in memory across time while the nonshared environment will be largely responsible for the discrepancy between longitudinally assessed memory scores.

Method

Participants

The current study involves 199 adopted probands assessed at age 9, 200 at age 10, 199 at age 12, and 189 at age 14. Additionally, 218 control families were assessed at age 9, 224 at age 10, 224 at age 12, and 210 at age 14. Of these adoptive and control subjects, data from 100 adopted siblings were collected at age 9, 96 at age 10, 79 at age 12, and 53 at age 14. For controls, there were 110 siblings at age 9, 111 at age 10, 91 at age 12, and 74 at age 14.

Measures

Memory was assessed using the Colorado Battery of Specific Cognitive Abilities (SCA; see DeFries et al., 1981) at 9, 10, 12, and 14 years. SCA memory is divided into two subtests: Names and Faces and Picture Memory. Names and Faces is a test of 16 faces paired with first names. After study, participants are immediately shown the faces and are requested to write the corresponding names. After a 15-minute delay, the faces are re-presented and the participants are again asked to write the appropriate names. Participants, therefore, receive scores for both immediate and delayed recall. The Picture Memory test is a test of immediate and delayed recognition memory of a group of 40 familiar objects. The Immediate Picture Memory test requires participants to complete a yes/no recognition test of 20 lure items and 20 target items immediately after study. The Delayed Picture Memory test consists of another yes/no recognition test on the remaining 20 target items, as well as another 20 lures.

Exploratory factor analyses were conducted to examine the covariance among the four SCA memory subtests (Names and Faces Immediate and Delayed, Picture Memory Immediate and Delayed). Using a varimax rotation, these analyses indicate that the Names and Faces measures load on one factor while Picture Memory measures load on a second factor. For example, at year 9, the Names and Faces factor accounted for 46% of the variance while the Picture Memory factor accounted for 36% of the variance. A similar pattern was obtained at 10, 12, and 14 years. Thus, memory measures appear to divide across recall versus recognition within measure, as opposed to dividing across immediate versus delayed memory. Thus, memory was assessed using a Names and Faces composite as well as a Picture Memory composite.

TABLE 5.1. Descriptive Statistics for Uncorrected
Picture Memory and Names and Faces Memory
at Years 9, 10, 12, and 14

Variable	Mean	SD	N
Picture Memory			
Year 9			
Male	16.48	6.46	238
Female	17.24	7.06	231
Average	16.86	6.77	469
Year 10			
Male	18.61	6.52	238
Female	19.71	6.77	231
Average	19.15	6.66	469
Year 12			
Male	22.50	5.72	238
Female	23.39	5.67	231
Average	22.94	5.71	469
Year 14			
Male	21.78	6.93	238
Female	23.05	6.55	231
Average	22.40	6.77	469
Names and Faces			
Year 9			
Male	6.06	3.83	238
Female	7.38	3.79	231
Average	6.71	3.86	469
Year 10			
Male	8.29	4.21	238
Female	9.87	3.96	231
Average	9.06	4.16	469
Year 12			
Male	6.06	3.77	238
Female	7.70	4.19	231
Average	6.87	4.06	469
Year 14			
Male	8.88	5.03	238
Female	11.46	6.10	231
Average	10.15	5.72	469

Note: Descriptive statistics are based on MANOVA results using
Listwise deletion

Results

Phenotypic Analyses

Means and standard deviations of memory measures by sex and age are
presented in table 5.1. Descriptively, there is a general increase in
memory score over the four assessment points. Additionally, it appears

TABLE 5.2. Multivariate Analysis of Variance
of Names and Faces and Picture Memory by Sex
at 9, 10, 12, and 14 Years

Effect	F	dftest	dfError	p
Between Subjects				
Intercept	5196.428	2	466	.000
SEX	18.512	2	466	.000
Within Subjects				
TIME	122.605	6	462	.000
TIME by SEX	1.141	6	462	.337

that females tend to perform slightly better than males on these memory tasks.

To examine these trends inferentially, a multivariate analysis of variance was conducted. Gender was the between-subject independent variable while Names and Faces and Picture Memory were examined longitudinally as two within-subject factors. Results of the omnibus multivariate test suggest highly significant sex and longitudinal effects, but do not suggest significant sex by time interactions (see table 5.2). Within-subject contrasts suggest that mean memory levels increase significantly across each measurement occasion, with the exception that the difference between Picture Memory at 12 and 14 was not significant.

Phenotypic Sibling Analyses

Given the gender and within-subject differences noted above, additional stepdown analyses were conducted to examine gender differences in memory at each age. First, the subject-level data used in the MANOVA were aggregated into family level data. Gender differences were then examined by conducting *t*-tests at each time point separately first using only the proband siblings (the sibling 1s). Next, a replication analysis was conducted using only the sibling 2s. These data are presented in table 5.3. Interestingly, these analyses suggest that females significantly outperform males on Names and Faces. Females also appear to outperform males on the Picture Memory subtests, but these effects are largely nonsignificant.

In sum, memory measures appear to index latent dimensions of recall (Names and Faces) and recognition (Picture Memory) memory as opposed to immediate vs. delay factors. There is also a general increase in the mean level of memory performance with age, with gender differences in Names and Faces memory performance, but not Picture Memory.

TABLE 5.3. Gender Differences on Year 9, 10, 12, and 14 Names and Faces and Picture Memory Subtests

	t	sig	Mean Diff
Sibling 1 s			
Year 9			
Names and Faces	−2.43	.02	−1.45
Picture Memory	−0.79	.43	0.73
Year 10			
Names and Faces	−3.13	.00	−1.84
Picture Memory	−1.75	.08	−1.72
Year 12			
Names and Faces	−3.08	.00	−1.73
Picture Memory	−1.34	.18	−1.07
Year 14			
Names and Faces	−3.65	.00	−3.15
Picture Memory	−0.76	.45	−0.75
Sibling 2 s			
Year 9			
Names and Faces	−1.26	.21	−0.76
Picture Memory	0.46	.65	0.45
Year 10			
Names and Faces	−1.99	.05	−1.21
Picture Memory	−1.11	.27	−1.10
Year 12			
Names and Faces	−2.07	.04	−1.21
Picture Memory	−1.55	.12	−1.39
Year 14			
Names and Faces	−1.97	.05	−2.02
Picture Memory	−1.97	.05	−2.44

Note: Mean differences are based on raw memory scores as presented in table 5.1

Univariate Genetic Analyses

Given the sex and time effects found in the phenotypic analyses, all variables were age and sex corrected (within measurement occasion) using a regression procedure. Univariate genetic and environmental influences upon Names and Faces Memory and Picture Memory were then estimated using both sibling (see table 5.4) and parent-offspring (see table 5.5) designs. Additionally, the full memory composite (the average of age- and sex-corrected Names and Faces and Picture Memory variables) was examined at each assessment occasion for comparison. The sibling and parent offspring results suggest that the effect of the shared environment is negligible. With the exception of a significant correlation between adoptive mother Names and Faces and child's year 9 Names and Faces, all other comparisons were at or near zero.

TABLE 5.4. Adoptive and Biological Sibling
Correlations for Names and Faces, Picture Memory,
and Memory Composite at 9, 10, 12, and 14 Years

Variable	Adoptive	Biological
Year 9		
Names and Faces	.03	.10
Picture Memory	−.14	.07
Memory Composite	−.12	.08
Year 10		
Names and Faces	.10	.20*
Picture Memory	.02	.31*
Memory Composite	.06	.29*
Year 12		
Names and Faces	−.06	.39*
Picture Memory	.04	.12
Memory Composite	−.07	.39*
Year 14		
Names and Faces	.06	.28*
Picture Memory	−.17	.21*
Memory Composite	−.09	.34*

Note: * = p < .05

TABLE 5.5. Biological, Adoptive, and Control Father-Offspring and
Mother-Offspring Correlations for Names and Faces, Picture Memory,
and Memory Composite for 9, 10, 12, and 14 Years

Variable	Biological		Adoptive		Control	
	F	M	F	M	F	M
Year 9						
Names and Faces	.16	.23*	.02	.16*	.02	.21*
Picture Memory	.11	.18*	−.15	.00	.26*	.08
Memory Composite	.15	.29*	−.06	.01	.16*	.14*
Year 10						
Names and Faces	.16	26*	.12	.10	.12	.10
Picture Memory	−.07	.11	−.09	.01	.14*	.19*
Memory Composite	.16	.22*	.05	.04	.16*	.12
Year 12						
Names and Faces	.09	.26*	−.07	.06	.20*	.23*
Picture Memory	−.09	.09	−.09	−.06	.25*	.04
Memory Composite	.15	.24*	−.10	−.04	.25*	.17*
Year 14						
Names and Faces	.09	.20*	−.04	.09	.10	.22*
Picture Memory	−.06	.29*	−.13	−.09	.20*	.07
Memory Composite	.00	.21*	−.07	−.05	.18*	.13

Note: * = p < .05

The sibling and parent-offspring data suggest a more complex genetic picture. Sibling intraclass correlations suggest that genetic effects are nonsignificant at age 9 for both Names and Faces and Picture Memory, but are significant at ages 10, 12, and 14 for Names and Faces and at ages 10 and 14 for Picture Memory. Parent-offspring results, in contrast, suggest that genetic effects are significant from year 9 onward for all variables with the exception of year 10 and year 12 Picture Memory. Interestingly, biological mothers show more significant relationships than do biological fathers, although the lack of prediction by biological fathers may be attributable to the smaller number of biological fathers (n = 39 to 45 using pairwise deletion) measured in the current study as compared to mothers (n = 190 to 205 using pairwise deletion).

Multivariate Genetic Analyses

Multivariate genetic analyses were then conducted to examine the covariance among recall (Names and Faces) and recognition (Picture Memory) memory using the sibling data at ages 9, 10, 12, and 14. First, the relationship between recall and recognition memory was examined. Names and Faces subtests at 9, 10, 12, and 14 were collapsed into a general recall memory factor while Picture Memory subtests were collapsed into a general recognition memory factor. These factors, as well as the residual variance on the individual memory measures, were then decomposed into genetic, shared environmental, and nonshared environmental effects using a latent factor cholesky analysis (see figure 5.1).

General effects are estimated by genetic (A1), shared environmental (C1), and nonshared environmental (E1) parameters (see figure 5.1). Specific effects on the Recognition Memory factor independent of Recall Memory are estimated through genetic (A2), shared environmental (C2), and nonshared environmental (E2) pathways. Finally, specific effects at the level of individual measures are estimated by genetic (a), shared environmental (c), and nonshared environment (e) parameters influencing each subtest, independent of that subtest's factor loading. The latent Cholesky model suggests substantial genetic overlap and zero genetic independence among memory factors as indicated by large loadings on A1 and the zero loadings on A2. Relatively little evidence for specific genetic effects for individual memory measures is suggested. Shared environmental effects appear to be uniformly small in effect while the nonshared environment (and error) is largely specific in effect.

To test the significance of these effects, the latent Cholesky model was compared to several submodels (see table 5.6). The full model provides a marginal, but acceptable fit to the data (X^2 = 326.98, df = 233, p = .00, $Akaike$ = −139.0). Dropping all genetic parameters (No Genetic)

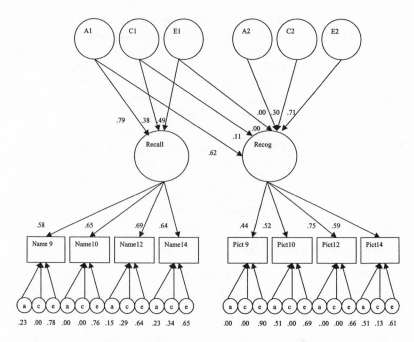

FIGURE 5.1. Latent Cholesky decomposition for recall and recognition memory factors. *Note*: All parameters are expressed as factor loadings, not as estimates of variance.

did not yield a significant decrease in chi-square, nor did dropping the specific latent genetic parameter, A2 (No General Indep). Dropping the genetic overlap (No Genetic correlation) yielded a significant decrease in model fit. Finally, it was possible to drop all specific genetic parameters on the individual memory measures (No Specific G) without a significant decrease in model fit.

Environmental submodels were also compared to the full common pathway model. Dropping all shared environmental parameters (No Shared Env) did not result in a significant decrease in chi-square. Thus, the shared environment was not necessary for the fit of the overall model. Taken together, these results suggest that the moderate overlap between recall and recognition memory from 9 to 14 years is almost completely due to genetic influences while the discrepancy between measures of memory ability is driven almost completely by the non-shared environment.

Given that there was little evidence for genetic specificity between recall and recognition memory, these measures were aggregated at each age to form memory composite scores at 9, 10, 12, and 14. Following this manipulation, a longitudinal latent Cholesky analysis was conducted to examine the covariance between memory measures in middle

TABLE 5.6. Latent Factor Cholesky Model for Recall
and Recognition Factors at Ages 9, 10, 12, and 14

Model	x2	df	p	Ak	X2cha	dfcha	pcha
Full	326.98	233	.00	−139.0			
No Genetic	336.61	244	.00	−151.4	9.63	11	NS
No General Indep	326.98	234	.00	−141.0	0.00	1	NS
No Genetic Correlation	333.83	234	.00	−134.2	6.85	1	*
No Specific G	328.97	241	.00	−153.0	0.01	8	NS
No Shared Env	330.25	244	.00	−157.8	3.27	11	NS

Note: * = p < .05

TABLE 5.7. Latent Factor Cholesky Model for Middle Childhood
and Early Adolescence Factors at Ages 9, 10, 12, and 14

Model	x2	df	p	Ak	X2cha	dfcha	pcha
Full	72.12	49	.02	−29.9			
No Genetic	81.54	56	.02	−30.5	9.42	5	NS
No General Indep	72.12	50	.02	−27.9	0.00	1	NS
No Genetic Correlation	79.61	50	.01	−20.4	7.49	1	*
No Specific G	73.33	53	.03	−32.7	1.21	4	NS
No Shared Env	73.29	56	.06	−38.7	1.17	5	NS

Note: * = p < .05

childhood and early adolescence. The model employed was similar
to the model presented in figure 5.1 with the exception that the composite memory scores at ages 9 and 10 were used to index a middle
childhood factor while memory scores at ages 12 and 14 were used
to index an early adolescence factor. These latent factors were then
examined using a latent Cholesky analysis.

These analyses are presented in table 5.7. The full model yields a
marginally acceptable fit to the data ($X^2 = 72.12$, $df = 49$, $p = .02$, $Ak =$
−29.9). The results of the longitudinal latent Cholesky analyses are
identical to the structural analyses presented in table 5.7. The only significant genetic parameter is the genetic correlation between middle
childhood and early adolescence factors (No Genetic Correlation). Specific genetic effects are nonsignificant (No General Indep, No Specific
G), as are shared environmental effects (No Shared Env).

The results of the analyses presented in tables 5.6 and 5.7 suggest
that the relationship among memory measures may be unidimensional
not only across recall and recognition but also across age. Thus, a
common pathway model was fit to the data to examine the extent to
which genes and environment shape the overlap and discrepancy among
the composite memory measures assessed from 9 to 14 years (see Neale

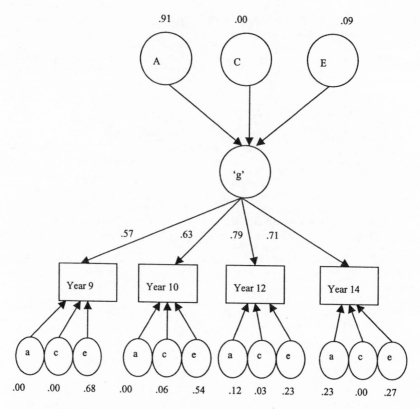

FIGURE 5.2. Common pathway model results: Memory composite at years 9, 10, 12, and 14. *Note*: Estimates for A, C, E, a, c, and e are expressed as variance components.

& Cardon, 1992). General effects are estimated by genetic (A), shared environmental (C), and nonshared environmental (E) parameters (see figure 5.2). Specific effects are estimated by genetic (*a*), shared environmental (*c*), and nonshared environmental (*e*) parameters influencing each subtest, independent of that subtest's g loading. The common pathway model mirrors the high degree of genetic overlap found in the latent Cholesky analyses as indicated by the moderate to large g factor loadings, the large "genetic g" estimate ($h^2 = -.91$), and the small estimates of specific genetic influence. Shared environmental effects appear to be uniformly small while the nonshared environment (and error) is largely specific in effect.

To test the significance of these effects, the full common pathway model was compared to several submodels (see table 5.8). The full model provides a more parsimonious fit to the data ($X^2 = 72.72$, *df* = 54, *p* = .05, *Akaike* = −35.3), than does the full latent cholesky model

TABLE 5.8. Common Pathway Model: Memory Composite
Years 9, 10, 12, and 14

Model	x2	df	p	Ak	X2cha	dfcha	pcha
Full	72.72	54	.05	−35.3			
No Genetic	83.23	59	.02	−34.8	10.51	5	NS
No General G	77.86	55	.02	−21.1	5.14	1	*
No Specific G	74.03	58	.08	−42.0	1.31	4	NS
No Shared Env	73.69	59	.09	−44.3	0.97	5	NS

Note: * = $p < .05$

using the same composite memory variables (table 5.7). Dropping all genetic parameters (No Genetic) did not yield a significant decrease in chi-square. Dropping genetic g (No General G), on the other hand, yielded a significant decrease in model fit. However, it was possible to drop all specific genetic parameters (No Specific G) without a significant decrease in model fit. These submodels further suggest that genetic effects are not only statistically significant but also largely shared across time.

Conclusions

In this chapter, we attempted to systematically examine memory measured through isomorphic tasks at ages 9, 10, 12, and 14. As hypothesized, the correlation between memory tests was driven almost completely by shared genetic factors while the discrepancy between memory tasks was influenced largely by the nonshared environment (and error). Chapter 2 (Bishop, Cherny, & Hewitt) suggests that this genetic variance associated with longitudinal stability in memory is, in fact, the same genetic variance operating in other tasks. What is interesting about the results of this chapter, however, is not only the degree of phenotypic overlap among memory measures but also the large magnitude of the genetic overlap in memory both across measurement occasion and across memory type.

However, the current results are based on a relatively limited measurement of memory at each age. Thus, the unitary factor of memory constructed for the current study may be one of many memory factors if a more comprehensive battery of memory measures was employed. Although behavioral genetics research is beginning to examine other measures of memory, these are largely based on standardized assessments such as the Stanford-Binet Intelligence Scale (Thorndike, Hagen, & Sattler, 1986) or the Wechsler Scales (Wechsler, 1972). Behavioral genetic methods have not been employed to examine individual differences in memory using theoretically driven and experimentally derived measures of human memory. Additionally, the discrepancy between twin and adoption data with respect to memory has been largely unexplored.

Despite these limitations, the results of the current study are part of an ever-increasing set of behavioral genetic studies that suggest genes operate in a unitary manner, resulting in the covariance of cognitive abilities across both task and age. Thus, far from being the result of independent processes, cognitive abilities are drawn together largely by shared genetic influences. Given this large degree of genetic covariance, molecular genetic studies that find links between DNA markers and memory (e.g., Nilsson et al., 1996) should be mindful that these DNA markers may also be linked with other cognitive abilities. In a similar vein, environmental researchers should look to identifying the nonshared environmental influences that account for the discrepancy between memory measures.

ACKNOWLEDGMENT

The authors would like to thank Robin Corley for his help in the data management and analysis of these data.

REFERENCES

Bouchard, T. J. Jr., Segal, N. L., & Lykken, D. T. (1990). Genetic and environmental influences on special mental abilities in a sample of twins reared apart. *Acta Geneticae Mededicae et Gemelloliguae, 39*, 193–206.

Cardon, L. R. (1994). Specific cognitive abilities. In J. C. DeFries, R. Plomin, & D. W. Fulker (Eds.), *Nature and nurture in middle childhood* (pp. 57–76). Oxford: Blackwell.

DeFries, J. C., Plomin, R., Vandenberg, S. G., & Kuse, A. R. (1981). Parent-offspring resemblance for cognitive abilities in the Colorado Adoption Project: Biological, adoptive, and control parents and one-year-old children. *Intelligence, 5*, 245–277.

Finkel, D., McGue, M., & Fox, P. (1991). A twin study of memory performance across the life span. *Behavior Genetics, 21*, 570–571. (abstract)

Finkel, D., Pedersen, N., & McGue, M. (1995). Genetic influences on memory performance in adulthood: Comparison of Minnesota and Swedish twin data. *Psychology and Aging, 10*, 437–446.

Finkel, D., Pedersen, N. L., McGue, M., & McClearn, G. E. (1995). Heritability of cognitive abilities in adult twins: Comparison of Minnesota and Swedish data. *Behavior Genetics, 25*, 421–432.

Neale, M. C. (1995). MX statistical modeling (3rd ed.) Richmond: Department of Psychiatry, Medical College of Virginia.

Neale, M. C., & Cardon, L. R. (1992). *Methodology for the genetic study of twins and families.* London: Kluwer.

Nilsson, L. G., Sikström, R., Adolfsson, K., Erngrund, K., Nylanbder, P. O., & Beckman, L. (1996). Genetic markers associated with high versus low performance on episodic memory tasks. *Behavior Genetics, 26,* 555–562.

Partanen, J., Brunn, K., & Markkanen, T. (1996). *Inheritance of drinking behavior.* Stockholm: Almquist & Wiksell.

Pedersen, N. L., Plomin, R., & McClearn, G. E. (1994). Is there G beyond g (is there genetic influence on specific cognitive abilities independent of genetic influences on general cognitive ability?). *Intelligence, 18,* 133–143.

Petrill, S. A. (1997). Molarity versus modularity of cognitive functioning? A behavioral genetic perspective. *Current Directions in Psychological Science, 6,* 96–99.

Petrill, S. A., Plomin, R., Berg, S., Johansson, B., Pedersen, N., Ahern, F., & McClearn, G. E., (1998). The genetic and environmental relationship between general and specific cognitive abilities in twins 80 years and older. *Psychological Science, 9,* 183–189.

Plomin, R., DeFries, J. C., McClearn, G. E., & Rutter, M. (1997). *Behavioral genetics* (3rd ed.). New York: Freeman.

Schoenfeldt, L. F. (1968). The heredity components of project TALENT two-day test battery. *Measures Evaluations and Guidance, 1,* 130–140.

Squire, L. R. (1992). Memory and the hippocampus: a synthesis from findings with rats, monkeys, and humans. *Psychological Review, 99,* 193–231.

Thapar, A., Petrill, S. A., & Thompson, L. A. (1994). The heritability of memory in the Western Reserve Twin Project. *Behavior Genetics, 24,* 155–160.

Thorndike, R. L., Hagen, E. P., & Sattler, J. M. (1986). *Guide for administering and scoring the fourth edition: Stanford-Binet Intelligence Scale.* Chicago, IL: Riverside.

Wechsler, D. (1972). *Wechsler's measurement and appraisal of adult intelligence* (5th ed.). Baltimore, MD: Williams & Wilkins.

NICHOLAS GIARDINO
RICHARD RENDE

6

Somatic Complaints from Early Childhood to Early Adolescence

Introduction

Somatic complaints are common in childhood. Recurrent pain, for example, may occur in as many as 30% of all children (McGrath & McAlpine, 1993). In addition, between 2 and 10% of children presenting at primary care clinics at any one time report suffering from physical symptoms, such as aches, pains, or fatigue, for which no diagnosable injury or disease is identified (Goodman & McGrath, 1991). However, relatively little research has been conducted on the phenomenon of somatic symptoms in children. Most of the work reported to date has addressed only the occurrence of somatic symptom complaints in relation to psychiatric problems, such as anxiety and depression. Several reports, for example, show that children complaining of abdominal pain or headaches show higher levels of anxiety and depression, as well as other medically unexplained somatic symptoms, than healthy children (Ernst, Routh, & Harper, 1984; Garber, Zeman, & Walker, 1990; Hodges, Kline, Barbero, & Woodruff, 1985; Walker & Greene, 1989; Wasserman, Whitington, & Rivera, 1988; Zuckerman, Stevenson, & Bailey, 1987). Parents of these children, especially mothers, also report increased emotional distress compared to parents of healthy children (Hodges et al., 1985; Walker & Greene, 1989; Zuckerman et al., 1987), suggesting a familial influence in the report of such symptoms.

Using the converse approach, others have reported that children diagnosed with psychiatric disorders, again typically depression and anxiety problems, are more likely to have significant somatic complaints, such as abdominal pain or headache (Jolly et al., 1994; Last, 1991; Livingston, Taylor, & Crawford, 1988; McCauley, Carlson, & Calderon, 1991).

Identifying a Phenotype of Somatic Complaints

While these associations are interesting and important findings, an often-neglected preliminary step in the investigation of the inter-relationship between somatic and psychological symptoms is the systematic characterization of the somatic symptoms themselves. Thus, if we want to examine the development of somatic symptoms in children with the hope of uncovering etiologic and maintaining factors, we may first need to better identify a specific phenotype of somatic complaints that is most amenable to such research. Among the features of somatic symptoms that we may wish to characterize is whether continuity in the report of somatic symptoms exists across development, and if so, whether these complaints remain homotypic or rather change with age or over time. Also important to an accurate phenotypic identification is whether the continuity (or discontinuity) of the presence or severity of somatic complaints over time applies mainly to a distinct subset of children (e.g., only those with extreme symptom reports) or whether the pattern of somatic symptom reports is observed similarly across the entire normal range.

More accurate identification of a phenotype for somatic complaints may have important implications for clinical nosology and research with problems such as somatization and emotional disorders. In the case of somatization disorder, for example, some have proposed and provided evidence to support a diagnostic conceptualization of somatization in terms of moderate, but persistent, somatic complaints, rather than the extreme and infrequently occurring presentation currently captured by the *DSM-IV* criteria (Escobar, Burnam, Karno, Forsythe, & Golding, 1987). Others, as well, present data to support a more dimensional profile of somatization, in which increasing levels of somatic symptoms are linearly associated with increasing psychosocial problems and distress (Katon et al., 1991). An examination of the longitudinal course of somatic symptoms over childhood and adolescence may further inform these important issues.

As mentioned above, little empirical work has been conducted with the aim of characterizing somatic symptom presentations in children, and especially in young children. One recent effort (Garber, Walker, & Zeman, 1991) has confirmed the presence of a somatization syndrome in a community sample of children and identified four symptom clusters of somatic complaints found in those with multiple symptoms.

Interestingly, these clusters, which include pseudoneurological, cardio-vascular, gastrointestinal, and pain/weakness factors, loosely parallel those identified in adult samples, although most clusters also include some unrepresentative symptoms. While this study probably represents the best effort to date to characterize somatic complaints in children, its cross-sectional design limits interpretations of age effects on symptom reporting and, of course, precludes assessment of symptom changes over time. Their finding of polysymptomatic conditions that have some tendency to form discrete clusters is important, however, both confirming other reports of somatic complaints syndromes in children (Achenbach, Conners, Quay, Verhulst, & Howell, 1989) and providing a link to the much more studied adult somatization disorder.

Data collected from the Colorado Adoption Project (CAP) are well suited to address several important issues regarding further identification of a suitable phenotype for the study of somatic symptom complaints in childhood and early adolescence. First, CAP provides data from an unselected normative sample. As such, these data allow for the characterization of somatic symptoms reports across the entire range of symptom frequency and severity, including the analysis of "extreme" groups. Second, taking advantage of the longitudinal design of CAP, we are able to assess changes in somatic symptoms across time, from early and middle childhood, and into late childhood, approaching early adolescence. Finally, drawing upon the genetically sensitive design of the CAP, we may examine associations between biological and adoptive siblings to study the role of genetic and environmental influences on somatic complaints.

In the following section we will describe the results of our analyses carried out to characterize somatic symptom complaints in children from the CAP. Specifically we report on the following: frequencies of specific somatic complaints in children across developmental periods; the continuity of somatic symptoms from early childhood to early adolescence; the utility of identifying individuals with somatic complaints in early childhood; sex differences in the characterization of somatic symptoms across childhood; and assessment of genetic and environmental influences through sibling analyses.

Method

Sample

Using data from the CAP, we examined the presence of several common somatic complaints in adopted and nonadopted probands and their siblings at 4, 7, 10, and 12 years of age. Participants included 217 adopted (114 male, 103 female) and 190 nonadopted control (102 male, 88 female) probands. In addition, siblings were included to yield 77 pairs of unrelated adoptive siblings (11 same-sex boys, 9 same-sex girls, 57 opposite-sex) and 93 pairs of biologically related siblings (37 same-sex boys, 24 same-sex girls, 32 opposite-sex).

Measures

Parents were asked to complete the Child Behavior Checklist (CBCL; Achenbach, 1991; Achenbach & Edelbrock, 1983) when their children were 4, 7, 10, and 12 years of age. At ages 7, 10, and 12, a teacher version of the CBCL was also administered. The CBCL is a widely used measure of childhood behavioral problems that asks parents to rate 118 problem items on a three-point Likert-type scale (0 = "not true," 1 = "sometimes true," 2 = "often true"). In our present study we report on the items that constitute the Somatic Complaints Scale of the CBCL, which includes dizziness, fatigue, aches and pains, headache, nausea, eye problems, rashes, stomachaches, and vomiting. Data are presented for the presence of each individual symptom, as well as a scale score representing the sum of all somatic symptom item scores. Where data were available from both mother and father, an average score for each item was calculated and used for all parent analyses.

Analysis Method

The continuity of somatic complaints over time was assessed by correlations of somatic complaints scale scores from year to year, and by comparing the odds of having somatic complaints at a given assessment age in probands who had those same complaints at an earlier age to those who did not. The resulting risk estimates are, thus, the ratios of the odds of reporting a somatic symptom at a given age, given that a proband had that somatic complaint earlier, over the odds of reporting a somatic symptom at that age, given that a proband did *not* report the symptom at the earlier age. In addition, odds ratios were calculated comparing the presence of later complaints in probands who had *any* somatic symptoms at an earlier age with those who did not, as well as those who had CBCL Somatic Complaints Scale scores in the upper 5%, and then again for the upper 10%, of the sample with those in the lower percentiles.

Sibling aggregation for the presence of somatic complaints was similarly determined at each assessment age by calculating sibling correlations for somatic complaints scale scores, and also pairwise odds ratios for biological and adoptive sibling pairs. Here the pairwise odds ratio for sibling pairs gives the ratio of odds that, given that one sibling reports any somatic complaints at a given age, the other sibling will also have one or more somatic symptoms, over the odds that, given that one sibling does *not* report any somatic symptoms, the other will have one or more somatic complaints (see Rende, Wickramaratne, Warner, &

Weissman, 1995; Rende, Warner, Wickramaratne, & Weissman, 1999; Rende & Weissman, 1999).

Results

Frequencies of individual somatic symptoms for all probands, as well as separately for boys and girls, are shown in table 6.1. Consistent with earlier reports (Achenbach et al., 1989; Apley, 1975; Garber et al, 1991; Oster 1972), somatic complaints were common in our sample. Thirty-five percent of all probands reported one or more somatic complaint at age 4. At age 12, 54% had at least one somatic symptom. Fatigue and stomachaches were the most common symptoms reported in early childhood, while headaches became more prominent in late childhood. No meaningful sex differences were observed in symptoms, with the exception of stomachaches, which were more common in female probands at ages 7–12. Somatic complaints scale scores increased with age for both male and female probands, $F(1,3) = 25.00$, $p < .001$. Adopted and control probands did not differ on somatic complaints scores at any assessment age.

Table 6.2 shows the frequency of children's somatic complaints as assessed by teacher report. While somatic complaints were not uncommon in teacher's reports, the frequency of all symptoms was significantly lower than parent-reported symptoms at every assessment age. On average, parents reported 1.2 symptoms per child at each assessment year, while teachers reported only 0.4.

Next we attempted to assess the continuity of somatic complaints from age to age by calculating correlations between CBCL somatic complaints scale scores at each assessment period (table 6.3). Scale score correlations between assessment periods were moderately strong for parent-reported symptoms across all ages, but only from age 10 to age 12 for teacher-report scales.

To try to better characterize the continuity of somatic complaints from age to age, we then computed risks for individual symptoms given the presence of those symptoms at earlier assessments (table 6.4). The risk of parent-reported symptoms was increased for all somatic complaints among those with symptoms at the previous assessment age. This was true between all adjacent assessment periods, and even across longer periods of time, such as from age 4 to age 12. The continuity of symptoms appeared less strong for teacher-reported complaints, although the very small number of symptoms recorded by teachers made meaningful calculations difficult. Nonetheless, among some of the more frequently reported complaints (e.g., aches and pains, headaches), continuity of symptoms was observed with teacher reports.

To further explore the presentation of somatic complaints over time, we next compared children who had any parent-reported somatic com-

TABLE 6.1. Proportion of Children with Parent-Reported Somatic Complaints at Each Assessment Year

| | Age 4 | | | Age 7 | | | Age 10 | | | Age 12 | | |
| | | % | | | % | | | % | | | % | |
	All	Male	Female	All	Male	Female	All	Male	Female	All	Male	Female
Any symptom	35.1	33.9	36.4	58.0	54.6	61.9*	50.9	47.3	55.1*	53.8	50.0	58.3
Dizziness	0.5	0.5	0.5	2.9	3.7	2.1	2.7	2.2	3.2	3.6	3.1	4.1
Fatigue	17.8	17.9	17.6	20.0	19.7	20.2	12.3	13.8	10.6	11.9	9.8	14.3
Aches & pains	7.1	6.3	8.0	18.3	16.1	20.7	13.8	13.3	14.3	19.1	16.5	21.9
Headaches	4.4	3.1	5.9	17.0	18.4	15.5	23.4	20.4	27.0	23.6	25.5	21.4
Nausea	1.5	1.4	1.6	12.2	10.1	14.5	8.0	6.7	9.5	8.3	6.7	10.2
Eye problems	1.0	0.5	1.6	7.1	8.7	5.2	7.5	6.2	9.0	9.8	8.0	11.7
Rashes	5.6	5.8	5.3	6.1	6.9	5.2	8.0	7.1	9.0	10.7	7.6	14.3
Stomachache	9.0	7.2	11.2	24.3	17.1	32.6*	19.1	13.8	25.4*	19.3	15.6	23.5
Vomiting	2.0	1.4	2.7	4.6	4.6	4.7	2.4	2.2	2.7	4.3	5.4	3.1
Constipation	8.5	8.5	8.5	13.4	14.3	12.5	8.7	9.8	7.4	5.5	5.8	5.1

*Females significantly different from males, p < .05

TABLE 6.2. Proportion of Children with Teacher-Reported Somatic Complaints at Each Assessment Year

	Age 7			Age 10			Age 12		
		%			%			%	
	All	Male	Female	All	Male	Female	All	Male	Female
Any symptom	21.9	18.6	25.5	20.1	18.1	22.5	18.0	16.5	19.6
Dizziness	1.0	1.4	0.5	1.5	1.7	1.3	2.6	2.5	2.7
Fatigue	10.0	9.4	10.5	8.6	8.5	8.9	6.8	8.2	5.3
Aches & pains	7.0	4.1	10.2*	6.1	3.3	9.6*	4.7	4.2	5.2
Headaches	7.0	3.6	10.7*	7.7	7.2	8.3	8.7	7.7	9.8
Nausea	5.4	2.1	9.0*	4.3	2.2	6.9*	6.2	4.9	7.5
Eye problems	4.9	5.2	4.6	5.0	5.6	4.2	7.7	7.1	8.4
Rashes	1.3	0.5	2.2	1.8	2.2	1.4	1.8	2.8	0.7
Stomachache	6.7	3.0	10.8*	4.9	3.3	6.8	6.1	4.8	7.5
Vomiting	0.3	0	0.6	0.6	1.1	0	2.9	2.8	3.0

*Females significantly different from males, $p < .05$

TABLE 6.3. Age-to-Age Correlations for Somatic Complaints Scale Scores from Parent and Teacher Reports

Age	4	7	10		Age	7	10
7	.373**						
10	.280**	.387**			10	.069	
12	.204**	.351**	.347**		12	−.039	.232**

**$p < .001$

plaints at age 4 with those who did not. Table 6.5 presents the odds ratios of reporting each somatic symptom given the presence or absence of any somatic complaints at age 4. Separate odds ratios were also calculated for male and female probands. Children who complained of one or more somatic symptom at age 4 were significantly more likely to report a variety of somatic symptoms at later assessment years, even up to eight years later at the age of 12. This continuity of somatic complaints over time was somewhat stronger in female probands. Interestingly, while a small gain in increased risk for reporting future symptoms was seen when these analyses were repeated using more extreme cutoffs to identify the at-risk group (e.g., top 5% split), adequate identification was achieved by a simple distinction of the presence or absence of *any* somatic complaints at age 4. Subsequent analyses described below further support this dichotomous conceptualization.

TABLE 6.4. Continuity of Individual Symptoms

A. Parent-Reported Symptoms

	Age 7–10				Age 10–12				Age 7–12			
	Yes %	No %	Odds ratio	95% CI	Yes %	No %	Odds ratio	95% CI	Yes %	No %	Odds ratio	95% CI
Any symptoms	60.6	34.1	2.98	1.96–4.51	69.0	35.9	3.98	2.60–6.08	63.3	41.4	2.44	1.62–3.69
Dizziness	20.0	2.1	11.47	2.09–62.81	20.0	2.6	9.20	1.73–48.97	9.1	3.8	2.56	0.31–21.45
Fatigue	27.0	7.4	4.64	2.38–9.03	33.3	9.0	5.05	2.45–10.39	30.4	7.7	5.63	2.95–10.75
Aches & pains	36.4	9.2	5.64	3.00–10.59	59.3	12.5	10.22	5.43–19.26	42.9	13.5	4.79	2.70–8.50
Headaches	51.6	17.3	5.09	2.88–9.00	47.3	16.7	4.48	2.68–7.48	50.8	18.0	4.70	2.68–8.27
Nausea	8.7	8.0	1.09	0.36–3.27	37.5	5.4	10.52	4.49–24.65	18.4	6.0	3.51	1.50–8.24
Eye problems	24.0	6.8	4.33	1.58–11.85	37.9	6.6	8.72	3.68–20.62	37.5	7.6	7.31	2.93–18.25
Rashes	32.1	5.9	7.49	3.02–18.56	41.4	8.2	7.89	3.44–18.10	23.1	9.7	2.81	1.06–7.47
Stomachache	40.7	12.2	4.99	2.90–8.60	38.4	15.9	3.30	1.88–5.79	37.9	13.4	3.93	2.30–6.73
Vomiting	15.8	1.7	11.16	2.56–48.70	40.0	3.5	18.46	4.64–73.45	10.5	3.3	3.42	0.71–16.51

B. Teacher-Reported Symptoms

	Age 7–10				Age 10–12				Age 7–12			
	Yes %	No %	Odds ratio	95% CI	Yes %	No %	Odds ratio	95% CI	Yes %	No %	Odds ratio	95% CI
Any symptoms	28.1	17.4	1.85	0.99–3.49	18.8	16.7	1.15	0.51–2.57	18.0	17.3	1.05	0.50 = 2.20
Dizziness	0	2.8	—	—	11.1	2.5	4.93	0.54–44.93	0	3.0	—	—
Fatigue	8.1	11.9	0.65	0.19–2.22	7.9	7.0	1.14	0.32–4.08	7.1	7.5	0.96	0.21–4.34
Aches & pains	27.2	14.7	2.17	0.81–5.83	12.5	3.1	4.53	1.25–16.46	6.3	3.7	1.75	0.21–14.93
Headaches	27.2	24.8	1.14	0.43–3.00	14.8	7.0	2.31	0.95–5.64	21.4	6.8	3.73	0.94–14.82
Nausea	23.5	8.2	3.47	1.06–11.40	5.3	6.2	0.84	0.11–6.76	11.1	4.9	2.41	0.28–21.00
Eye problems	18.8	7.7	2.75	0.73–10.32	14.3	7.7	2.01	0.54–7.47	11.1	7.4	1.38	0.17–11.60
Rashes	0	3.0	—	—	0	2.1	—	—	0	1.3	—	—
Stomachache	13.6	20.6	0.61	0.17–2.12	7.5	5.2	1.49	0.46–4.89	12.5	4.5	3.01	0.60–15.11
Vomiting	100	2.1	—	—	0	2.7	—	—	0	2.2	—	—

For each assessment age pair, percentages of probands for whom symptoms were reported at the later age are presented separately for those who did (yes) and did not (no) report that symptom at the earlier age. Additionally, the risk of reporting symptoms at the later age are given the presence of that symptom at the earlier age are given.

TABLE 6.5. Somatic Symptoms at Ages 4, 7, 10, and 12 for Probands with (yes) or without (no) Any Somatic Symptom Complaints at Age 4

	Age 4			Age 7						Age 10						Age 12					
	Yes n	%	No n	Yes n	%	No n	%	Odds ratio	95% CI	Yes n	%	No n	%	Odds ratio	95% CI	Yes n	%	No n	%	Odds ratio	95% CI
Dizziness	2	1.57%	0	5	4.42%	7	2.67%	1.69	0.52–5.43	4	3.39%	5	1.92%	1.79	0.47–6.79	3	2.59%	10	3.83%	0.67	0.18–2.49
Frigue	73	57.48%	0	40	35.40%	38	14.50%	3.23	1.93–5.42	26	22.03%	21	8.08%	3.22	1.73–6.00	25	21.55%	23	8.81%	2.84	1.54–5.26
Aches & pains	29	22.83%	0	37	32.74%	34	12.98%	3.25	1.91–5.54	26	22.03%	30	11.54%	2.15	1.21–3.83	32	27.59%	43	16.48%	1.91	1.13–3.21
Headaches	18	14.17%	0	30	26.55%	36	13.74%	2.26	1.31–3.90	41	34.75%	49	18.85%	2.27	1.39–3.71	39	33.62%	54	20.69%	1.93	1.19–3.15
Nausea	6	4.72%	0	21	18.58%	24	9.16%	2.25	1.20–4.25	17	14.41%	15	5.77%	2.74	1.32–5.69	14	12.07%	19	7.28%	1.75	0.85–3.63
Eye problems	4	3.15%	0	13	11.50%	14	5.34%	2.29	1.04–5.05	7	5.93%	21	8.08%	0.74	0.31–1.79	11	9.48%	27	10.34%	0.89	0.43–1.87
Rashes	23	18.11%	0	14	12.39%	13	4.96%	2.69	1.22–5.92	12	10.17%	17	6.54%	1.61	0.74–3.48	19	16.38%	23	8.81%	2.02	1.05–3.88
Stomachache	37	29.13%	0	46	40.71%	46	17.56%	3.26	1.99–5.33	33	27.97%	40	15.38%	2.12	1.25–3.58	30	25.86%	46	17.62%	1.63	0.96–2.75
Vomiting	8	6.30%	0	9	7.96%	9	3.44%	2.42	0.94–6.28	5	4.24%	5	1.92%	2.24	0.64–7.89	6	5.17%	12	4.60%	1.13	0.42–3.10
Females Only																					
Dizziness	1	1.61%	0	2	3.45%	2	1.75%	2.00	0.27–14.57	3	5.36%	1	0.87%	6.45	0.66–63.50	2	3.51%	5	4.35%	0.82	0.15–4.34
Fatigue	33	53.23%	0	21	36.21%	16	14.04%	3.48	1.64–7.38	15	26.79%	4	3.48%	10.15	3.18–32.37	15	26.32%	11	9.57%	3.38	1.43–7.95
Aches & pains	15	24.19%	0	21	36.21%	16	14.04%	3.44	1.62–7.30	17	30.36%	9	7.83%	5.09	2.09–12.35	18	31.58%	22	19.13%	1.95	0.94–4.04
Headaches	11	17.74%	0	13	22.41%	15	13.16%	1.89	0.83–4.30	22	39.29%	25	21.74%	2.28	1.14–4.57	18	31.58%	22	19.13%	1.95	0.94–4.04
Nausea	3	4.84%	0	12	20.69%	12	10.53%	2.20	0.92–5.26	10	17.86%	7	6.09%	3.32	1.19–9.27	9	15.79%	10	8.70%	2.01	0.77–5.27
Eye problems	3	4.84%	0	4	6.90%	5	4.39%	1.60	0.41–6.20	3	5.36%	13	11.30%	0.44	0.12–1.63	4	7.02%	16	13.91%	0.46	0.15–1.45
Rashes	10	16.13%	0	8	13.79%	6	5.26%	2.83	0.93–8.58	5	8.93%	10	8.70%	1.01	0.33–3.11	14	24.56%	14	12.17%	2.38	1.05–5.43
Stomachache	21	33.87%	0	29	50.00%	28	24.56%	3.04	1.56–5.93	19	33.93%	24	20.87%	1.90	0.93–3.89	18	31.58%	24	20.87%	1.76	0.86–3.61
Vomiting	5	8.06%	0	4	6.90%	4	3.51%	2.02	0.49–8.38	3	5.36%	2	1.74%	3.14	0.51–19.37	2	3.51%	4	3.48%	1.02	0.18–5.74
Males Only																					
Dizziness	1	1.52%	0	3	5.45%	5	3.38%	1.65	0.38–7.15	1	1.61%	4	2.76%	0.58	0.06–5.28	1	1.69%	5	3.42%	0.49	0.06–4.25
Fatigue	40	60.61%	0	19	34.55%	22	14.86%	3.02	1.48–6.19	11	17.74%	17	11.72%	1.62	0.71–3.71	10	16.95%	12	8.22%	2.28	0.93–5.61
Aches & pains	14	21.21%	0	16	29.09%	18	12.16%	2.96	1.38–6.35	9	14.52%	21	14.48%	1.00	0.43–2.32	14	23.73%	21	14.38%	1.81	0.85–3.86
Headaches	7	10.61%	0	17	30.91%	21	14.19%	2.71	1.30–5.64	19	30.65%	24	16.55%	2.23	1.11–4.47	21	35.59%	32	21.92%	1.95	1.01–3.78
Nausea	3	4.55%	0	9	16.36%	12	8.11%	2.22	0.88–5.60	7	11.29%	8	5.52%	2.18	0.75–6.30	5	8.47%	11	6.16%	1.39	0.45–4.33
Eye problems	1	1.52%	0	9	16.36%	9	6.08%	3.02	1.13–8.07	4	6.45%	8	5.52%	1.25	0.36–4.31	7	11.86%	11	7.53%	1.62	0.59–4.39
Rashes	13	19.70%	0	6	10.91%	7	4.73%	2.47	0.79–7.70	7	11.29%	7	4.83%	2.51	0.84–7.49	5	8.47%	9	6.16%	1.39	0.45–4.33
Stomachache	16	24.24%	0	17	30.91%	18	12.16%	3.32	1.56–7.07	14	22.58%	16	11.03%	2.35	1.07–5.18	12	20.34%	22	15.07%	1.43	0.66–3.11
Vomiting	3	4.55%	0	5	9.09%	5	3.38%	2.86	0.80–10.29	2	3.23%	3	2.07%	1.58	0.26–9.68	4	6.78%	8	5.48%	1.25	0.36–4.31

TABLE 6.6. Risk of Multiple Somatic Complaints
at Age y Given One or More Symptoms at Earlier
Assessment Age x

Age x	Age y	Odds Ratio	95% CI
4	7	4.01	2.50–6.43
4	10	2.53	1.57–4.09
4	12	1.90	1.19–3.06
7	10	3.17	1.91–5.26
7	12	3.06	1.87–4.98

As described above, an important distinction in determining risk for somatization-like behavior appears to be that one report multiple types of unexplained somatic symptoms. Therefore we next examined the progression and continuity of multiple symptoms over time. First, we compared the number of somatic symptoms at ages 7, 10, and 12 in probands who were rated as having one or more somatic symptoms at earlier ages to those who had reported none (see table 6.6). Risk of multiple somatic complaints in middle childhood was significantly and strongly associated with presentation of at least one somatic complaint at an earlier age. And, although this risk decreased with increasing number of years between assessments, it was notable as well across the eight-year span from age 4 to age 12. Furthermore, as 56% of those who had more than one somatic complaint at age 4 retained more than more symptom at age 12, compared to only 24% who had either one or no symptoms at age 4, it appears that the magnitude of continuity for children with symptoms at age 4 does not change with increasing age, but rather some individuals acquire somatic symptoms at later ages. This view is additionally supported by data showing that, while the ability to predict multiple somatic complaints at age 12 increases if probands are examined for the presence of any symptoms at age 7 (OR = 3.06), the continuity of multiple symptoms from age 7 to age 12 is no stronger than that shown from age 4.

Finally, in order to assess genetic and environmental contributions to somatic complaints, correlations on CBCL somatic complaints scale scores and adjusted pairwise odds ratios for the presence of somatic symptoms were calculated for adoptive and biological sibling pairs. Sibling correlations calculated with scale scores derived from parent-reported data showed moderate similarity between siblings, but no significant differences between adoptive and biological pairs were observed (table 6.7). Correlations of teacher-reported scores, on the other hand, were stronger for biological siblings, suggesting possible genetic influence.

Odds-ratio calculations reveal similar patterns of sibling aggregation, but are less discrepant for parent and teacher reports. While younger

TABLE 6.7. Siblings' Correlations on CBCL Somatic Complaints Scale Scores

Age		4	7	10	12
Parent Report					
Adoptive siblings	All pairs	.27	.19	.10	.25
	Opposite sex pairs	.28	.19	.26	.26
Biological siblings	All pairs	.16	.07	.41	.18
	Opposite sex pairs	.10	.22	.59	.33
Teacher Report					
Adoptive siblings	All pairs		−.02	.23	−.10
	Opposite sex pairs		.00	.38	−.10
Biological siblings	All pairs		.66	.61	.13
	Opposite sex pairs		.37	.60	−.09

siblings of probands with somatic complaints at age 4 showed increased risk for somatic complaints at the same age, no significant differences between biological and adoptive siblings were observed with parent-reported data. Evidence for genetic influence continued to be absent at later assessment ages (see table 6.8), and sibling aggregation for somatic complaints decreased in early adolescence. Similar results were found using teacher's reports. At age 7, for example, pairwise odds ratios were 2.48 for adoptive siblings (95% CI = 0.73 – 8.46) and 2.44 for biological siblings (95% CI = 0.82 – 7.25). When these analyses were once again repeated using more a more extreme symptom severity criterion to define somatic complaints (e.g., top 5% of symptom reporters), a substantial drop-off in the number of concordant pairs was seen for both biological and adoptive siblings. For example, using parent-reported data at age 4, there were 11 pairs of biological siblings and 7 pairs of adoptive siblings who both had somatic complaints. When either a top 5% or decile split was used to categorize siblings, however, only one pair of siblings in each group remained concordant. Thus, the utility of a simple symptoms/no symptoms distinction was again supported.

As one of the potential problems in the sibling adoption design is that an uneven proportion of same-gender versus mixed-gender sibling pairs typically exist between adoptive and control probands (Plomin, DeFries, & Fulker, 1988), we also calculated sibling correlations and pairwise odds ratios for the presence of any somatic complaints at ages 4, 7, 10, and 12 separately for opposite-sex adoptive and same- and opposite-sex biological sibling pairs (there were too few same-sex adoptive siblings for meaningful calculations). With parent-reported data, comparisons of opposite-sex adoptive and biological siblings revealed no strong or consistent systematic bias due to the greater proportion of same-sex siblings among biological probands, although there was a

TABLE 6.8. Adjusted Pairwise Odds Ratios for the Presence of Any Somatic Complaints in Siblings

	Adopted Siblings			Biological Siblings		
Age	Odds ratio	95% CI	Pearson χ^2	Odds ratio	95% CI	Pearson χ^2
4	2.24	0.76–7.51	2.28	2.40	0.92–6.46	3.31
7	2.68	1.14–7.01	5.14*	2.52	1.07–5.97	4.56*
10	1.26	0.50–3.26	0.26	1.04	0.41–2.89	0.21
12	1.78	0.71–4.46	1.53	1.60	0.60–4.29	0.90

*$p < .05$

trend for greater aggregation in biological siblings with increasing age. Among biological siblings there were no consistent differences in aggregation between same-sex and opposite-sex pairs.

Comparisons of sibling correlations and odds ratios from teacher's reports, on the other hand, did reveal a bias for biological sibling similarity. In almost all cases, aggregation among opposite-sex biological siblings was significantly lower than for same-sex pairs, and thus may be a source of bias for biological sibling correlations and risk estimates, as they contain a greater proportion of same-sex sibling pairs than the adoptive sample. At all assessment ages, risk estimates for opposite-sex biological siblings were less than half of those computed for all biological sibling pairs (see also table 6.7). It is also important to note that, due to the small number of somatic symptoms reported by teachers, many calculations, especially those assessing sex differences, were either not possible or were accompanied by very large confidence intervals.

Discussion

In this chapter we attempted to better identify and characterize a profile of somatic complaints from early childhood to early adolescence using data from the CAP. Consistent with earlier reports (Achenbach et al., 1989; Apley, 1975; Garber et al., 1991; Oster, 1972), somatic complaints were common in children, especially when using parental reports. More than one-third of all probands in our study reported one or more somatic complaints at age 4; by age 12, more than half had at least one somatic symptom. Fatigue and stomachache were the most common symptoms in early childhood, while headache and other aches and pains gained prominence in later childhood years. Also consistent with the reports, no significant sex differences in the frequency or distribution of most somatic symptoms were observed, although stomachaches were reported more for female probands. Also, somatic complaints, in general, increased with age for both sexes. Garber et al. (1991), using cross-sectional data, reported that somatization may peak in late childhood and early adolescence. Our longitudinal data are consistent with their finding, although we did not yet follow children into middle and late adolescence, so we cannot be sure if a peak was in fact reached.

Despite a general trend for increased symptom reporting in late childhood, we also found evidence for symptom continuity from one assessment age to the next, as well as across longer periods of time. For almost all somatic complaints, the presence of a symptom at one assessment period significantly increased the risk of experiencing that same symptom at the next assessment age. Furthermore, symptom continuity could be observed over multiple assessment periods, in most cases even from age 4 to age 12. Using a community-based sample, Hotopf, Carr, Mayou, Wadsworth, & Wessely (1998) reported that 10% of children reporting abdominal pain at age 7 continued to do so at ages 11 and 15. These authors also report that persistent abdominal complaints in childhood modestly predict medically unexplained physical symptoms in adulthood, but that they are more strongly associated with adult anxiety and mood disorders. It is unknown whether the same holds true for other types of childhood somatic complaints.

In attempting to identify a useful categorization of somatic symptoms in early childhood that would predict later somatic complaints, we find several of our findings supporting the utility of distinguishing children early on by a simple symptom/no symptoms dichotomy, rather than by extreme scores on somatic symptoms. First, children who report any somatic complaints at age 4 showed significantly higher risk for several later symptoms. While the risk of reporting future symptoms does increase somewhat for those with more extreme somatic complaints in early childhood, the predictive value of a more simple dichotomous distinction is remarkable, given that increased risk is sustained not only over an eight-year period but, more important, over an eight-year period of rapid change across multiple developmental periods. Second, children who report one or more somatic complaint at age 4 are nearly twice as likely to report multiple symptoms at age 12. Again, while the risk of multiple complaints at later ages is slightly higher when more extreme group categories are used at age 4, the simpler binary distinction achieves impressive discrimination given the ages studied. Finally, sibling pairs concordant for symptom status become all but absent when applying more extreme criteria.

While not specifically addressing the complex syndrome of somatization disorder, these results are supportive of efforts to reconceptualize somatization in terms of chronic somatic complaints that may not necessarily number into the extreme range required by DSM, but still discriminate those at high risk for disability. Escobar et al. (1987, 1998), for example, noting the low prevalence of somatization disorder, have proposed a more inclusive notion of somatization with a required number of somatic symptoms well below the diagnostic criteria of recent and current DSM somatization disorder. Compared to prevalences of standard somatization disorder of 0.03 and 0.5% in the community and primary care setting, respectively, 4.4 and 22% meet criteria for the abridged construct. More important, the success of these more inclusive criteria in identifying individuals with psychopathology and physical disability was not improved upon by the addition of the standard diagnosis (Escobar et al., 1998).

Consistent with previous twin (Torgensen, 1986) and adoption (van den Oord, Boomsma, & Verhulst, 1994) studies, we found no strong support in our data for a genetic contribution to a childhood somatic complaints syndrome. Sibling aggregation of somatic complaints was found, however, for both biological and adoptive siblings, indicating a role for shared environmental influences on this behavioral profile, especially in early and middle childhood. Several potential mechanisms have been proposed to explain the development and maintenance of somatization, many of which would influence siblings similarly. Some of these include classical and operant conditioning, social learning, response to traumatic or stressful life events or conditions, and even physician behaviors. Several studies have found that recurrent abdominal pain in childhood is associated with ill health and neuroticism in parents (Hodges, Kline, Barbero, & Flanery, 1984; Hotopf et al., 1998; Walker, Garber & Greene, 1993; Walker & Greene, 1989). Thus, parental anxiety and preoccupation with illness may, in turn, reinforce attentiveness and preoccupation with somatic complaints in their children (Hotopf et al., 1998). While we did not examine psychosocial factors associated with somatic complaints in our present analyses, many of them have been assessed in the CAP and will be the subject of future work.

Our lack of strong findings for a genetic influence on somatic complaints is inconsistent, though, with twin studies of related problem behaviors in children. Hewitt, Silberg, Neale, Eaves, & Erickson (1992), for example, reported high heritability for the CBCL Internalizing Scale, especially in boys. Similarly, twin studies of internalizing problems, such as anxiety and depression, also show significant genetic influences (Eaves et al., 1997; Eley, 1997; Rende, Plomin, Reiss, & Hetherington, 1993; Thapar & McGuffin, 1994; Wierzbicki, 1987). Contrary to these twin studies, however, a recent report of sibling adoption and parent-offspring analyses from the CAP (Eley, Deater-Deckard, Fombonne, Fulker, & Plomin, 1998) showed little evidence for heritability and a modest effect of shared environment on depression and anxiety symptoms in children age 7 to 12. Likewise, one twin study that presented separate analyses for children and adolescents showed that only shared environment significantly contributed to twin-pair similarity for parent-reported depressive symptoms in 8- to 11-year-olds (Thapar & McGuffin, 1994).

Finally, it is important to note that our conclusions in this chapter are drawn in large part from data obtained by parental reports of their children's somatic complaints. It is possible, and even likely, that parent reports introduce at least some bias to these measures, such as inflation in symptom continuity across assessment ages. In an attempt to address these concerns, all analyses were replicated using teacher reports, which are unlikely to be subject to these same potential biases. While a similar, but weaker, general trend in patterns of somatic complaints obtained by teacher reports was observed, the relatively small number of symptoms reported by teachers necessarily reduces confidence in these findings. Also, the use of teacher reports may help reduce the problem of common rater biases, but also may introduce other problems, such as the possibility that teachers are not well suited to report internalizing symptoms in their students. In this case self-report would, of course, be desirable, but, unfortunately, is not part of the large CAP assessment protocol for childhood years.

In summary, this chapter has provided an overview of the problem of somatic complaints in childhood, as well as a prospective analysis of somatic symptoms from early childhood to early adolescence. We have made a case for dichotomous risk profile for somatization characterized by the presence or absence of any somatic complaints in early childhood. Furthermore, we have demonstrated the utility of assessing individual symptoms that constitute syndrome classifications in order to better define a phenotype for empirical research. These results provide a foundation for further informed investigation of CAP data to better understand etiologic and maintaining processes in somatic complaints in childhood and adolescence and the role somatic complaints in other behavioral problems.

ACKNOWLEDGMENT

This work was supported in part by a Mentored Research Scientist Award (R. Rende) from the National Institute of Mental Health (MH01559-01).

REFERENCES

Achenbach, T. M. (1991). *Integrative guide for the 1991 CBCL/4-18, YRS, and TRF profiles.* Burlington: University of Vermont.
Achenbach, T. M., & Edelbrock, C. (1983). *Manual for the Child Behavior Checklist and Revised Child Behavior Profile.* Burlington: University of Vermont, Department of Psychology.
Achenbach, T. M., Conners, C. K., Quay, H. C., Verhulst, F. C., & Howell, C. T. (1989). Replication of empirically derived syndromes as a basis for taxonomy of child/adolescent psychopathology. *Journal of Abnormal Child Psychology, 17,* 299–323.
Apley, J. (1975). *The child with abdominal pain.* Oxford: Blackwell.
Eaves, L. J., Silberg, J. L., Meyer, J. M., Maes, H. H., Simonoff, E., Pickles, A., Rutter, M., Neale, M. C., Reynolds, C. A., Erikson, M. T., Heath, A. C., Loeber, R., Truett, K. R., & Hewitt, J. K. (1997). Genetics and develop-

mental psychopathology: 2. The main effects of genes and environment on behavioral problems in the Virginia Twin Study of Adolescent Behavioral Development. *Journal of Child Psychology and Psychiatry, 38*, 965–980.

Eley, T. C. (1997). Depressive symptoms in children and adolescents: Etiological links between normality and abnormality: A research note. *Journal of Child Psychology and Psychiatry, 38*, 861–866.

Eley, T. C., Deater-Deckard, K., Fombonne, E., Fulker, D. W., & Plomin, R. (1998). An adoption study of depressive symptoms in middle childhood. *Journal of Child Psychology and Psychiatry, 39*, 337–345.

Ernst, A. R., Routh, D. K., & Harper, D. C. (1984). Abdominal pain in children and symptoms of somatization disorder. *Journal of Pediatric Psychology, 9*, 77–85.

Escobar, J. I., Burnam, M. A., Karno, M., Forsythe, A., & Golding, J. M. (1987). Somatization in the community. *Archives of General Psychiatry, 44*, 713–718.

Escobar, J. I., Waitzkin, H., Silver, R. C., Gara, M., & Holman, A. (1998). Abridged somatization: A study in primary care. *Psychosomatic Medicine, 60*, 466–472.

Garber, J., Walker, L., & Zeman, J. (1991). Somatization symptoms in a community sample of children and adolescents: Further validation of the children's somatization inventory. *Psychological Assessment: A Journal of Consulting and Clinical Psychology, 3*, 588–595.

Garber, J., Zeman, J., & Walker, L. (1990). Recurrent abdominal pain in children: Psychiatric diagnoses and parental psychopathology. *Journal of the American Academy of Child and Adolescent Psychiatry, 29*, 648–656.

Goodman, J. E., & McGrath, P. J. (1991). The epidemiology of pain in children and adolescents: A review. *Pain, 46*, 247–264.

Hewitt, J. K., Silberg, J. L., Neale, M. C., Eaves, L. J., & Erickson, M. (1992). The analysis of parental ratings of children's behavior using LISREL. *Behavior Genetics, 22*, 293–317.

Hodges, K., Kline, J. J., Barbero, G., & Flanery, R. (1984). Life events occurring in families of children with recurrent abdominal pain. *Journal of Psychosomatic Research, 28*, 185–188.

Hodges, K., Kline, J. J., Barbero, G., & Woodruff, C. (1985). Anxiety in children with recurrent abdominal pain and their parents. *Psychosomatics, 26*, 859–866.

Hotopf, M., Carr, S., Mayou, R., Wadsworth, M., Wessely, S. (1998). Why do children have chronic abdominal pain, and what happens to them when they grow up? Population based cohort study. *British Medical Journal, 316*, 1196–1200.

Jolly, J. B., Wherry, J. N., Wiesner, D. C., Reed, D. H, Rule, J. C., & Jolly, J. M. (1994). The mediating role of anxiety in self-reported somatic complaints of depressed adolescents. *Journal of Abnormal Child Psychology, 22*, 691–702.

Katon. W., Lin, E., Von Korff, M., Russo, J., Lipscomb, P., & Bush, T. (1991). Somatization: A spectrum of severity. *American Journal of Psychiatry, 148*, 34–40.

Last, C. G. (1991). Somatic complaints in anxiety disordered children. *Journal of Anxiety Disorders, 5*, 125–138.

Livingston, R., Taylor, J. L., & Crawford, S. L. (1988). A study of somatic complaints and psychiatric diagnosis in children. *Journal of the American Academy of Child and Adolescent Psychiatry, 27,* 185–187.

McCauley, E., Carlson, G. A., & Calderon, R. (1991). The role of somatic complaints in the diagnosis of depression in children and adolescents. *Journal of the American Academy of Child and Adolescent Psychiatry, 230,* 631–635.

McGrath, P., & McAlpine, L. M. (1993). Psychological perspectives on pediatric pain. *Journal of Pediatrics, 122,* 52–58.

Oster, J. (1972). Recurrent abdominal pain, headache and limb pains in children and adolescents. *Pediatrics, 50,* 429–436.

Plomin, R., DeFries, J. C., & Fulker, D. W. (1988). *Nature and nurture during infancy and early childhood.* Cambridge: Cambridge University Press.

Rende, R., Plomin, R., Reiss, D., & Hetherington, E. M. (1993). Genetic and environmental influences on depression in adolescence: Etiology of individual differences and extreme scores. *Journal of Child Psychology and Psychiatry, 34,* 1387–1398.

Rende, R., Warner, V., Wickramaratne, P., & Weissman, M. M. (1999). Sibling aggregation for psychiatric disorders in offspring at high and low risk: Ten-year follow-up. *Psychological Medicine, 28,* 1291–1298.

Rende, R., & Weissman, M. M. (1999). Sibling aggregation for psychopathology in offspring of opiate addicts: Effects of parental comorbidity. *Journal of Clinical Child Psychology, 28,* 342–348.

Rende, R., Wickramaratne, P., Warner, V., & Weissman, M. M. (1995). Sibling resemblance for psychiatric disorders in offspring at high and low risk for depression. *Journal of Child Psychology and Psychiatry, 36,* 1353–1363.

Thapar, A., & McGuffin, P. (1994). A twin study of depressive symptoms in childhood. *British Journal of Psychiatry, 165,* 259–265.

Torgensen, S. (1986). Genetics of somatoform disorders. *Archives of General Psychiatry, 43,* 502–505.

Van den Oord, E. J. C. G., Boomsma, D. I., & Verhulst, F. C. (1994). A study of problem behaviors in 10- to 15-year-old biologically related and unrelated international adoptees. *Behavior Genetics, 24,* 193–205.

Walker, L. S., Garber, J., & Greene, J. W. (1993). Psychosocial correlates of recurrent childhood pain: a comparison of pediatric patients with recurrent abdominal pain, organic illness and psychiatric disorders. *Journal of Abnormal Psychology, 102,* 248–258.

Walker, L. S., & Greene, J. W. (1989). Children with recurrent abdominal pain and their parents: More somatic complaints, anxiety, and depression than other patient families? *Journal of Pediatric Psychology, 14,* 231–243.

Wasserman, A. L., Whitington, P. F., & Rivera, F. P. (1988). Psychogenic basis for abdominal pain in children and adolescents. *Journal of the American Academy of Child and Adolescent Psychiatry, 27,* 179–184.

Wierzbicki, M. (1987). Similarity of monozygotic and dizygotic child twins in level and lability of subclinically depressed mood. *American Journal of Orthopsychiatry, 57,* 33–40.

Zuckerman, B., Stevenson, J., & Bailey, V. (1987). Stomachaches and headaches in a community sample of preschool children, *Pediatrics, 79,* 677–682.

Teacher and Mother Ratings
of Attention Problems

Introduction

Attention problems are highly prevalent, with rates of occurrence depending on the definition used, and make up to 50% of child psychiatric cases (Cantwell, 1996). Previous behavior genetic studies of attention problems have analyzed either twin, family, or adoption data, and reported heritability estimates ranging from .55 to .98 (Cantwell, 1996; LaBuda, Gottesman, & Pauls, 1993; Waldman, Levy, & Hay, 1995). Adoption studies can provide a direct test for the presence of shared environmental influences by examining the similarity between unrelated siblings growing up in the same family; however, the power will be relatively low in small samples. Some researchers have questioned whether results based on twin or adoptees can be generalized due to their unique situation (Levy, Hay, McLaughlin, Wood, & Waldman, 1996; Vandell, 1990) while a number of studies reported no differences to the general population (Kendler, 1993). Previous studies reported that adoptees show an increased incidence of attention problems, as high as 13 to 21% in some samples (Deutsch et al., 1982).

A number of studies, particularly family studies on attention problems, focused on data mainly from boys (e.g., Biederman et al., 1996) while Faraone, Biederman, Keenam, and Tsuang (1991) reported on a sample of girls only. Twin studies usually recruited same-sex pairs

(Gjone, Stevenson, & Sundett, 1996; Hay, Levy, McLaughlin, & Wood, 1996; Thapar, Hervas, & McGuffin, 1995) while one study (Heath et al., 1996; Hudziak et al., 1998) reported on data from twin girls only. Data from same-sex twin pairs are sufficient to test for sex differences in the amount of genetic and environmental variation to the phenotype of interest; however, it is necessary to collect data for opposite-sex twins and brother-sister sib pairs to resolve whether they involve the same or different sets of genetic and/or environmental factors. Additionally, most twin studies have been cross-sectional, meaning that changes in the contribution of genetic and environmental factors to the developmental course of attention problems could not be tested.

Different criteria are proposed in *DSM-IV* (American Psychiatric Association, 1994) for disorders that are mainly (1) inattention problems, (2) those mainly concerned with hyperactivity/impulsivity, or (3) those of the combined type. Pelham, Evans, Gnagy, and Greenslade (1992) factor-analyzed teacher questionnaire data on boys in special education classes, resulting in two factors reflecting inattention and impulsivity, backing such diagnostic distinction from a data analytic point of view. Similarly, Healey et al. (1993) suggested that hyperactivity might result from response inhibition while inattention problems were associated with visual search patterns. Achenbach (1996), in response to requests from clinicians and researchers, suggested dividing the attention-problem scale of the Teacher Rating Form (TRF; Achenbach, 1991b), publishing the results of a factor analysis from a large referred sample.

While inattention problems seem to be constant over time, hyperactivity seems to decline with age (Hart, Lahey, Loeber, Applegate, & Frick, 1995; Lahey et al., 1994). While probably related to age and sex, these differences between aspects of attention problems occur similarly in all ethnic groups (Lahey et al., 1994). Hay et al. (1996) reported a high genetic correlation between hyperactivity and inattention, but this was not replicated by Heath et al. (1996). In a report using an all-female sample, their data suggested a genetic correlation of .50, indicative of partially distinct etiologies. The discrepancies between these preliminary results might be due to differences in age and sex in the sample compositions. Supporting differences in diagnoses with data on similar or differing genetic and environmental pathways—that is, comparing the magnitude of genetic and environmental correlations—could in turn have consequences on the timing and the strategies of clinical interventions. This has consequences on whether a family or classroom intervention (due to the presence of shared environmental correlations) or an individual program (due to the presence of nonshared environmental correlations) is to be preferred. It is also hypothesized that genetic factors play a larger role for inattention than hyperactivity, due to its greater persistence. An important question, to both clinicians and researchers, is that of (1) the validity of screening instruments and, related to that, (2) the application of findings from the general population to a clinical sample. Generally, the Child Behavior Checklist (CBCL; Achenbach, 1991a) and the matching instruments for other informants, the TRF and the Youth-Self Report (YSR; Achenbach, 1991c) have shown a good correspondence to clinically derived diagnosis. Barkley (1988) attests the attention scale's concurrent validity to the Conners scales, the Werry-Weiss-Peters activity rating scale, and the DISC-P, as well as its adequate discriminant validity (see also reports by Biederman, Faraone, Doyle, et al., 1993; Jensen, Salzberg, Richters, & Watanabe, 1993; Nadder et al., 1998).

Inattention and hyperactivity problems show continuous variation—that is, the clinical expression of ADHD is not of a qualitatively different diagnostic class but the extreme of a normal distribution. Gjone et al. (1996) tested this by exploring differential heritabilities for varying cut-off points with CBCL data (increasing by half standard-deviations above the mean) and did not detect a change of parameter estimates. Levy, May, McStephen, Wood, & Waldman (1997), who used a *DSM-III-R*–based maternal rating scale, came to the same conclusion, namely that ADHD is the extreme of a distribution, rather than a discrete disorder unrelated to the rest of the population.

Another potential problem is that of the agreement between different informants. In the literature, the agreement between parents and teachers has been reported as moderate (Achenbach, McConaughy, & Howell, 1987; McConaughy, Stanger, & Achenbach, 1992). Fergusson, Horwood, and Lynskey (1993) report an agreement of $r = .40$ between teachers and parents, while Stevenson in his twin study (1992) only obtained a correlation of .25. These low correlations might reflect the children's different behavior in the home and the classroom situation. While a good agreement between informants is usually desirable, a low correlation between teachers and parents might be indicative of a less severe problem (*DSM-IV* requires symptoms to be present in two or more settings for a diagnosis).

This difference between raters might be of practical interest in pointing toward differences in severity. Biederman, Faraone, Milberger, and Doyle (1993) concluded that even though the agreement between parents and teachers on single items was low, there was a high probability of teachers endorsing a sufficient number of items for a diagnosis when the same endorsement level was observed in parents.

Factor analyses of the Attention Problems scale of the Teacher Report Form (TRF) on a large clinically referred sample (Achenbach, 1996) showed two subscales, roughly corresponding to the *DSM-IV* classifications of mainly inattentive and mainly hyperactive-impulsive. We performed factor analyses of TRFs (Schmitz, 1997) for children from the normal range, at 7 and 12 years of age. These analyses replicated Achenbach's (1996) results, indicating that are there no differences in factor structure for the TRF Attention Problems scale between a general population and a clinical sample. Thus the measurement properties of the questionnaire were shown to be assessing the same latent constructs across very different samples.

Using teacher-reported data and adoption data will complement results reported on twin studies with mainly maternal ratings (Goodman & Stevenson, 1989a,b; Stevenson, 1992; Eaves et al., 1993; Thapar, Hervas, & McGuffin, 1995; Gjone, Stevenson, & Sundett, 1996; Zahn-Waxler, Schmitz, Fulker, Robinson, & Emde, 1996; Levy et al., 1997). In summary, the twin study results can be mainly characterized by high estimates of heritability (between .60 and .95); and no differential heritabilities—that is, severely affected children did not show a more heritable trait. Rhee, Waldman, Hay, and Levy (1999) reported differences in genetic and environmental influences on ADHD for boys and girls in that the shared environment played a role for girls and there were dominance effects for boys. Most of those studies were cross-sectional, and it is unclear whether differences in parameter estimates across studies are due to age differences. Sibling contrasts are postulated as an explanation of the low DZ within-pair correlations (e.g., Thapar et al., 1995), and the few teacher reports resulted in lower estimates of heritabilities than parental (mainly maternal) ratings.

Most of the studies published prior to 1995 used either *DSM-III* or *DSM-III-R* criteria. There are a few behavior genetic studies using *DSM-IV* approaches that differentiate ADHD problems into those of a more inattentive or a more hyperactive nature, as will be done for this chapter for the teacher ratings. Results reported by Heath et al. (1996) on a sample of 13-year-old girls and their diagnostic interviews (self-report) showed a heritability for inattention of .88 and for hyperactivity of .89, with a genetic correlation of .52. Hay et al. (1996) used a sample of boys and girls, ages 4 to 12 years, for whom they obtained maternal reports (questionnaires). Some shared environmental influences unique to hyperactivity were reported and a genetic correlation of .90 between hyperactivity and inattention.

Sherman, Iacono, and McGue (1997) reported lower parameter estimates, using teacher ratings and maternal interviews for 11- and 12-year-old boys. Teacher ratings resulted in a heritability of .39 for inattention and .69 for hyperactivity, and significant shared environmental variance for inattention. Maternal ratings resulted in higher estimates of heritability of .69 for inattention and .91 for hyperactivity. Using family data, Young (1998) reported low estimates for familial transmission for both inattention (transmissible factors accounted for 13% of the variance) and hyperactivity (26%), and genetic correlations between the ADHD subscales of .55 to .77. These estimates increased slightly when the sample was trichotomized into unaffected, slightly affected, and definitively affected. However, the results of this family study of transmissible factors accounted for just slightly more than a third of the variance—significantly less than what twin studies reported.

The present chapter will address the following questions: what is the estimated magnitude of genetic and environmental influences when adoption, rather than twin, data are analyzed? What are parameter estimates when teacher, rather than parental, ratings are used? Are these estimates different for the inattention versus hyperactivity aspects? To what extent do mother and teacher ratings correlate?

Methods

Sample and Measures

Annually, the teachers completed the TRF (Teacher Report Form; Achenbach, 1991b), a screening instrument with 112 items regarding problem behavior, constructed similarly to the Child Behavior Checklist (CBCL/4–18; Achenbach, 1991a) used for the mother ratings. For each of these items the teacher and the mother is asked to rate the child as either never showing that behavior (score of 0), occasionally showing that behavior (score of 1), or definitively showing that behavior (score of 2). The Attention Problems scale is constructed by the summation of the teacher's responses over 20 items; the factor-analytically-derived subscales of Inattention and Hyperactivity contain 14 and 10 items, respectively, with an overlap of four items (Achenbach, 1996). Separate analyses that excluded the overlapping items showed a slight attenuation of the correlations but did not change any conclusions. Factor analyses performed on the current sample at ages 7 and 12 replicated the reported factor structure with internal consistencies of .90 and above. Sample sizes of individual teacher ratings range from 501 for age 7 to 351 at age 12. The maternal scale is the result of summation over 11 items and does not have subscales; sample sizes for mother ratings range from 501 at age 11 to 564 at age 10.

Analyses

Descriptive statistics will be used to describe the data in terms of mean scores, separately for each sex, and correlations across time. Within-pair correlations are then computed, which give an indication of genetic and environmental influences on behaviors. These influences are then formally estimated and tested with structural equation modeling techniques with the Mx program (Neale, 1997).

TABLE 7.1. Means and Standard Deviations, by Sex

	Age 7	Age 8	Age 9	Age 10	Age 11	Age 12	Averaged
Mother Ratings							
Attention Problems	***		**	*	(*)	*	***
Boys	3.14 ± 3.09		2.80 ± 2.93	2.74 ± 2.71	2.56 ± 2.83	2.68 ± 2.98	2.86 ± 2.62
Girls	2.30 ± 2.47		2.22 ± 2.55	2.21 ± 2.52	2.11 ± 2.45	2.12 ± 2.80	2.18 ± 2.15
Teacher Ratings							
Attention Problems	***	***	***	***	***	***	***
Boys	7.39 ± 8.41	7.68 ± 8.06	7.89 ± 8.26	8.36 ± 8.50	6.89 ± 7.43	7.70 ± 7.81	8.13 ± 7.12
Girls	4.65 ± 6.25	4.64 ± 6.89	5.09 ± 6.86	4.27 ± 5.76	4.05 ± 5.74	4.09 ± 6.00	5.39 ± 5.75
Inattention	***	***	**	***	***	***	***
Boys	4.97 ± 6.25	5.26 ± 5.94	5.40 ± 6.09	5.80 ± 6.26	4.88 ± 5.56	5.44 ± 5.85	5.84 ± 5.34
Girls	3.28 ± 4.83	3.38 ± 5.35	3.78 ± 5.50	3.19 ± 4.61	3.18 ± 4.76	3.20 ± 4.99	3.75 ± 4.33
Hyperactivity	***	***	***	***	***	***	***
Boys	4.45 ± 4.79	4.54 ± 4.80	4.54 ± 4.71	4.73 ± 4.87	3.78 ± 4.30	4.24 ± 4.40	4.47 ± 4.10
Girls	2.63 ± 3.60	2.49 ± 3.67	2.61 ± 3.49	2.20 ± 3.09	1.92 ± 2.84	1.84 ± 2.83	3.01 ± 3.22

Note: Mean differences significant with $p < .10$ (*), $p < .05$ *, $p < .01$ **, or $p < .001$ ***

TABLE 7.2. Attention Problems: Correlations Across Time

	Year 7	Year 8	Year 9	Year 10	Year 11	Year 12
Year 7	1.00	.56	.56	.52	.47	.43
Year 8		1.00	.57	.58	.47	.43
Year 9	.61		1.00	.60	.55	.55
Year 10	.58		.71	1.00	.58	.42
Year 11	.58		.64	.69	1.00	.46
Year 12	.48		.59	.64	.70	1.00

Notes: 1. Teacher ratings are in the upper, mother ratings in the lower triangle.
2. All correlations are significant at the $p < .001$ level.
3. Sample sizes ranged from 246 to 365 for teacher, and 462 to 547 for mother ratings.
4. Scores were sex-corrected and square root–transformed.

TABLE 7.3. Inattention: Correlations Across Time

	Year 7	Year 8	Year 9	Year 10	Year 11	Year 12
Year 7	1.00	.54	.57	.54	.44	.40
Year 8		1.00	.58	.59	.48	.44
Year 9			1.00	.59	.57	.54
Year 10				1.00	.54	.41
Year 11					1.00	.51
Year 12						1.00

Notes: 1. Teacher ratings only.
2. All correlations are significant at the $p < .001$ level.
3. Sample sizes ranged from 247 to 366.
4. Scores were sex-corrected and square root–transformed.

Results

Table 7.1 shows the means and standard deviations for the Attention Problems scale for mother and teacher ratings, as well as the Inattention and Hyperactivity subscales for each of the ages of 7 through 12 from teacher ratings, separately for boys and girls.

As was to be expected from the literature, boys were rated as showing a higher mean on all aspects of attention problems. Teacher-mother agreement on the overall Attention Problems scale ranged from .37 at age 7 to .49 at age 11. Mother ratings in particular show a decline of the mean scores at older ages.

Data transformations were performed prior to further analyses for two reasons: problem behavior scores in a nonclinical population show a J-shaped distribution and therefore a square root transformation was applied. Additionally, the data were sex-corrected due to the mean and variance differences for boys and girls. Tables 7.2 through 7.4 show the correlations across time for the Attention Problems scale and its

TABLE 7.4. Hyperactivity: Correlations Across Time

	Year 7	Year 8	Year 9	Year 10	Year 11	Year 12
Year 7	1.00	.56	.52	.51	.49	.44
Year 8		1.00	.55	.57	.47	.45
Year 9			1.00	.56	.51	.54
Year 10				1.00	.56	.42
Year 11					1.00	.43
Year 12						1.00

Notes: 1. Teacher ratings only.
2. All correlations are significant at the $p < .001$ level.
3. Sample sizes ranged from 249 to 368.
4. Scores were sex-corrected and square root–transformed.

TABLE 7.5. Sample Sizes (Pairs)

	Age 7	Age 8	Age 9	Age 10	Age 11	Age 12
Teacher Ratings						
Adopted sibs	51	49	46	33	31	28
Nonadopted sibs	62	49	57	49	49	29
Mother Ratings						
Adopted sibs	70		70	73	62	69
Nonadopted sibs	73		86	85	67	68

subscales from teacher ratings. Correlations are shown for the whole sample, since there were no differences between boys and girls regarding the magnitude of these correlations. All correlations indicate attention problems are rated as fairly stable across years and similar for mother and teacher ratings; this is all the more impressive, since a different teacher rated the children in each grade. These correlations were similar in magnitude for adopted and nonadopted children; the few statistically significant differences that were detected went both ways— that is, no group showed more consistent stability than the other.

Similar to the results reported by Hay et al. (1996), inattention and hyperactivity are highly correlated throughout the grade-school period, with rs ranging between .82 and .86. This suggests that, at least during this period, the distinction between "mainly hyperactive" and "mainly inattentive" might not be useful with respect to teacher ratings.

To assess the etiology of individual differences, genetically informative samples such as the current one are needed. In the case of the CAP sample, differences between adopted (genetically unrelated) and nonadopted (sharing 50% of their segregating genes, on average) siblings are investigated. The numbers of sibling pairs available for teacher ratings are shown in the top half of table 7.5, and for mother ratings in

TABLE 7.6. Within-Pair Correlations,
Teacher Ratings

	Adopted	Nonadopted
Attention Problems		
Age 7	.20	.15
Age 8	.29*	.14
Age 9	.10	.25
Age 10	.04	.54***
Age 11	−.12	.44***
Age 12	.25	.37*
Averaged	.17	.38***
Inattention		
Age 7	.19	.17
Age 8	.31*	.07
Age 9	.18	.16
Age 10	.09	.46***
Age 11	−.12	.45***
Age 12	.26	.42*
Averaged	.21	.36***
Hyperactivity		
Age 7	.22	.02
Age 8	.25	.25
Age 9	.13	.27*
Age 10	.00	.45***
Age 11	−.18	.29*
Age 12	.25	.23
Averaged	.20	.30**

$*p < .05$, $**p < .01$, $***p < .001$.
For sample sizes, see table 7.5.

the bottom half. Usually, there are a few more sibling pairs available for the nonadopted sample than the adopted one. There is a marked difference between the adopted and nonadopted samples in that the adopted sibling sample consists mainly of brother-sister pairs while the nonadopted sibling sample contains slightly more same-sex than opposite sex pairs. However, the current sample is too small to take these composition differences into account.

Within-pair correlations were calculated as an initial check on the genetic and environmental etiology of attention problems. These correlations are presented in table 7.6 for teacher and in table 7.7 for mother ratings, respectively. Correlations for ratings averaged over the grade-school years, increasing the reliability of ratings, are shown as well.

While correlations of teacher ratings for the ages 9 through 11 seem to indicate genetic influences, due to the higher within-pair correlations for the nonadopted siblings, this picture is not that clear for the other

TABLE 7.7. Within-Pair Correlations,
Mother Ratings

	Adopted	Nonadopted
Attention Problems		
Age 7	.13	.17
Age 9	−.01	.14
Age 10	.10	.24*
Age 11	.06	.29*
Age 12	−.04	.20
Averaged	.06	.33***

$*p < .05$, $**p < .01$, $***p < .001$.
For sample sizes, see table 7.5.

three age points. Analyses incorporating sibling contrast effects, as suggested by some reports in the twin literature (Eaves et al., 1997; Thapar et al., 1995), however, showed nonsignificant results regarding the contrasting of the siblings behavior. This is not too surprising, considering that these teacher ratings were done three years apart, on average, therefore less open to comparisons between the children as might happen when a mother rates both her twins at the same time. Additionally, the members of a sibling pair might have had different teachers—information that is not readily available for this sample. Comparing within-pair correlations separately for those sibling pairs who are of the same sex and brother-sister pairs did not reveal significant differences in the magnitude of these correlations. Larger sample sizes are needed to obtain sufficient power to detect possible sex differences. Within-pair correlations of maternal ratings are higher for nonadopted than adopted siblings at all age points, particularly pronounced so for the averaged score.

Based on these within-pair correlations, covariance matrices were generated as data input for structural equation modeling. The results from fitting univariate behavior genetic models to mother and averaged teacher ratings are shown in tables 7.8 and 7.9. As expected, parameter estimates for the genetic variance estimate of attention problems derived from teacher ratings is lower than that from maternal ratings (e.g., Sherman et al., 1997). However, the magnitude of heritability estimates resulting from mother ratings ($h^2 = .64$) is considerably lower than the .80 to .90+ range reported from twin studies. Estimated heritabilities for the two aspects of attention problems, inattention and hyperactivity, are similar in magnitude.

In fitting models to all the time points obtained from mother ratings, it was possible to reduce the model to one or two genetic factors common to all ratings (results not shown). This implies a genetic con-

TABLE 7.8. Parameter Estimates, Averaged
Teacher Ratings

Scale	h^2	c^2	e^2	χ^2_3	p
Attention	.39	.16	.45	3.74	.29
Inattention	.35	.19	.46	8.21	.04
Hyperactivity	.12	.19	.68	0.90	.83

h^2 denotes variance due to additive genetic influences, c^2 variance due to shared environmental influences, and e^2 variance due to non-shared environmental influences, including measurement error; χ^2_3 and its associated p-value indicate the goodness of fit.

TABLE 7.9. Parameter Estimates, Mother Ratings

Scale	h^2	c^2	e^2	χ^2_3	p
Year 7	.13	.12	.75	1.00	.80
Year 9	.31	.00	.69	1.55	.67
Year 10	.42	.08	.50	3.34	.34
Year 11	.40	.05	.55	2.00	.57
Year 12	.46	.00	.54	6.23	.10
Averaged	.64*	.06	.30	6.78	.08

*$p < .05$. h^2 denotes variance due to additive genetic influences, c^2 variance due to shared environmental influences, and e^2 variance due to nonshared environmental influences, including measurement error; χ^2_3 and its associated p-value indicate the goodness of fit.

tinuity to attention problems during the grade-school years. Teacher ratings did not replicate this; the reasons for this need to be explored further.

Discussion

This study replicates the findings that boys are rated as having more attention problems, by both mothers and teachers. Separating the overall score into the aspects of inattention and hyperactivity follows the overall picture in that boys seem to show more of these behaviors than girls. For both sexes, the behaviors were rated as relatively stable during the grade-school years.

The teacher-mother agreement is moderate but slightly higher than the average reported in the literature (Achenbach et al., 1987). Possible reasons for a disagreement between parent and teacher ratings can be that a teacher bases his ratings on his experience from a lot of observations while mothers have a different reference frame. On the other hand, it is possible that children exhibit behaviors to a differing degree at school and in the home and that therefore the observed correlation reflects a real difference. A further factor that might influence the correlation between the two raters is that the two scales are based on a different number of items, with teachers rating nearly twice as many aspects of a child's behavior than mothers do.

The slightly higher across-time correlations for mothers could be due in part to rater bias, meaning that the observed continuity is partly mediated by the mother's perception of her children. However, the difference between between mother and teacher correlations needs to be more pronounced for rater bias to be considered and is negligible in the current study.

There is a marked difference between the adopted and nonadopted samples in that the adopted sibling sample consists mainly of brother-sister pairs while the nonadopted sibling sample contains slightly more same-sex than opposite-sex sibling pairs. For the current analyses this was not of importance, since no statistically significant differences could be detected in the within-pair correlations for same and opposite-sex sibling pairs (correlations available upon request).

The heritability estimates obtained from teacher ratings were lower than those obtained from parental ratings in the twin literature; however, estimates obtained from mother ratings ($h^2 = .64$) also are lower than those reported from twin ratings. This raises the possibility that heritability estimates from twin studies might have been inflated due the presence of nonadditive genetic effects, which tend to increase the MZ-correlation comparative to the DZ-correlation.

The high observed correlation between the inattention and hyper-activity aspects of attention problems (average $r = .84$) is in contrast to Heath et al. (1996), who reported a moderate correlation of $r = .50$. This needs to be further investigated, particularly with respect to co-occurring problems to both subscales.

A limitation to our results is the small sample size of sibling pairs on which they are based, particularly for teacher ratings at older ages. A possible explanation for the difference in decline of number of sibling pairs for teacher versus mother ratings could be that a teacher rating was obtained for one sibling but not the other, while usually mothers completed the questionnaires for both children.

REFERENCES

Achenbach, T. M. (1991a). *Manual for the Child Behavior Checklist/4–18 and 1991 Profile*. Burlington: University of Vermont, Department of Psychiatry.
Achenbach, T. M. (1991b). *Manual for the Teacher's Report Form and 1991 Profile*. Burlington: University of Vermont, Department of Psychiatry.

Achenbach, T. M. (1991c). *Manual for the Youth Self-Report and 1991 Profile.* Burlington: University of Vermont, Department of Psychiatry.

Achenbach, T. M. (1993). *Empirically based taxonomy: How to use syndromes and profile types derived from the CBCL/4–18, TRF, and YSR.* Burlington: University of Vermont, Department of Psychiatry.

Achenbach, T. M. (1996). Subtyping ADHD: A request for suggestions about relating empirically based assessment to DSM-IV. *The ADHD Report, 4,* 5–9.

Achenbach, T. M., McConaughy, S. H., & Howell, C. T. (1987). Child/ adolescent behavioral and emotional problems: Implications of cross-informant correlations for situational specificity. *Psychological Bulletin, 101,* 213–232.

American Psychiatric Association. (1994). *Diagnostic and statistical manual of mental disorders* (4th ed.). Washington, DC: American Psychiatric Association.

Barkley, R. A. (1988). Child behavior rating scales and checklists. In M. Rutter, A. H. Tuma, & I. S. Lann (Eds.), *Assessment and diagnosis in child psychopathology* (pp. 113–155). New York/London: Guilford Press.

Biederman, J., Faraone, S., Mick, E., Wozniak, J., Chen, L., Ouellette, C., Marrs, A., Moore, P., Garcia, J., Mennin, D., & Lelon, E. (1996). Attention-deficit hyperactivity disorder and juvenile mania: An over-looked comorbidity? *Journal of the American Academy of Child and Adolescent Psychiatry, 35*(8), 997–1008.

Biederman, J., Faraone, S. V., Doyle, A., Lehman, B. K., Kraus, I., Perrin, J., & Tsuang, M. T. (1993). Convergence of the Child Behavior Checklist with structured interview-based psychiatric diagnoses of ADHD children. *Journal of Child Psychology and Psychiatry, 34*(7), 1241–1251.

Biederman, J., Faraone, S. V., Milberger, S., & Doyle, A. (1993). Diagnoses of attention-deficit hyperactivity disorder from parent reports predict diagnoses based on teacher reports. *Journal of the American Academy of Child and Adolescent Psychiatry, 32*(2), 315–317.

Cantwell, D. P. (1996). Attention deficit disorder: A review of the past 10 years. *Journal of the American Academy of Child and Adolescent Psychiatry, 35*(8), 978–987.

Deutsch, C. K., Swansson, J. M., Bruell, J. H., Cantwell, D. P., Weinberg, F., & Baren, M. (1982). Overrepresentation of adoptees in children with the attention deficit disorder. *Behavior Genetics, 12*(2), 231–238.

Eaves, L. J., Silberg, J., Hewitt, J. K., Meyer, J., Rutter, M., Simonoff, E., Neale, M., & Pickles, A. (1993). Genes, personality, and psychopathology: A latent class analysis of liability to symptoms of attention-deficit hyper-activity disorder in twins. In R. Plomin & G. E. McClearn (Eds.), *Nature, nurture, and psychology* (pp. 285–303). Washington, DC: American Psychological Association.

Eaves, L. J., Silberg, J. L., Meyer, J. M., Maes, H. M., Simonoff, E., Pickles, A., Rutter, M., Neale, M. C., Reynolds, C. A., Erikson, M. T., Heath, A. C., Loeber, R., Truett, K. R., & Hewitt, J. K. (1997). Genetics and develop-mental psychopathology: 2. The main effects of genes and environment on behavioral problems in the Virginia Twin Study of Adolescent Behavioral Development. *Journal of Child Psychology and Psychiatry, 38*(8), 965–980.

Faraone, S. V., Biederman, J., Keenan, K., & Tsuang, M. T. (1991). A family-genetic study of girls with DSM-III attention deficit disorder. *American Journal of Psychiatry, 148*, 112–117.

Fergusson, D. M., Horwood, L. J., & Lynskey, M. T. (1993). The effect of conduct disorder and attention deficit in middle childhood on offending and scholastic ability at age 13. *Journal of Child Psychology and Psychiatry, 34*(6), 899–916.

Gjone, H., Stevenson, J., & Sundett, J. M. (1996). Genetic influence on parent-reported attention-related problems in a Norwegian general population twin sample. *Journal of the American Academy of Child and Adolescent Psychiatry, 35*(5), 588–598.

Goodman, R., & Stevenson, J. (1989a). A twin study of hyperactivity—II. The aetiological role of genes, family relationships and perinatal adversity. *Journal of Child Psychology and Psychiatry, 30*, 691–709.

Goodman, R., & Stevenson, J. (1989b). A twin study of hyperactivity—I. An examination of hyperactivity scores and categories derived from Rutter teacher and parent questionnaires. *Journal of Child Psychology and Psychiatry, 30*, 671–689.

Hart, E. L., Lahey, B. B., Loeber, R., Applegate, B., & Frick, P. J. (1995). Developmental change in attention-deficit hyperactivity disorder in boys: A four-year longitudinal study. *Journal of Abnormal Child Psychology, 23*(6), 729.

Hay, D. A., Levy, F., McLaughlin, M., & Wood, K. (1996). A developmental behaviour genetic approach to attention deficit hyperactivity disorder. *Abstracts of the XIVth Biennial Meetings of the International Society for the Study of Behavioural Development*, 488 (abs).

Healey, J. M., Newcorn, J. H., Halperin, J. M., Wolf, L. E., Pascualvaca, D. M., Schmeidler, J., & O'Brien, J. D. (1993). The factor structure of ADHD items in DSM-III-R: Internal consistency and external validation. *Journal of Abnormal Child Psychology, 21*(4), 441–453.

Heath, A. C., Hudziak, J., Reich, W., Madden, P. A. F., Slutske, W. S., Bierut, L., & Bucholz, K. K. (1996). The inheritance of hyperactivity and inattention in girls: Results from the Missouri Twin Study. *Behavior Genetics, 26*(6), 587 (abs).

Hudziak, J. J., Heath, A. C., Madden, P. F., Reich, W., Buchholz, K. K., Slutske, W., Bierut, L. J., Neuman, R. J., & Todd, R. D. (1998). Latent class and factor analysis of DSM-IV ADHD: A twin study of female adolescents. *Journal of the American Academy of Child and Adolescent Psychiatry, 37*(8), 848–857.

Jensen, P. S., Salzberg, A. D., Richters, J. E., & Watanabe, H. K. (1993). Scales, diagnoses, and child psychopathology: I. CBCL and DICA relationships. *Journal of the American Academy of Child and Adolescent Psychiatry, 32*(2), 397–406.

Kendler, K. S. (1993). Twin studies of psychiatric illness. *Archives of General Psychiatry, 50*, 905–915.

LaBuda, M. C., Gottesman, I. I., & Pauls, D. L. (1993). Usefulness of twin studies for exploring the etiology of childhood and adolescent psychiatric disorders. *American Journal of Medical Genetics (Neuropsychiatric Genetics), 48*, 47–59.

Lahey, B. B., Applegate, B., Biederman, J., Greenhill, L., Hynd, G. W., Barkley, R. A., Newcorn, J., Jensen, P., Richters, J., Garfinkel, B., Kerdyck, L., Frick, P. J., Ollendick, T., Perez, D., Hart, E. L., Waldman, I., & Shaffer, D. (1994). DSM-IV field trials for attention deficit hyperactivity disorder in children and adolescents. *American Journal of Psychiatry*, *151*(11), 1673–1685.

Levy, F., Hay, D., McLaughlin, M., Wood, C., & Waldman, I. (1996). Twin-sibling differences in parental reports of ADHD, speech, reading and behaviour problems. *Journal of Child Psychology and Psychiatry*, *37*(5), 569–578.

Levy, F., Hay, D. A., McStephen, M., Wood, C., & Waldman, I. (1997). Attention deficit hyperactivity disorder (ADHD): A category or a continuum? Genetic analysis of a large scale twin study. *Journal of the American Academy of Child and Adolescent Psychiatry*, *36*(6), 737–744.

McConaughy, S. H., Stanger, C., & Achenbach, T. M. (1992). Three-year course of the behavioral/emotional problems in a national sample of 4- to 16-year-olds: I Agreement among informants. *Journal of the American Academy of Child and Adolescent Psychiatry*, *31*(5), 932–940.

Nadder, T. S., Silberg, J. L., Eaves, L. J., Maes, H. H., & Meyer, J. M. (1998). Genetic effects on ADHD symptomatology in 7- to 13-year-old twins: Results from a telephone survey. *Behavior Genetics*, *28*(2), 83–99.

Neale, M. C. (1997). *Mx: Statistical modeling* (4th ed.). Richmond: Department of Psychiatry, Medical College of Virginia.

Pelham, W. E., Evans, S. W., Gnagy, E. M., & Greenslade, K. E. (1992). Teacher ratings of DSM-III-R symptoms for the disruptive behavior disorders: Prevalence, factor analyses, and conditional probabilities in a special education sample. *School Psychology Review*, *21*(2), 285–299.

Rhee, S. H., Waldman, I. D., Hay, D. A., & Levy, F. (1999). Sex differences in genetic and environmental influences on DSM-III-R attention-deficit/hyperactivity disorder. *Journal of Abnormal Psychology*, *108*(1), 24–41.

Schmitz, S. (1997). Quantitative genetic analyses of the TRF Attention Scale and its subscales. *Behavior Genetics*, *27*(6), 605 (abs).

Sherman, D. K., Iacono, W. G., & McGue, M. K. (1997). Attention-deficit hyperactivity disorder dimensions: A twin study of inattention and impulsivity-hyperactivity. *Journal of the American Academy of Child and Adolescent Psychiatry*, *36*(6), 745–753.

Stevenson, J. (1992). Evidence for a genetic etiology in hyperactivity in children. *Behavior Genetics*, *22*(3), 337–344.

Thapar, A., Hervas, A., & McGuffin, P. (1995). Childhood hyperactivity scores are highly heritable and show sibling competition effects: Twin study evidence. *Behavior Genetics*, *25*(6), 537–544.

Vandell, D. L. (1990). Development in twins. *Annals of Child Development*, 7, 145–174.

Waldman, I. D., Levy, F., & Hay, D. (1995). Multivariate genetic analyses of the overlap among DSM-III-R disruptive behavior disorder symptoms. *Behavior Genetics*, *25*(3), 293–294 (abs).

Young, S. E. (1998). *Family factors underlying attention deficit hyperactivity disorder and conduct disorder: A study of comorbidity.* Unpublished doctoral dissertation, University of Colorado, Boulder.

Zahn-Waxler, C., Schmitz, S., Fulker, D., Robinson, J. L., & Emde, R. (1996). Behavior problems in five-year-old MZ and DZ twins: Estimates of genetic and shared environmental influences. *Development and Psychopathology, 8,* 103–122.

Adopted and Nonadopted Adolescents' Adjustment

Introduction

Persons under the age of 18 adopted by nonrelatives constitute approximately 2–3% of the population of the United States (Brodzinsky, 1993). Although the majority of interfamilial adoptions have positive outcomes (Hoopes, 1982), there is some evidence to suggest that adoptees might be at an increased risk of maladjustment, especially when the adoptees are selected from clinical samples (e.g., Rogeness et al., 1988). Population-based cross-sectional studies have also sometimes found poorer levels of emotional and educational adjustment among adoptees, though findings from these studies have been less consistent, and have often indicated small or negligible behavioral differences between adopted and nonadopted persons (Haugaard, 1998).

There is some indication that these differences begin to emerge in middle childhood (Brodzinsky, 1993). Although adopted infants and preschool children do not differ significantly from their nonadopted peers, externalizing problems and conduct problems become more pronounced among adoptees in middle childhood. It is unclear, however, whether adopted children continue to exhibit greater levels of maladjustment during adolescence. Aside from a few longitudinal studies that have addressed this issue, little is known about adoptees' adjustment during the adolescent years. It is the aim of this chapter to review current research and to present data that explore adoptees' emotional, social, and scholastic adjustment in the transition from early to late adolescence.

Adoption as a Risk Factor for Psychiatric Morbidity and Maladjustment

In order to evaluate the psychological outcomes of adoption, three types of investigations have been conducted. These include (1) studies of the prevalence and type of psychiatric morbidity among adopted children in clinical populations, (2) population-based cross-sectional studies that compare adopted and nonadopted children's adjustment, and (3) longitudinal studies that examine adoptees' adjustment developmentally.

Studies of Clinical Populations

Studies of clinical populations have been primarily concerned with two issues: whether the number of adopted children referred to psychiatric clinics corresponds to that expected on the basis of the number of adopted children in the general population, and whether adopted children are more likely than nonadopted children to present specific disorders.

On the first issue, research suggests that there is a clear overrepresentation of adopted children within psychiatric settings. Epidemiological studies have reported that, in comparison to 2–3% of adopted children in the general population, approximately 3–15% of children attending psychiatric facilities are adopted (Brodzinsky, 1993; Deutsch et al., 1982; Kim, Davenport, Joseph, Zrull, & Wolford, 1988; Lipman, Offord, Boyle, & Racine, 1993; Rogeness et al., 1988; Sullivan, Wells, & Bushnell, 1995). For example, a retrospective investigation of child and adolescent psychiatric morbidity found that adopted children and adolescents represented 9.8% and 15%, respectively, of the overall adolescent inpatient population of an Ohio psychiatric hospital (Kim et al., 1988). Similarly, Rogeness et al. (1988) reported that 8.7% of children admitted to an inpatient psychiatric facility were adopted. These percentages are equivalent to an approximate fourfold increase in the risk of psychiatric morbidity among adoptees (e.g., Deutsch et al., 1982; Haugaard, 1998; Mech, 1973).

Clinical population studies have also suggested some specificity in the type of symptomatology that adopted children present. Externalizing and acting-out behaviors such as aggression, disobedience, overactivity, and conduct problems are reported to be more common among adopted children, and more frequently diagnosed in this group, than behaviors of a withdrawn or emotional type (Brodzinsky, 1993; Deutsch et al., 1982; Kim et al., 1988; Sullivan et al., 1995). In an investigation of the prevalence of attention deficit disorder among adoptees, Deutsch et al. (1982) found that adoption was related to an eightfold increase in the diagnosis of this disorder, a much greater risk than that associated with psychiatric morbidity more generally. Similarly, Sullivan et al. (1995), using data from the Christchurch Psychiatric Epidemiological Survey, found that adoptive status was a significant predictor of childhood conduct disorder, substance use, and antisocial personality disorder, but was not predictive of other psychiatric conditions such as depression, dysthymia, phobic disorder, and obsessive compulsive disorder.

Findings from clinical population studies must be interpreted with some caution, however. Small sample sizes and lack of suitably selected comparison groups are methodological limitations of these studies. Furthermore, factors that have been shown to predict poor emotional adjustment in children, such as prenatal complications, history of abuse and neglect, number of adoptive placement disruptions, and the nature of adoption (i.e., extra or intrafamilial), are often not well documented in studies using clinical populations (Hersov, 1990).

Those children who were adopted later in life, and as a result of early stressful experiences of neglect and abuse, are not surprisingly much more likely to present maladjusted behaviors than are children who were adopted as infants and raised in a loving and caring environment (Verhulst, Althaus, & Verluis-den-BiemanHerma, 1992). The failure to discriminate between these diverse early experiences in clinical investigations is, therefore, likely to result in an overestimation of psychiatric morbidity among adoptees (Haugaard, 1998). Greater psychiatric morbidity among adoptees in clinical samples may also be inflated by referral bias. For example, Warren (1992) found that adoptive parents were more likely than nonadoptive parents to seek help for minor behavioral problems. When more severe conditions were examined, however, the percentage of adopted and nonadopted youths seeking psychiatric treatment did not differ.

The greatest limitation of clinical population studies, however, is their poor generalizability to the wider population of adopted children (Haugaard, 1998).

Population-Based Cross-Sectional Studies

Population-based cross-sectional studies have addressed many of these methodological shortcomings by examining the outcomes of adoption in general population samples, and by comparing homogeneous groups of adopted children with nonadopted children from comparable birth or family backgrounds. It is not surprising that these studies have suggested a less dramatic conclusion. Indeed, the behavioral differences that are observed among adopted and nonadopted persons in the general population are often negligible and much smaller than those estimated by clinical comparison studies (Brodzinsky, 1993; Haugaard, 1998; Sharma, McGue, & Benson, 1998).

In order to evaluate the outcomes of adoption in samples of the general population, two different approaches have been employed and these yield very different results.

Studies that have compared adopted children with nonadopted children from similar birth circumstances, who remained with their biological parents, often in disadvantageous environments, have found better indices of psychological and scholastic adjustment among adoptees (Hodges & Tizard, 1989a,b). For example, a prospective comparison of children raised in institutional care up to age 8, and subsequently either adopted or restored to their biological parents, suggested improved levels of behavioral adjustment and cognitive development among adoptees in adolescence (Hodges & Tizard, 1989a,b). When alternatives to adoption are considered, research has also shown that adopted children are more likely than children brought up in permanent foster homes or residential care to develop more intimate and meaningful attachments and to experience positive feelings of belonging toward their adoptive family (Triseliotis & Hill, 1990). What has emerged from this research, therefore, clearly indicates that adoption is a positive experience for the lives of many children, and that adopted children are generally well adjusted (Hoopes, 1982).

Studies that compared adopted children with nonadopted children raised in intact biological two-parent families have, however, yielded mixed results. While some studies have found that adopted and nonadopted children do not differ on measures such as temperament, problem behavior, and attachment (e.g., Plomin & DeFries, 1985; Singer, Brodzinsky, Ramsay, Steir, & Everett, 1985), others have indicated better (Sharma, McGue, & Benson, 1996, 1998) or worse (e.g., Verhulst & Verluis-den-BiemanHerma, 1995) adjustment among adoptees.

The general pattern that emerges from these studies, however, suggests that few behavioral differences are apparent between adopted and nonadopted children during infancy and early childhood. For example, a study of 100 6- to 18-month-old infants indicated comparable patterns of attachment in adopted and nonadopted mother-infant pairs (Singer et al., 1985). Similarly, a study of infants and preschool children found comparable indices of cognitive development, psychomotor activity, and temperament characteristics in adopted and nonadopted children (Brodzinsky, 1993).

In contrast to these findings, studies of middle childhood suggest that it is at this time that adopted children, and especially adopted boys, begin to exhibit greater vulnerability to the development of conduct problems, externalizing behaviors, and poor school performance (e.g., Brodzinsky, Schechter, Braff, & Singer, 1984; Verhulst & Verluis-den-BiemanHerma, 1995).

In a study of 260 6- to 11-year-old adopted and nonadopted children, parents rated adoptees as less socially competent and more externalizing and internalizing on the Child Behavior Profile, a widely used instrument for the measurement of child behavioral problems (Brodzinsky et al., 1984). Teachers also rated adopted children as performing less well academically and exhibiting lower levels of scholastic adjustment than nonadopted children. The behavioral differences that were observed between adopted and nonadopted children were however slight and did not indicate clinical levels of pathology. A large-scale epidemiological study of 2,148 10- to 15-year-old interracial adoptees in the Netherlands (Verhulst, Althaus, & Verluis-den-BiemanHerma, 1989) also reported increased levels of externalizing and delinquent behaviors among adoptees—behaviors that were more accentuated among adopted boys as compared to adopted girls.

Overall, these investigations suggest that beginning in middle childhood, adopted children exhibit somewhat greater adjustment problems than nonadopted children, and that the majority of these problems fall in the category of externalizing behaviors (Brodzinsky, 1993).

When adoptees move into the adolescent years, the pattern of adjustment observed is less clear (Brodzinsky, 1993). Adolescence has been described as a particularly difficult transition for adopted children. Factors such as increased awareness of one's own biological roots and unknown genealogical links have, for example, been associated with adolescents' search behavior, problematic identity formation, low self-esteem, and feelings of anxiety (see Hoopes, 1990, for a review).

Studies comparing adopted and nonadopted adolescents have, however, yielded inconclusive results (Brodzinsky, 1993). In a recent investigation conducted in the United States, Sharma et al. (1998) compared 881 12- to 18-year-old adopted adolescents to their nonadopted siblings, and to a population norm on measures of psychological and scholastic adjustment (Achenbach Youths Self-Report). When compared to their nonadopted siblings, though the differences observed were small, adopted youths reported greater levels of illicit drug use, delinquent behavior, and poorer levels of scholastic adjustment (Sharma et al., 1998). However, a subsample of adopted adolescents, which excluded adolescents who had been referred for psychological help, denoted better adjustment on indices of withdrawn behavior among adopted boys, when compared to the population norms of the Achenbach Youths Self-Report. Although adopted girls exhibited better adjustment on indices of withdrawn behavior and social problems, they fared worse on measures of delinquent and externalizing behaviors, when compared to the Youth Self-Report norms.

Discrepant findings were reported in a study of a cohort of 1,265 New Zealand children: while no differences in indices of depression and anxiety were found between adopted and nonadopted adolescents, adopted adolescents reported lower levels of self-esteem (Fergusson, Lynskey, & Horwood, 1995).

Although these cross-sectional studies provide important information about the adjustment of adopted adolescents, they do not address directly important issues pertinent to developmental changes that might be apparent in adopted, as compared to nonadopted, persons' psychological, emotional, and scholastic functioning.

Longitudinal Investigations

A few longitudinal investigations have addressed these issues and have provided a good basis for the study of developmental trajectories in the adjustment of adopted and nonadopted persons.

A prospective longitudinal study conducted in Sweden examined emotional and scholastic adjustment in children who were initially registered for adoption, but who were subsequently brought up by their socially disadvantaged biological parents, placed in permanent foster homes, or adopted (Bohman & Sigdvarsson, 1985). Overall, adopted children fared better on both emotional and scholastic measures than did children brought up by their socially disadvantaged biological parents or children placed in foster care. However, when adopted children were followed up at ages 10 to 11 and compared to randomly selected classmates of the same sex, findings suggested elevated problem behavior and lower scholastic achievement among adopted boys, though not among adopted girls. The increased risk of maladjustment observed among adoptees at age 11 was, however, not observed at age 15, when no apparent differences in adjustment were found between adopted and nonadopted youths. Similarly, at age 23, no significant differences were observed in the frequency of criminal and drug-related offenses among adopted and nonadopted persons, thus suggesting similar patterns of adjustment for adopted and nonadopted persons in adolescence and adulthood.

The findings from this study largely corroborate data from the National Child Development Study conducted in the United Kingdom (Maughan & Pickles, 1990), a longitudinal prospective investigation of all children born in the United Kingdom in the week of 3–9 March 1958 (see Maughan & Pickles, 1990). As part of this investigation, two groups of illegitimately born children, one group adopted at birth and one group brought up by their biological parents, were compared at ages 11, 15, and 23 with a group of legitimate nonadopted children. Overall, adoptees exhibited better indices of adjustment than did illegitimate children who had remained with their biological parents, though they did show slightly poorer adjustment when compared to their legitimate nonadopted counterparts. The developmental patterns observed in this study suggested elevated rates of problem behavior among adoptees around middle childhood but improved indices of adjustment in adolescence. One exception was, however, observed for emotional problems at age 16, when adoptees had begun to exhibit greater indices of unhappiness and anxiety that were greater than those observed for both legitimate and illegitimate nonadopted adolescents. The authors suggested that the emergence of emotional problems in adolescence might be reflective of growing anxieties related to the experience of coming to terms with the adoptive experience.

Similar increased levels of internalizing and withdrawn behaviors among adoptees were reported in a follow-up study of 1,538 inter-country adoptees aged 14 to 18 years in the Netherlands, previously studied at ages 10 to 15 (Verhulst & Verluis-den-BiemanHerma, 1995). When compared to a sample of adolescents from the general population, adoptees had greater withdrawn, internalizing, and delinquent behavioral problems, which increased with age. The increase in problem behaviors observed among adoptees was in contrast to the decrease observed for a sample of nonadopted adolescents from the general population. Unlike Bohman and Sigdvarsson's (1985) findings of comparable indices of behavioral problems for adopted and non-adopted persons in adolescence, the findings from this study indicated that adopted and nonadopted persons differed increasingly with age (Verhulst & Althaus, 1988).

Plomin and DeFries (1985), using data from the Colorado Adoption Project (CAP), compared 490 adopted and nonadopted infants matched for age, sex, and socioeconomic status of family. Comparing adopted and nonadopted children at 12 and 24 months of age revealed only neg-ligible differences. The few meaningful differences that were observed, moreover, favored adopted children: adopted children showed greater task-oriented attention, less emotionality, and fewer difficult tempera-ment and problem behaviors. A follow-up of these same children, however, suggested that, in contrast to the lack of differences observed in infancy and early childhood, at ages 7 and 9 adopted children had begun to exhibit slightly more problems than their nonadopted comparisons (Rhea & Corley, 1994). Adopted children were rated by their parents and teachers as more externalizing, less attentive, and more emotional and active than nonadopted children. Furthermore, adopted children's self-reports indicated lower levels of appropriate classroom conduct and lower scholastic self-esteem. At ages 7 and 12, adopted children also dis-played somewhat lower indices of cognitive verbal ability when com-pared to their nonadopted counterparts (Wadsworth, DeFries, & Fulker, 1993). The differences observed between adopted and nonadopted persons were however small, with effect sizes ranging between .2 and .4.

As a follow-up to Plomin and DeFries (1985), Rhea and Corley (1994), and Wadsworth et al. (1993), the present investigation examines differences between adopted and nonadopted children in CAP as they make their transition from early to late adolescence. The questions that we will aim to answer include: Do adopted children differ on a range of behavioral and scholastic achievement measures when compared to their nonadopted counterparts? If so, do adopted children fare better or worse? Is there some specificity in the problems that adopted children present? And, are the behavioral differences observed between adopted and nonadopted youths more accentuated at particular times during the transition to adolescence? CAP offers a particularly valuable set of data to address these issues, in that it is one of the few large-scale prospective longitudinal adoption studies in which all children were adopted very early on after birth, adoptive and nonadoptive families were carefully matched, and independent reports from parents and teachers were collected using standardized instruments of child and adolescent adjustment.

Methods

Sample

Birth parents of adopted-away children were recruited by social workers in collaboration with two large adoption agencies in the Rocky Mountain Region of Denver, Colorado. All of the adopted children were relinquished at birth and placed in foster care for an average of 29 days prior to being placed in a permanent adoptive home. Psychological and emotional adjustment was, therefore, unlikely to have been influenced by preadoption factors other than possible prenatal factors. Of these children, 90% were of Caucasian origin, and the remaining 10% were Hispanic or Asian. Following the adoption procedures, families of adopted children were contacted when the child was on average 7 months old. These families were subsequently matched to nonadoptive families—families selected from hospital records—on the basis of sex of the proband; number of children in the household; and father's age, educational, and occupational status. The ethnic background of the nonadopted children closely matched that of the adopted children (95% Caucasian and 5% Hispanic and Asian). The analyses discussed in this chapter are based on a sample of adopted and nonadopted adolescents for whom data were available on measures of psychological and educational adjustment from ages 9 to 16. Depending on the measures, the sample ranged from 142–200 adopted adolescents and 170–233 nonadopted adolescents.

Measures

Parent's ratings and adolescents' self-reports of problem behavior and emotional adjustment were administered in the laboratory when the adolescents were aged 12 and 16 and during telephone interviews at ages 9, 10, 11, 13, 14, and 15. At these ages, teachers were also mailed questionnaires that assessed adolescents' emotional adjustment and scholastic competence. Ratings of problem behavior were obtained from parents and teachers on the Child Behavior Checklist (CBCL; Achenbach & Edelbrock, 1983). The CBCL consists of 118 items that yield two second-order factors—Externalizing Behavior and Internalizing Behavior, and a Total Problem score. The externalizing factor includes items that tap into delinquent and aggressive behaviors (e.g., "disobeys at home," "cruel to others," and "destroys others' things"), while the internalizing factor includes items that underlie behaviors of a withdrawn and depressed nature (e.g., "withdrawn," "won't talk," and "feels worthless").

In addition to problem behavior, teachers reported indices of scholastic competence and classroom conduct on the teacher version of the CBCL. Grade and class performance of reading and arithmetic were rated by teachers on a 5-point scale (Far Below Grade, Somewhat Below Grade, At Grade Level, Somewhat Above Grade, and Far Above Grade). Similar ratings were used to measure indices of appropriate scholastic conduct and hard work (Much Less, Somewhat Less, About Average, Somewhat More, and Much More).

Adolescents' self-perceived competence was assessed using the Harter's Self-Perception Profile (Harter, 1982), which measures five dimensions of Perceived Self-Competence: Social Acceptance, Athletic Competence, Scholastic Competence, Physical Appearance, Behavior Conduct, and a Total Self-Worth score.

Data Preparation and Analyses

Because aggregation increases reliability of measurement and decreases measurement error, the measures employed to assess psychological, emotional, and scholastic adjustment were averaged across age to form two composite aggregates. The mean score of ages 9, 10, 11, and 12 formed the early adolescence composite, and the mean score of ages 13, 14, 15, and 16 formed the late adolescence composite. Due to extensive missing data for teachers' ratings at age 16, the late adolescence composite for teachers included measures for ages 13–15 only.

Aggregated scores of behavioral, emotional, and scholastic adjustment were examined using a 2 (adoptive status, 1 = adopted, 2 = non-adopted) × 2 (gender, 0 = males, 1 = females) × 2 (age, 1 = early adolescence composite, 2 = late adolescence composite) mixed plot factorial analysis of variance (ANOVA). The reliability and longitudinal stability of parents and teachers' reports of problem behavior were examined using Pearson's correlations.

Results

Problem Behavior

Parents' and teachers' interrater agreement on the two second-order factors of the CBCL—externalizing and internalizing—were computed using Pearson correlations (table 8.1). For externalizing behaviors, the correlations between parents' and teachers' reports were very high, and their reports were therefore combined. Both measures were consistent across time for parents ($r = .775$, $p < .001$) and for teachers ($r = .678$, $p < .001$), as was the composite ($r = .767$, $p < .001$). For internalizing behaviors, however, the correlations between parents' and teacher's reports were low (table 8.1). Furthermore, while parental reports showed good consistency across time ($r = .712$, $p < .001$), teachers' reports indicated poor longitudinal stability ($r = .237$, $p < .001$). Analyses of variance for this measure were therefore conducted independently for parents' and teachers' reports. These findings agree with a large body of evidence from meta-analytic studies, which finds greater interrater reliability for undercontrolled behaviors, as opposed to overcontrolled behaviors (Achenbach, McCounaghy, & Howell, 1987).

Parents' and teachers' reports of adopted and nonadopted adolescents' problem behavior were compared using the $2 \times 2 \times 2$ repeated measures ANOVA, and independent comparisons were made for the externalizing and the internalizing scales of the CBCL (table 8.2).

As indicated in table 8.2 and illustrated in figure 8.1, adopted adolescents were rated by their parents and teachers as more externalizing than their nonadopted counterparts ($F[1, 405] = 16.49$, $p < .001$). A significant adoptive status by age interaction ($F[1,405] = 11.10, p < .001$) was also observed, and simple effect tests indicated that adopted youths exhibited increasingly more externalizing behaviors in late adolescence, as compared to early adolescence ($F[1,405] = 10.41$, $p < .001$). This

TABLE 8.1. Correlations for Parents' and Teachers' Ratings of CBC Externalizing and Internalizing Behaviors, and Total Problem Score in Early (9–12) and Late (13–16) Adolescence

| Externalizing | 9–12 | .786** | Externalizing | 13–15 | .979** |
| Internalizing | 9–12 | .224** | Internalizng | 13–15 | .276** |

**Correlation is significant at the 0.01 level (two tailed). $N = 316$–401.

TABLE 8.2. Means, Standard Deviations, and Effect Sizes for Teachers' and Parents' Ratings of Adopted and Nonadopted Adolescents' Externalizing and Internalizing Behavior

Measure	Age	Adopted Males		Adopted Females		Nonadopted Males		Nonadopted Females		Effect Size[a]
		Mean	SD	Mean	SD	Mean	SD	Mean	SD	
Combined externalizing[b]	9–12	9.07	6.39	6.89	5.10	7.93	5.27	4.86	3.89	0.27
Combined externalizing[b]	13–16	9.47	8.79	7.40	6.91	7.40	6.91	4.58	4.39	0.40
Parent internalizing[b]	9–12	5.86	4.56	5.92	4.80	5.66	4.38	6.09	4.93	0.00
Parent internalizing[b]	13–16	5.95	5.25	7.00	5.47	4.60	4.35	5.38	4.61	0.29
Teacher internalizing[c]	9–12	5.96	4.24	5.83	4.87	5.91	5.41	6.20	5.39	0.02
Teacher internalizing[c]	13–15	5.27	4.85	6.30	6.23	4.83	5.65	3.94	3.62	0.26

[a]Effect size averaged across sex. [b]N = 192 adopted, 217 nonadopted. [c]N = 142 adopted, 170 nonadopted.

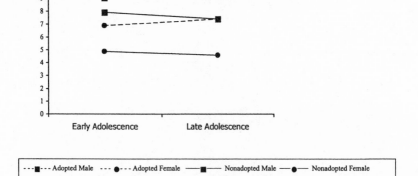

FIGURE 8.1. Parents'- and teachers'-rated externalizing behavior for adopted and nonadopted youths.

increase, which was more accentuated for adopted females $(F[1,407] = 14.56, p < .001)$, was not observed for nonadopted adolescents $(F[1,405] = 1.71, p > .05)$, indicating that adopted and nonadopted adolescents differed increasingly apart with age.

Internalizing behavioral ratings for adopted and nonadopted adolescents are presented in table 8.2 and figures 8.2a and 8.2b. As illustrated in figure 8.2a, parents' ratings indicated similar levels of internalizing problems for adopted and nonadopted adolescents $(F[1,405] = 2.91, p > .05)$, as did teachers' ratings in figure 8.2b—$(F[1,308] = .177, p > .05)$. However, an age-by-adoptive-status interaction was observed for both parents $(F[1,405] = 16.89, p < .001)$ and teachers $(F[1,308] = 4.45, p < .05,$ for teachers), and simple effect analyses indicated that parents rated adopted youths as exhibiting greater internalizing problems in late adolescence, as compared to early adolescence $(F[1,405] = 4.47, p < .05)$. Although teachers did not report this increase $(F[1,308] = .98, p > .05)$, both parents $(F[1,407] = 13.60, p < .001)$ and teachers $(F[1,308] = 4.47, p < .05)$ agreed in their report of a significant decrease for the same measure for nonadopted youths. In other words, as with the externalizing measure, adopted and nonadopted youths differed increasingly with age.

Scholastic Adjustment

Teachers' reports of adopted and nonadopted adolescents' scholastic performance and classroom conduct indicated that adopted adolescents experience somewhat more difficulties in adjusting to school than nonadopted adolescents (table 8.3).

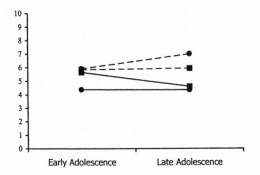

FIGURE 8.2a. Parent-rated internalizing behavior for adopted and nonadopted males and females.

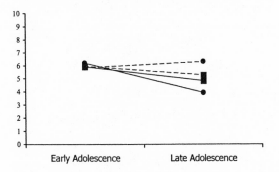

FIGURE 8.2b. Teacher-rated internalizing behavior for adopted and nonadopted males and females.

As indicated in table 8.3, adoptees were rated by their teachers as performing less well academically—arithmetic grade ($F[1, 281] = 13.76$, $p < .001$); arithmetic class performance ($F[1,275] = 13.83$, $p < .001$); reading grade ($F[1,299] = 15.93$, $p < .001$); reading class performance ($F[1,301] = 16.03$, $p < .001$); working less hard ($F[1,308] = 12.58$, $p < .001$); and exhibiting less appropriate classroom conduct ($F[1,309] = 18.59$, $p < .001$)—than their nonadopted counterparts. While no significant age-by-adoptive-status interactions were found, significant increases in scholastic performance—arithmetic grade ($F[1,281] = 8.69$, $p < .01$); arithmetic class performance ($F[1,275] = 9.36$, $p < .01$); reading grade ($F[1,299] = 4.36$, $p < .05$); reading class performance ($F[1,301] = 11.17$, $p < .001$); hard work ($F[1,308] = 8.06$, $p < .01$); and appropriate classroom conduct ($F[1,309] = 45.69$, $p < .001$)—were observed for both adopted and nonadopted adolescents from early to late adolescence.

TABLE 8.3. Means, Standard Deviations, and Effect Sizes for Teachers' Ratings of Adopted and Nonadopted Adolescents' Reading and Mathematics Grade and Class Performance, Appropriate School Conduct, and Amount of Hard Work

Measure	Age	Adopted Males		Adopted Females		Nonadopted Males		Nonadopted Females		Effect Size[a]
		Mean	SD	Mean	SD	Mean	SD	Mean	SD	
Reading grade[b]	9–12	3.40	0.9	3.51	0.7	3.61	0.8	3.98	0.7	0.34
Reading grade[b]	13–15	3.46	0.9	3.51	0.8	3.81	0.8	3.96	0.7	0.46
Maths grade[c]	9–12	3.46	0.7	3.26	0.6	3.62	0.7	3.68	0.6	0.39
Maths grade[c]	13–15	3.52	0.8	3.37	0.8	3.73	0.8	3.84	0.7	0.38
Reading performance[d]	9–12	3.42	0.9	3.47	0.7	3.59	0.8	3.84	0.7	0.31
Reading performance[d]	13–15	3.48	0.9	3.53	0.8	3.82	0.7	4.04	0.7	0.48
Maths performance[e]	9–12	3.51	0.8	3.26	0.7	3.64	0.7	3.76	0.7	0.38
Maths performance[e]	13–15	3.57	0.8	3.41	0.9	3.78	0.8	3.90	0.8	0.38
Good conduct[f]	9–12	3.11	0.8	3.68	0.6	3.37	0.8	3.97	0.7	0.32
Good conduct[f]	13–15	3.46	1.0	3.82	0.7	3.85	0.9	4.30	0.7	0.42
Hard work[g]	9–12	3.21	0.7	3.45	0.7	3.31	0.7	3.76	0.8	0.24
Hard work[g]	13–15	3.31	1.1	3.43	0.7	3.48	0.9	4.05	0.8	0.39

[a]Effect size averaged across sex. [b]N = 138 adopted, 165 nonadopted. [c]N = 132 adopted, N = 153 nonadopted. [d]N = 140 adopted, N = 165 nonadopted. [e]N = 143 adopted, N = 170 nonadopted. [f]N = 143 adopted, N = 169 nonadopted. [g]N = 129 adopted, N = 150 nonadopted.

Youths' Self-Reported Perceived Self-Competence

Adopted and nonadopted adolescents' perceptions of self-competence were compared for each of the five dimensions of the Harter's Self-Perception Profile: Social Acceptance, Behavior Conduct, Athletic Competence, Scholastic Competence, Physical Appearance, and Total Self-Worth (table 8.4).

Table 8.4 and figure 8.3 show that, in accordance with parents' and teachers' reports of internalizing behavior, adopted and nonadopted adolescents gave similar ratings of total self-worth ($F[1,419] = .03, p > .05$), self-perceived attractiveness ($F[1,419] = 2.47, p > .05$), and athletic competence ($F[1,419] = 1.67, p > .05$). In comparison to nonadopted youths, however, adoptees gave lower self-ratings of scholastic competence ($F[1,419] = 8.54, p < .01$) and appropriate conduct ($F[1,419] = 12.45, p < .001$), though adoptees rated themselves as more socially confident and outgoing ($F[1,419] = 11.48, p < .001$).

A significant adoptive status by age interaction was observed for youths' ratings of conduct ($F[1,419] = 7.97, p < .01$). Simple effect analyses revealed that adopted youths' appropriate conduct decreased from early to late adolescence ($F[1,421] = 6.48, p < .01$). This decrease, which was not observed for nonadopted persons, indicated that, as with externalizing and internalizing behaviors, adopted and nonadopted adolescents differed increasingly apart with age. There were no other significant adoptive-status-by-age interactions. However, a sex-by-adoptive-status interaction ($F[1,419] = 5.35, p < .05$) was observed for youths' ratings of conduct, and simple effect analyses revealed that adopted females reported a greater decrease in appropriate conduct, with increasing age as compared to adopted males ($F[1,311] = 10.93, p < .01$).

Discussion

Previous reports of children in CAP have suggested comparable indices of adjustment for adopted and nonadopted persons in infancy and early childhood (Plomin & DeFries, 1985), but somewhat greater levels of behavioral problems among adoptees in middle childhood (Rhea & Corley, 1994). The present study aimed to bring the picture forward to the adolescent years, and to answer questions concerning the psychological, emotional, and scholastic adjustment of these same children as they made their transition from early to late adolescence. We addressed these questions by examining (1) whether adopted and nonadopted adolescents differed on measures of psychological and scholastic functioning, (2) whether some specificity was apparent in the type of problems that adopted and nonadopted adolescents experience, and (3) whether behavioral differences observed between adopted and nonadopted

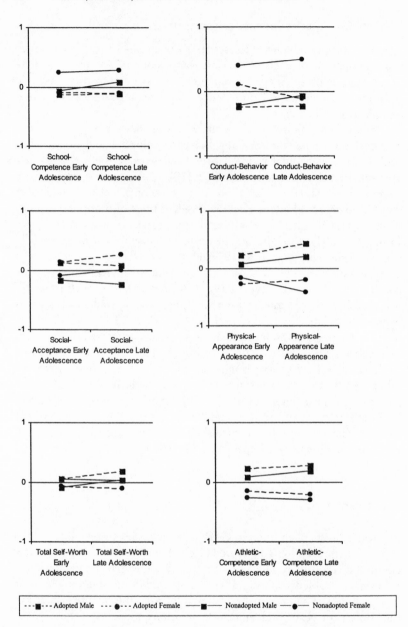

FIGURE 8.3. Adopted and nonadopted adolescents' self-report of perceived competence.

TABLE 8.4. Means, Standard Deviations, and Effect Sizes for Adopted and Nonadopted Adolescents' Reports of Self-Perceived Competence Using the Harter's Self-Competence Profile

Measure	Age	Adopted Males		Adopted Females		Nonadopted Males		Nonadopted Females		Effect Size[a]
		Mean	SD	Mean	SD	Mean	SD	Mean	SD	
Social acceptance	9–12	16.16	2.5	16.20	2.3	15.35	2.8	15.58	2.8	0.26
Social acceptance	13–15	16.73	2.4	17.17	2.0	15.92	2.4	16.53	2.4	0.30
Perceived attractiveness	9–12	15.88	2.5	14.53	2.9	15.48	2.2	14.84	2.6	0.03
Perceived attractiveness	13–15	16.34	2.3	14.42	2.8	15.65	2.6	13.73	3.3	0.23
Athletic self-esteem	9–12	13.34	2.5	12.32	2.2	12.95	2.7	12.01	2.8	0.13
Athletic self-esteem	13–15	13.95	2.7	12.48	2.6	13.66	2.8	12.20	3.4	0.09
Scholastic self-esteem	9–12	14.76	2.6	14.48	2.5	14.92	2.5	15.74	2.5	0.19
Scholastic self-esteem	13–15	15.08	2.7	15.10	2.8	15.62	2.5	16.20	2.9	0.28
Good conduct	9–12	14.27	2.3	15.10	2.2	14.31	2.2	15.81	2.3	0.14
Good conduct	13–15	13.99	2.7	14.34	2.8	14.46	2.5	16.09	2.8	0.37
Total self-worth	9–12	91.68	10.1	90.29	9.4	90.09	10.1	91.53	10.4	0.02
Total self-worth	13–15	93.93	9.5	90.91	9.06	92.37	9.44	92.93	11.6	0.02

[a]Effect size averaged across sex. $N = 200$ adopted, $N = 123$ nonadopted.

adolescents were more accentuated at particular times during the transition to adolescence.

Emotional and Scholastic Adjustment

Do adopted and nonadopted adolescents differ for measures of emotional and scholastic adjustment? On the first issue, comparisons of adopted and nonadopted adolescents indicated small but significant differences for measures of problem behavior and scholastic adjustment. Although for each of these measures adopted adolescents showed slightly poorer indices of adjustment, adoptive status seldom accounted for more than 4% of the overall variance of these measures, and adopted adolescents' scores were within the normal variation range. The results from this study therefore suggest that adoptees are no more likely than nonadopted adolescents to experience clinical levels of psychopathology and educational morbidity, and that adoptees are comparably well adjusted.

These findings disagree with studies of adoptees from clinical populations that have indicated a fourfold increase in the risk of psychiatric morbidity among adopted adolescents (e.g., Kim et al., 1988; Rogeness et al., 1988). This discrepancy in findings is likely to be associated with methodological inconsistencies. All of the children in CAP, approximately 400 in total, were adopted very early after birth and were carefully matched to nonadopted children on the basis of family structure and social economic status. In contrast, clinical population studies often rely on small sample sizes and unmatched comparison groups and fail to discriminate between adolescents who were adopted in childhood following abuse, neglect, and several adoption disruptions from adolescents who were adopted at birth with relatively little or no preadoption complications. Adoptees are a highly heterogeneous group, with a few of its members experiencing severe maladjustment and the majority experiencing none. Much of this heterogeneity is accounted for by factors such as adverse preadoption experiences, and the failure to control for these can inflate mean differences and lead to inaccurate estimates of clinical morbidity (Haugaard, 1998). The large differences in adjustment reported in studies of clinical populations are, therefore, likely to be representative of a few extreme cases but largely unrepresentative of the wider population of adoptees (Haugaard, 1998; Sharma et al., 1998).

A further factor that could account for the divergent findings between the current study and studies of adoptees in clinical samples is referral bias. Warren (1992) found that adopted adolescents were more likely than nonadopted persons to be referred to health professionals for relatively minor problems. Consequently, the estimated risk of psychopathology and educational morbidity reported in studies of adopted adolescents in clinical settings might be exaggerated by this bias.

Accordingly, when compared to studies of adoptees from the general population that employed similarly stringent methodological criteria, the results from this study find considerable support. In accordance with our findings, Brodzinsky et al. (1984) and Sharma et al. (1998) reported comparably slight differences between adopted and nonadopted persons' emotional and scholastic adjustment that were not indicative of an increased risk of psychopathology and educational morbidity among adoptees.

Types of Problems

Is there some specificity in the type of problems that adopted and non-adopted adolescents experience? The second question that we addressed in this study was concerned with whether adopted adolescents are more susceptible to specific problem behaviors, as opposed to poor adjustment more generally. A strong body of evidence has suggested that adoptees show greater vulnerability to behaviors of an externalizing, as opposed to an internalizing nature (e.g., Deutsch et al., 1982; Sullivan et al., 1995). In line with this research, the largest differences between adopted and nonadopted adolescents in this study were found for measures of externalizing and conduct problems. Specifically, while adopted and nonadopted adolescents did not differ on measures of internalizing behaviors, adoptees were rated by their parents and teachers as exhibiting increased levels of externalizing behaviors and conduct problems. Similarly, while adoptees' self-reports of perceived competence indicated lower appropriate behavior conduct and scholastic competence, they suggested comparable indices of self-worth and greater social confidence and social orientation. These findings are consistent with those reported by Sharma et al. (1996, 1998) who, in a study of 881 adopted and nonadopted adolescents, found greater externalizing behavior but lower indices of social problems and higher prosocial behavior among adoptees.

The greater vulnerability to externalizing and conduct problems observed among adopted persons is poorly understood. However, a few hypotheses that aim to explain this specific pattern of vulnerability have been advanced. Among these are poor parent-child bonding and the experience of coming to terms with adoption (Brodzinsky, 1993). The problem with these hypotheses is that, although they offer an explanation for increased problem behavior among adoptees, they tell us very little as to why adopted adolescents are more prone to experiencing externalizing as opposed to internalizing-type behaviors (Fergusson et al., 1995). The limited empirical evidence available furthermore indicates that factors such as poor child-parent bonding and the experience of coming to terms with adoption are more often associated with feelings of anxiety and depression than they are to aggression and disobedience (Hoopes, 1990).

A more viable explanation for this pattern of specificity, and one that has received greater empirical support, suggests that adoptees are at a heightened genetic risk for externalizing problems (Cadoret, 1984). For example, Crowe (1974) reported greater problem behavior among adopted-away offsprings of mothers who scored high on measures of antisocial behavior, and Bohman, Cloninger, Sigvardsson, & Von Knorring (1982) found considerable genetic influence on measures of antisocial personality and delinquent behaviors in a sample of Swedish adoptees and their biological parents.

This explanation finds little support in the current study, however. While comparisons of adopted and nonadopted children's biological parents in CAP suggest slightly higher indices of hysteria and depression among biological parents of adoptees, these differences are small and largely accounted for by differences in age (e.g., Plomin & DeFries, 1985; Rhea & Corley, 1994). Furthermore, factors such as parent-child bonding, the experience of coming to terms with adoption, and genetic vulnerability are unlikely to account for the positive indices of social confidence and outgoing behavior observed among adoptees in the study of Sharma et al. (1996, 1998) and the present study. Further research is, therefore, needed to identify factors implicated in the etiology of this specificity effect.

The Transition to Adolescence

Are differences observed between adopted and nonadopted adolescents accentuated at specific times during the transition to adolescence? The final question addressed in this study explored whether adopted and nonadopted adolescents are more likely to differ apart at particular times during the transition to adolescence, and whether specific problems are more likely to become accentuated during early as compared to late adolescence. Consistent with the findings reported by Verhulst and Verluis-den-BiemanHerma (1995), but contrary to those reported by Bohman and Sigvarsson (1985), results from longitudinal comparisons in the current study indicated increasingly diverging patterns of adjustment for adopted and nonadopted adolescents with increasing age. This divergence was accentuated by a meaningful increase in conduct, externalizing, and internalizing behaviors among adopted adolescents that contrasted the decrease observed for the same behaviors in nonadopted adolescents.

These results are consistent with the increased indices of emotional problems observed in adopted adolescents by Maughan and Pickles

(1990) in the National Child Development Study. In line with their findings, differences in internalizing behavior between adopted and nonadopted persons in the present study only emerged at ages 13–16, suggesting that late adolescence might be a particularly vulnerable developmental stage for adopted youths. The overall increase in problem behavior observed for adopted persons in adolescence has been interpreted by some authors as a consequential outcome of emerging anxieties concerning lost genealogical links and uncertainties related to the identity of biological parents (Hoopes, 1990). Although not addressed in the present study, one alternative explanation is that the transition to adolescence—characterized by increased freedom from parental supervision, growing responsibilities, and pubertal maturation—might represent a more stressful experience for adopted as compared to nonadopted persons.

Longitudinal comparisons of adopted and nonadopted adolescents self-perceived competence and scholastic adjustment did not replicate the divergent patterns of development observed for measures of problem behavior. Rather, an age-related increase in academic performance observed for both adopted and nonadopted adolescents indicated comparable indices of improvement for this measure. These findings are in contrast to the age-related decrease in indices of social competence and scholastic adjustment reported by Verhulst and Verluis-den-BiemanHerma (1995) in a study of 1,538 intercountry adoptees in the Netherlands. As the age of the adoptees in the sample and the mearing instruments employed to assess adjustment closely matched those in the current study, this discrepancy in findings is likely to be related to differences in ethnicity. While the large majority of adopted adolescents in CAP were from ethnic backgrounds similar to those of their adoptive families and their nonadopted comparisons, those studied by Verhulst and Verluis-den-BiemanHerma (1995) were not. One possible explanation for the inconsistent findings, therefore, is that intercountry adoptees might experience greater problems than intercountry adoptees in both areas of social and scholastic adjustment.

Summary and Conclusion

Consistent with previous research on adopted adolescents from general-population samples, the findings from this study suggest that although adoptees show less favorable outcomes in some areas of adjustment, the differences observed between adopted and nonadopted adolescents are negligible and not representative of an increased risk of psychiatric and educational morbidity. These results further indicate that in areas of social adjustment and social competence, adoptees fare better than their nonadopted counterparts.

Although the results from this study support previous research indicating that adopted adolescents are more likely to differ from their nonadopted counterparts on measures of externalizing behaviors, longitudinal comparisons indicate that adoptees become increasingly more vulnerable to internalizing problems in late adolescence with increasing age. The comparable increase and greater susceptibility to problem behaviors observed among adoptees in late adolescence further suggest that this developmental period might be somewhat more stressful for adopted, as compared to nonadopted, adolescents. Though the present study compared adopted and nonadopted persons only up to the aolescent years, research has suggested that the increase in problem behavior observed among adoptees during adolescence reflects transitory rather than long-lasting difficulties (Collishaw, Maughan, & Pickles, 1998). The present study also failed to consider mediating factors implicated in the adjustment of adolescents following adoption. The developmental pathways responsible for the increase in behavioral problems among adoptees in late adolescence remain elusive. Further research examining these issues would, therefore, be timely.

In conclusion, given that the differences observed between adopted and nonadopted adolescents were negligible and remained small during late adolescence, the findings from this study imply that parents, social workers, clinicians, and educators should focus on the needs of individual hildren rather than on adoptive children as a group.

REFERENCES

Achenbach, T. M., & Edelbrock, C. (1983). *Manual for the Child Behavior Checklist and revised Child Behavior Profile.* Burlington: University of Vermont, Department of Psychology.

Achenbach, T. M., McCounaghy, S. H., & Howell, C. T. (1987). Child/adolescent behavioral and emotional problems: Implications for cross-informant correlations for situational specificity. *Psychological Bulletin, 101,* 213–232.

Bohman, M., Cloninger, C. R., Sigvardsson, S., & Von Knorring, A. L. (1982). Predisposition to petty criminality in Swedish adoptees I: Genetic and environment heterogeneity. *Archives of General Psychiatry, 39,* 1233–1241.

Bohman, M., & Sigvardsson, S. (1985). A prospective longitudinal study of adoption. In A. R. Nicol (Ed.), *Longitudinal studies in child psychology and psychiatry.* New York: John Wiley.

Brodzinsky, D. M. (1993). Long-term outcomes of adoption. *The future of children, adoption, 3,* 153–165.

Brodzinsky, D. M., Schechter, D. E., Braff, A. M., & Singer, L. M. (1984). Psychological and academic adjustment in adopted children. *Journal of Consulting and Clinical Psychology, 52,* 582–590.

Cadoret, R. J. (1984). Biological perspectives of adoptee adjustment. In D. Brodzinsky & M. Schechter, (Eds.), *The psychology of adoption* (pp. 25–41). New York: Oxford University Press.

Collishaw, S., Maughan, B., & Pickles, A. (1998). Infant adoption: Psychosocial outcomes in adulthood. *Social Psychiatry and Psychiatric Epidemiology*, *33*, 2, 57–65.

Crowe, R. R. (1974). An adoption study of antisocial personality. *Archives of General Psychiatry*, *31*, 785–791.

Deutsch, C. K., Swanson, J. M., Bruell, J. H., Cantwell, D. P., Weinberg, F., & Baren, M. (1982). Overrepresentation of adoptees in children with attention deficit disorder. *Behavior Genetics*, *12*, 231–238.

Fergusson, M., Lynskey, M., & Horwood, L. J. (1995). The adolescent outcome of adoption: A 16-year longitudinal study. *Journal of Child Psychology and Psychiatry*, *36*, 597–615.

Harter, S. (1982). The perceived self-competence scale for children. *Child Development*, *56*, 87–97.

Haugaard, J. J. (1998). Is adoption a risk factor for the development of adjustment problems? *Clinical Psychology Review*, *81*, 47–69.

Hersov, L. (1990). The Seventh Jack Tizard Memorial Lecture: Aspects of adoption. *Journal of Child Psychology and Psychiatry*, *31*, 493–510.

Hodges, J., & Tizard, B. (1989a). IQ and behavioral adjustment of ex-institutional adolescents. *Journal of Child Psychology and Psychiatry*, *30*, 53–75.

Hodges, J., & Tizard, B. (1989b). Social and family relationships of ex-institutional adolescents. *Journal of Child Psychology and Psychiatry*, *30*, 77–97.

Hoopes, J. L. (1982). *Prediction in child development: A longitudinal study of adoptive and non-adoptive families*. New York: Child Welfare League of America.

Hoopes, J. L. (1990). Adoption and identity formation. In D. Brodzinsky & M. Schechter (Eds.), *The psychology of adoption* (pp. 145–166). New York: Oxford University Press.

Kim, W. J., Davenport, C., Joseph, J., Zrull, J., & Wolford, E. (1988). Psychiatric disorder and juvenile delinquency in adopted children and adolescents. *Journal of the American Academy of Child and Adolescent Psychiatry*, *27*, 111–115.

Lipman, E. L., Offord, D. R., Boyle, M. H., & Racine, Y. A. (1993). Follow-up of psychiatric and educational morbidity among adopted children. *Journal of the American Academy of Child and Adolescent Psychiatry*, *32*, 1007–1012.

Maughan, B., & Pickles, A. (1990). Adopted and illegitimate children growing up. In L. N. Robins & M. Rutter (Eds.), *Straight and devious pathways from childhood to adulthood* (pp. 36–62). New York: Cambridge University Press.

Mech, E. V. (1973). Adoption: A policy perspective. In B. M. Caldwell & H. N. Ricciuti (Eds.), *Reviews of child development research: Vol. III. Child development and social policy* (pp. 467–508). Chicago: University of Chicago Press.

Plomin, R., & DeFries, J. C. (1985). *Origins of individual differences in infancy: The Colorado Adoption Project*. Orlando, FL: Academic Press.

Rhea, S., & Corley, R. P. (1994). Applied issues. In J. C. DeFries, R. Plomin, & D. Fulker (Eds.), *Nature and nurture during middle childhood* (pp. 295–310). Oxford: Blackwell.

Rogeness, G. A., Hoppe, S. K., Macedo, C. A., Fischer, C., & Harris, W. R. (1988). Psychopathology in hospitalized, adopted children. *Journal of the American Academy of Child and Adolescent Psychiatry, 27,* 628–631.

Sharma, A. R., McGue, M. K., & Benson, P. (1996). The emotional and behavioral adjustment of United States adopted adolescents: I. A comparison study. *Children and Youth Services Review, 18,* 77–94.

Sharma, A. R., McGue, M. K., & Benson, P. (1998). The psychological adjustment of United States adopted adolescents and their nonadopted siblings. *Child Development, 69,* 791–802.

Singer, L. M., Brodzinsky, D. M., Ramsay, D., Steir, M., & Everett, W. (1985). Mother-infant attachment in adoptive families. *Child Development, 56,* 1543–1551.

Sullivan, P. F., Wells, J. E., & Bushnell, J. A. (1995). Adoption as a risk factor for mental disorders. *Acta Psychiatrica Scandinavica, 92,* 119–124.

Triseliotis, J., & Hill, M. (1990). Contrasting adoption, foster care, and residential rearing. In D. Brodzinsky & M. Schechter (Eds.), *The psychology of adoption* (pp. 107–120). New York: Oxford University Press.

Verhulst, F. C., & Althaus, M. (1988). Persistence and change in behavioral/emotional problems reported by parents of children aged 4–14. *Acta Psychiatrica Scandinavica, 77,* (Suppl.), 399.

Verhulst, F. C., Althaus, M., & Verluis-den-BiemanHerma, H. J. M. (1989). Problem behavior in international adoptees: I An epidemiological study. *Journal of the American Academy of Child and Adolescent Psychiatry, 29,* 94–103.

Verhulst, F. C., Althaus, M. & Verluis-den-BiemanHerma, H. J. M. (1992). Damaging backgrounds: Later adjustment of international adoptees. *Journal of the American Academy of Child and Adolescent Psychiatry, 33,* 518–524.

Verhulst, F. C., & Verluis-den-BiemanHerma, H. J. M. (1995). Developmental course of problem behaviors in adolescent adoptees. *Journal of the American Academy of Child and Adolescent Psychiatry, 34,* 151–159.

Wadsworth, S. J., DeFries, J. C., & Fulker, D. W. (1993). Cognitive abilities of children at 7 and 12 years of age in the Colorado Adoption Project. *Journal of Learning Disabilities, 26,* 611–615.

Warren, S. B. (1992). Lower threshold for psychiatric treatment for adopted adolescents. *Journal of the American Academy of Child and Adolescent Psychiatry, 31,* 512–527.

Stability and Change in Internalizing Problems in the Transition to Early Adolescence

Genetic and Environmental Influences

Introduction

Internalizing problems in childhood and adolescence reflect a broad array of emotions, including feelings of sadness, loneliness, depression, anxiety, and somatic problems (Achenbach, 1991). Much research has focused on describing the prevalence, predictors, and stability of depressive illness in referred and clinically diagnosed groups of children (Harrington, Rutter, & Fombonne, 1996). Another series of studies has focused on understanding the nature and correlates of individual differences in internalizing symptoms in unselected community samples (O'Connor, Neiderhiser, Reiss, Hetherington, & Plomin, 1998; Stanger, Achenbach, & McConaughy, 1993). The two approaches complement each other since the etiology of the extremes may be similar to the etiology of the full range of internalizing problems in the population (Deater-Deckard, Reiss, Hetherington, & Plomin, 1997; Eley, 1997; Rende, Plomin, Reiss, & Hetherington, 1993). This approach is one component within the field of developmental psychopathology that emphasizes exploration of the links between the normal and abnormal in order to learn more about development. The present study focused on the etiology of stability and change in several aspects of internalizing problems (loneliness, depression, and anxiety) in a community-based sample, the Colorado Adoption Project (CAP).

Research indicates that individual differences in depressive symptoms can emerge in childhood (Harrington et al., 1996). Like adults, children can experience persistent feelings of sadness, despair, loneliness, and hopelessness, as well as fearfulness and anxiety (Asher et al., 1984; Cassidy & Asher, 1992; Eley, 1997). Childhood episodes of major depression have been documented in the clinical literature. Furthermore, individual differences in internalizing symptoms in unselected samples of children suggest that variation in subclinical depressive and anxious symptoms is substantial and may be an important factor to consider in terms of liability for the more extreme clinical manifestations of depressive illness (Eley, 1997).

These individual differences in internalizing symptoms in childhood are correlated with a host of family, school, and peer-group factors including parental mental illness, being rejected or neglected by peers, and major loss events, to name but a few (Asher & Wheeler, 1985; Cassidy & Asher, 1992; Eley, Deater-Deckard, Fombonne, Fulker, & Plomin, 1998; Harrington et al., 1996; Rubin, Hymel, & Mills, 1989). Familiality for depressive symptoms is considerable, with both genetic and environmental factors being implicated in the etiology of depression. However, the findings with respect to childhood depressive symptoms are complicated. Although genetic studies consistently find evidence for a genetic contribution to depression in adolescence (Eley, 1997) and adulthood (McGuffin, Owen, O'Donovan, Thapar, & Gottesman, 1994), the evidence is more mixed for prepubescent samples—for example, genetic variance has been found in twin studies but not adoption studies (Eley et al., 1998). It is also important, however, that researchers examine genetic and environmental correlates of stability and change in depression and loneliness. Indeed, the years leading up to, during, and following the pubertal transition may be a particularly important time with respect to the development of internalizing symptoms. There is evidence that loneliness is higher during adolescence compared to other parts of the life span (Parkhurst & Asher, 1992; Peplau & Perlman, 1982). It is also possible that depressive symptoms fluctuate more during this time compared to externalizing problems because of new social cognitive processes that emerge during the teenage years (Damon, 1977; Elkind, 1985). It is plausible that the gene-environment processes underlying changes in depressive symptoms may become more clearly defined during and following this pubertal transition. To this end, the main goal of this chapter is to explore genetic and environmental sources of variance in intraindividual changes in self-reported and parent-reported internalizing symptoms in the transition to early adolescence.

Stability and Change in Internalizing Problems

Longitudinal studies have shown that children's internalizing problems are modestly to moderately stable over time (see O'Connor et al., 1998).

However, the stability correlations are far from unity, suggesting that intraindividual change may be considerable. The longitudinal clinical research indicates that children and adolescents with depression remain at risk for depressive episodes in adulthood, but that many adults who are depressed had shown a variety of psychopathologies, or no symptoms at all, during childhood and adolescence (Harrington et al., 1996). This suggests that childhood and adolescent internalizing problems are heterogeneous, perhaps comprising a variety of subtypes—for instance, some of which are strongly associated with adult depression, some linked to later anxiety disorders, and so forth. This heterogeneous syndrome can be more fully understood using a developmental change framework. It may be that certain patterns of change in depressive symptoms distinguish different syndromes of depressive symptoms as they relate to subsequent psychopathology.

There has been surprisingly little research on intraindividual change and stability in internalizing problems in childhood, or in any age group, for that matter. Therefore, it is difficult to make predictions about the causes of these changes in depressive symptoms. Nonetheless, there are several hypotheses that could account for change and stability. First, it may be that any instability over time is a result of poor measurement and is not a meaningful phenomenon. This is a legitimate concern, in light of the poor agreement found between different informants (Eley et al., 1998) and given that much of the previous research on internalizing symptoms in childhood has been based on parents' and teachers' reports—informants who may not be aware of the child's internal states. Second, assuming acceptable validity and reliability of measurement, intraindividual fluctuations in internalizing problems over time may arise from developmental processes that are themselves critically important to our understanding of the etiology of depressive and anxiety disorders. In light of the lack of research examining intraindividual change, this second possible explanation has not been rigorously tested.

Using Behavioral Genetics to Examine Stability and Change

Quantitative genetics is a valuable tool for testing both explanations for the intraindividual change in internalizing symptoms in childhood. In addition to providing *static* estimates of genetic and environmental variance in adjustment problems (i.e., based on single assessments or the average scores across several assessments), these models can be used to estimate genetic and environmental sources of variance in dynamic intraindividual changes in depressive symptoms, as well as sources of variance in the stable components of these problems. In these dynamic models, intraindividual change and stability that is attributable to systematic additive *genetic* and *shared environmental* factors (i.e., environmental factors that lead to sibling similarity in depressive symptoms) is estimated separately from variance that is due to nonshared environmental factors (i.e., nongenetic factors that lead to sibling differences) and measurement error. To the degree that intraindividual change in internalizing symptoms in childhood is due to random measurement error, little if any of the change in depressive symptoms will be predicted from shared genetic or environmental factors—that is, all or nearly all of the variance will be accounted for by nonshared environmental sources of variance. In contrast, finding evidence for systematic genetic or shared environmental variance in intraindividual change, or any source of variance mediating the stability of symptoms over time, would suggest that the change and stability in internalizing problems are predictable.

The examination of intraindividual change in body mass provides an example of how genetic models of change can further our understanding of the etiology of depressive symptoms in the transition to early adolescence. The role of additive genetic influences on weight and body mass is well established, with heritability estimates in the .7 range (Grilo & Pogue-Geile, 1991). However, the sources of variance for individual differences in weight at any point in time are, in theory, independent of the sources of variance for intraindividual change in weight over time. Thus, an individual's weight at any point in time may not be a reliable predictor of change in that individual's weight over time. That is, the causes of "weight" as tested using static genetic models are not necessarily the causes of "change in weight" as tested by the dynamic genetic models.

More recent genetic research has found that the degree and direction of intraindividual variations in body mass among adults is influenced by genetic factors, even after known environmental factors and initial weight level have been statistically controlled (Austin et al., 1997). In other words, it appears that there are genetic factors that predict weight gain and loss over time. More important, the examination of intraindividual variation in weight change has led to new discoveries with respect to gene-environment processes. For instance, the level of physical activity appears to moderate the magnitude of the genetic influence on intraindividual change in weight among adult males (Heitmann et al., 1997).

Although there has been very little research examining gene-environment processes in intraindividual change in internalizing problems, one combined twin/step-family/adoptive sibling study examined increases and decreases in symptoms across two time points (O'Connor et al., 1998). In this large sample of 10- to 18-year-olds, change in internalizing symptoms was accounted for by nonshared environmental factors that were specific to each child within the same family; there was no evidence for a genetic or shared environmental contribution to intraindividual change in internalizing problems over this four-year period. Stability over time ($r = .59$) was accounted for by both genetic and nonshared environmental factors, suggesting that the gene-environment processes underlying continuity in internalizing symptoms are clearly in place by adolescence.

Given the lack of research examining intraindividual variations and stability in internalizing problems, and in light of evidence that change over time can be substantial, our aim was to examine genetic and environmental sources of variation in change and stability in internalizing symptoms during the transition to early adolescence using an adoptive sibling design.

Method

Sample

Participants included same- and opposite-sex (59%) pairs of unrelated adoptive siblings and related nonadoptive siblings from CAP (DeFries, Plomin, & Fulker, 1994), an ongoing longitudinal study of 490 families. Four annual assessments were conducted when each child was 9, 10, 11, and 12 years old. Only those pairs with complete data for the four assessments were included in the current study, resulting in a range of sample sizes: 39 adoptive pairs for parents' reports and 64 unrelated adoptive pairs for children's self-reports, and 45 nonadoptive pairs for parents' reports, and 69 nonadoptive pairs for children's self-reports.

The adoptive families in the CAP were contacted via adoption agencies in Colorado from 1975 to 1982, when the child was 7 months old on average (average age at placement was 29 days). The families were European American (95% of sample) and middle class. Selective placement was negligible. A comparison sample of nonadoptive families with biological siblings was recruited from the same area using birth records. The adoptive and comparison families were matched on five factors: the first child's sex, family size, paternal age, occupational status, and years of education. For more details about the sample and procedures, see Plomin & DeFries (1985).

Measures

Children reported their own depressive symptoms when they were 9, 10, 11, and 12 years old, using a six-item adapted version of the Kandel Depressive Mood Inventory (Kandel & Davies, 1982). The items pertain to difficulties in sleeping, lethargy, feelings of sadness and hopelessness, worrying, and anxiety. This scale has been validated and is a reliable screening instrument for assessing individual differences in depressive symptoms among children. Children also completed an adapted version of the Asher Loneliness Questionnaire (Asher, 1985). This included eight items pertaining to the child's feelings of isolation, loneliness, and sadness.

Parents completed annual assessments using the Child Behavior Checklist (Achenbach, 1991). For this study, we used the 14-item Anxious/Depressed subscale. The CBCL is a reliable and validated instrument for individual differences in emotional problems. See Eley et al. (1998) and McGuire and Clifford (chapter 10, this volume) for detailed descriptions of all of these instruments.

Analyses

The adoptive sibling and matched nonadoptive sibling design allows for the estimation of possible genetic and environmental sources of variance in children's depressive symptoms. Because adoptive siblings are genetically unrelated but are reared in the same home, their similarity is a direct estimate of *shared environmental* influences—environmental factors that lead to sibling similarity. In contrast, modest adoptive and biological sibling similarity is a strong indicator of *nonshared environmental* influences—environmental factors that result in sibling differentiation. In addition, by comparing the two groups of siblings—unrelated adoptive siblings and related full-siblings who share (on average) 50% of their genes—all of whom are reared together from infancy, the presence of genetic influences or *heritability* can also be detected. Genetic and environmental sources of variance and co-variance were derived using structural equation modeling (Neale & Cardon, 1992). This model is described in detail in the "Results" section, which follows.

Results

We analyzed the etiology of stability and change in children's self-reported loneliness, depressive symptoms, and parent-rated CBCL depression and anxiety separately. First, we examined mean differences in children's internalizing symptoms as a function of adoptive status, birth order, gender, and age for the entire sample of sibling pairs. Second, we estimated the phenotypic stability for each measure using bivariate Pearson's correlations for those with complete longitudinal data. Finally, we estimated genetic and environmental sources of variance in change in internalizing problems, and sources of genetic and environmental covariance mediating the stability of internalizing problems.

TABLE 9.1. Means and Standard Deviations for Children's Reports of Depressive Symptoms (Kandel) at 9, 10, 11, and 12 Years by Gender, Adoptive Status, and Birth Order

| | Adoptive | | Biological | |
	Boys	Girls	Boys	Girls
Older Siblings	$n = 109$	$n = 97$	$n = 113$	$n = 97$
Year 9	14.1 (5.1)	15.6 (5.1)	13.4 (4.9)	13.4 (4.9)
Year 10	12.6 (4.5)	13.6 (4.9)	12.3 (5.0)	12.5 (4.2)
Year 11	11.4 (4.3)	11.6 (4.9)	11.1 (4.1)	11.6 (4.5)
Year 12	12.1 (4.9)	11.1 (4.0)	11.8 (4.6)	11.9 (4.1)
Younger Siblings				
Year 9	16.6 (5.8)	17.0 (5.9)	14.9 (4.4)	14.7 (4.0)
Year 10	14.4 (4.7)	13.6 (5.1)	13.1 (4.8)	13.2 (4.3)
Year 11	12.9 (6.1)	13.3 (5.8)	14.1 (5.2)	12.8 (4.8)
Year 12	13.9 (4.6)	12.4 (5.1)	12.6 (4.7)	11.8 (4.0)

TABLE 9.2. Means and Standard Deviations for Parents' Reports of Child Depression/Anxiety (CBCL) at 9, 10, 11, and 12 Years by Gender, Adoptive Status, and Birth Order

| | Adoptive | | Biological | |
	Boys	Girls	Boys	Girls
Older Siblings	$n = 83$	$n = 77$	$n = 92$	$n = 72$
Year 9	3.7 (3.2)	3.2 (3.5)	3.1 (3.4)	3.6 (3.4)
Year 10	3.5 (3.5)	3.2 (3.5)	3.3 (3.6)	3.1 (2.7)
Year 11	3.0 (3.0)	3.3 (3.7)	2.6 (2.9)	2.7 (2.8)
Year 12	3.2 (3.2)	3.0 (3.2)	2.6 (2.8)	2.9 (2.8)
Younger Siblings				
Year 9	2.8 (3.1)	3.4 (4.6)	1.9 (2.1)	2.1 (2.8)
Year 10	2.5 (2.5)	2.8 (3.2)	2.2 (2.5)	3.0 (3.0)
Year 11	2.1 (2.5)	3.5 (4.7)	2.4 (2.9)	3.5 (4.0)
Year 12	2.3 (2.4)	3.5 (4.1)	2.7 (3.3)	3.1 (3.3)

Descriptive Analyses of Internalizing Problems

Means and standard deviations for the measures of internalizing problems are shown in table 9.1 and table 9.2. The descriptive statistics and means analyses for the Asher Loneliness Questionnaire data are presented elsewhere in this volume—there were no gender, birth-order, or adoptive-status differences in children's reports of loneliness, although levels of loneliness decreased over time (McGuire & Clifford, chapter 10, this volume).

For children's self-reported Kandel depressive symptoms (see table 9.1), three analysis of variance (ANOVA) models were tested. First, a 2 (older sibling gender) by 2 (adoptive status) by 4 (time) mixed model was examined for older siblings' reports of depression on the Kandel. There was a main effect for time, $F = 36.79$, $df = 3,410$, $p < .01$. This main effect was subsumed by a two-way interaction between time and adoptive status, $F = 2.79$, $df = 3,410$, $p < .01$. Post-hoc analyses revealed that there was only one adoptive status mean difference, at year 9, $F = 7.02$, $df = 1,412$, $p < .01$, with adopted older siblings reporting slightly higher levels of depressive symptoms than nonadoptive older siblings.

Second, a 2 (younger sibling gender) by 2 (adoptive status) by 4 (time) mixed model was examined for younger siblings' reports of depression on the Kandel. There was a significant main effect for time, $F = 13.01$, $df = 2, 133$, $p < .01$; post-hoc analyses showed that the younger siblings' Kandel scores were higher at 9 years compared to 10, 11, and 12 years of age, $t = 3.08$ to 3.63, $p < .01$.

Third, a 2 (birth order) by 2 (adoptive status) by 4 (time) mixed model was examined for older and younger siblings' reports of depression; to examine birth-order effects and interactions. As in the first ANOVA, there was a main effect of time, and a two-way interaction between time and adoptive status (see above). However, there was also a main effect of birth order, $F = 10.13$, $df = 1, 130$, $p < .01$, with younger siblings having higher scores than older siblings at 9 and 11 years, $t = -2.32$ and -2.89, $p < .05$.

Next, we examined mean differences in parents' reports of children's depressive and anxious symptoms on the CBCL (see table 9.2). Again, three analyses of variance (ANOVA) models were tested. First, a 2 (older sibling gender) by 2 (adoptive status) by 4 (time) mixed model was examined for parents' reports of older siblings' depression/anxiety. There was a main effect for time, $F = 3.95$, $df = 3,318$, $p < .01$. Post-hoc analyses revealed that older siblings' CBCL scores were lower at 11 years than at 9 and 10 years of age, $t = 2.29$ and 2.89, $p < .05$.

Second, a 2 (younger sibling gender) by 2 (adoptive status) by 4 (time) mixed model was examined for parents' reports of the younger siblings' CBCL depression/anxiety symptoms. There were no significant main effects or interactions. Third, a 2 (birth order) by 2 (adoptive status) by 4 (time) mixed model was examined for older and younger siblings' CBCL depression/anxiety scores. There was a main effect of birth order, $F = 10.39$, $df = 1,82$, $p < .01$. This main effect was subsumed by a two-way interaction with time, $F = 3.61$, $df = 3, 80$, $p < .05$. Older siblings had higher CBCL scores than younger siblings at 9 and 10 years of age, $t = 3.42$ and 3.52, $p < .01$.

TABLE 9.3. Stability of Children's Reports
of Loneliness (Asher) Over Time ($n = 133$ pairs
of siblings) for Older Siblings (below diagonal)
and Younger Siblings (above diagonal)

Year	9	10	11	12
9	—	49	34	42
10	48	—	58	54
11	38	54	—	57
12	17	29	31	—

Note: Decimals excluded, all significant at two-tailed $p < .05$

TABLE 9.4. Stability of Children's Reports
of Depression (Kandel) Over Time ($n = 132$ pairs
of siblings) for Older Siblings (below diagonal)
and Younger Siblings (above diagonal)

Year	9	10	11	12
9	—	33	39	18
10	42	—	26	26
11	29	46	—	41
12	10+	10+	25	—

Note: Decimals excluded, + not significant, otherwise all significant
at two-tailed $p < .05$

Stability over Time: Bivariate Correlations

Next, we examined the stability in children's symptoms over a three-year period (from 9 years of age to 12 years of age). Bivariate correlations over time were estimated separately for older and younger siblings. The correlations for children's self-reported loneliness are shown in table 9.3. For both older and younger siblings, consecutive year-to-year stability was moderate ($r = .31$ to $.58$), with lagged correlations being somewhat more modest ($r = .17$ to $.54$). A similar pattern emerged for children 's self-reported depressive symptoms (see table 9.4). Again, consecutive year-to-year stability was modest to moderate ($r = .25$ to $.46$), with lagged correlations being slightly lower ($r = .10$ to $.39$). Finally, we estimated the stability correlations for parents' reports of their children's depressive and anxious symptoms on the CBCL (see table 9.5). These correlations were somewhat higher than those found for children's self-reports, with substantial consecutive as well as lagged stability correlations ($r = .52$ to $.80$).

TABLE 9.5. Stability of Parents' Reports of Child
Depression/Anxiety (CBCL) Over Time ($n = 84$
pairs of siblings) for Older Siblings (below diagonal)
and Younger Siblings (above diagonal)

Year	9	10	11	12
9	—	77	75	67
10	56	—	80	73
11	55	70	—	76
12	52	66	67	—

Note: Decimals excluded, all significant at two-tailed $p < .05$

Genetic Analysis of Stability and Change

None of these stabilities approached the reliability of measurement, suggesting modest to substantial amounts of change over time, particularly for children's self-reported feelings of loneliness and depressive symptoms. The final set of analyses examined the genetic and environmental sources of variance in the stability and changes in internalizing symptoms over time. The gender of each child was statistically controlled in these analyses. We used a longitudinal bivariate ACE model to estimate sources of genetic and environmental covariance between consecutive annual assessments of internalizing problems. Thus, we examined genetic and environmental mediation of children's internalizing problems from 9 to 10 years of age, from 10 to 11 years of age, and from 11 to 12 years of age. These parameters were estimated using a bivariate extension of the univariate ACE model (Neale & Cardon, 1992). In the bivariate model, the covariances between a scores at one age and scores at another age are decomposed into mediating (i.e., common or overlapping) and unique environmental and genetic components through the estimation of six latent variables: additive genetic variance (A), shared environmental variance (C), and nonshared environmental variance (E) that mediate or are common across both assessments, as well as A, C, and E that are unique to each assessment. This model is represented graphically in figure 9.1.

Before examining stability and change, for descriptive purposes we estimated the univariate genetic and environmental variance parameters for all three measures (excluding year 12, as the univariate estimates for the final assessment are not derived in this longitudinal bivariate model). The univariate parameter estimates for children's self-reported loneliness (Asher) are reported elsewhere—significant additive genetic variance accounted for one-half to two-thirds of the variance at ages 10 and 11 years, and shared environmental variance was modest (McGuire & Clifford, chapter 10, this volume). For children's self-reported

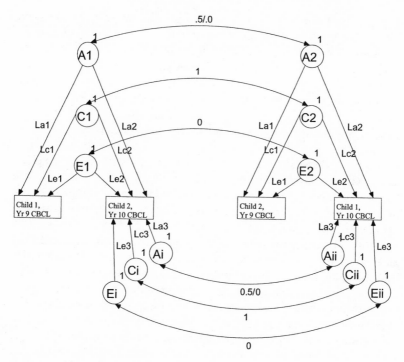

FIGURE 9.1. Bivariate ACE model for longitudinal genetic analyses. This example is the model specification for nonadoptive biological siblings' data at years 9 and 10 on parent-rated CBCL depression/anxiety. The covariation between year 9 and year 10 scores is decomposed into additive genetic (A) [fixed at .5 for nonadoptive biological siblings, and .0 for adoptive unrelated siblings], shared environmental (C) [fixed at 1 for all siblings], and nonshared environmental (E) [fixed at 0 for all siblings] sources of variance and covariance. Parameters with the same name (e.g., Lal, La2) were constrained to be equal. The parameters Lal/La2, Lcl/Lc2, and Lel/Le2 provide estimation of additive genetic, shared environmental, and nonshared environmental mediation respectively of stability over time. The remaining parameters La3, Lc3, and Le3 provide estimates of the unique genetic, shared environmental, and nonshared environmental variance, respectively, for change over time.

depressive symptoms, sibling intraclass correlations were slightly higher among biological siblings than adoptive siblings, ranging from −.09 to .04 for adoptive siblings, and from .05 to .20 for nonadoptive siblings. Additive genetic influence was significant at several ages (year 9 = .49, $p < .05$, year 10 = .09, *ns*, year 11 = .49, $p < .05$), and shared environment was negligible and nonsignificant (year 9 = .04, years 10 and 11 = .00). A different pattern emerged for parents' ratings of their children on the CBCL. Sibling intraclass correlations suggested substantial shared environmental variance, ranging from .49 to. 77 for adoptive siblings, and from .34 to .51 for nonadoptive siblings. Heritability was negligible (year 9 = .19, *ns*, year 10 = .00, year 11 = .01, *ns*), with nearly all sibling similarity being accounted for by shared environment (year 9 = .46, year 10 = .61, year 11 = .59, all significant at *p* < .05).

Next, we examined the etiology of stability and change in internalizing symptoms over time, using the bivariate ACE model (described above). The parameters, presented in table 9.6 along with the model fit indices, represent the sources of covariance that mediate the stability of these symptoms over time (Lal and La2, Lcl and Lc2, Le1 and Le2), as well as those sources of variance that account for change over time (La3, Lc3, and Le3, discussed below). For example, the second line in table 9.6 shows the results of the bivariate model for children's reports of loneliness from year 10 to year 11. The two parameters assessing genetic contributions to stability (Lal and La2) were significant (.76 and .78). The paths for shared environmental mediation (Lcl and Lc2) and nonshared enviromental mediation (Le1 and Le2) were not significant. This suggests that stability in loneliness from 10 to 11 years of age was mediated entirely by genetic factors. In order to simplify the description of these results, we estimated the proportion of each one-year lagged stability correlation that was mediated by genetic, shared environmental, and nonshared environmental sources of covariance (see table 9.7).

For both child self-reported measures (Asher loneliness and Kandel depressive symptoms), nearly all of the stability was accounted for by genetic mediation. An exception to this pattern was found for Asher loneliness for years 9 to 10, and Kandel depression for years 10 to 11, where the majority of stability was accounted for by nonshared environmental mediation. There was little evidence to suggest shared environmental mediation of the stability of loneliness and depressive symptoms over time in the transition to adolescence. In contrast, nearly all of the stability in parent-rated CBCL depression and anxiety symptoms was accounted for by shared environmental mediation, with the remaining covariance over time accounted for by nonshared environmental mediation. There was no evidence that additive genetic factors accounted for stability in parent-rated child depressive/anxious symptoms.

Finally, we examined sources of variation contributing to change in internalizing symptoms over time. These are estimated from parameters La3, Lc3, and Le3 (see figure 9.1, and table 9.6). For children's self-reported loneliness (Asher), the majority of the variance in change over time, for all three one-year lagged models (i.e., 9 to 10, 10 to 11, and 11 to 12 years), was accounted for by nonshared environmental variance. Only one model (9 to 10 years) showed evidence for a genetic contribution to change (La3 = −.59, *p* < .05), and there was little evidence to suggest that shared environmental factors were contributing

TABLE 9.6. Additive Genetic (A), Shared Environmental (C), and Nonshared Environmental (E) Parameter Estimates and Fit Indices from Bivariate ACE Structural Equation Model (see figure 9.1)

	La1	Lc1	Le1	La2	Lc2	Le2	La3	Lc3	Le3	x^2(p)	df	GFI
Asher Loneliness:												
Year 9 to year 10	.47	-.36*	-.81	.38	-.23	-.27	.59*	—	.61	5.1 (.96)	12	.98
Year 10 to year 11	.76*	.11	-.63	.78*	.13	.08	—	.24	.56	7.02 (.80)	12	.98
Year 11 to year 12	-.80	-.27	-.54	-.57*	-.18	.11	—	.15	.78	10.5 (.57)	12	.97
Kandel Depression:												
Year 9 to year 10	.70*	.19	.69	.43	.20	.05	—	—	.88	16.5 (.22)	13	.96
Year 10 to year 11	-.31	.00	.95	-.40	.00	.29	.59*	—	-.63	7.2 (.82)	12	.97
Year 11 to year 12	.70*	.02	-.71	.50	-.18	.00	—	—	-.85	6.9 (.91)	13	.97
CBCL depression/anxiety:												
Year 9 to year 10	-.44	.68*	.59	.01	.77*	.23	—	—	.60	16.3 (.24)	13	.92
Year 10 to year 11	—	-.78*	.63	—	-.76*	.20	—	—	.61	12.1 (.67)	15	.94
Year 11 to year 12	-.11	-.77*	.63	-.09	-.72*	.23	—	—	-.65	25.9 (.01)	13	.90

Note: *two-tailed $p < .05$—constrained at 0. Note that the statistical significance of nonshared environment parameters (Le1, Le2, Le3) was not tested, given that error variance is included in these estimates and that the null hypothesis that error variance is = 0 is not tenable.

145

TABLE 9.7. Genetic (A), Shared Environmental (C), and Nonshared Environmental (E) Mediation of Bivariate Model Longitudinal Correlations r, as the Percentage of Model Stability Correlation

		Percentage of Stability Correlation (r)		
	r	A	C	E
Asher Loneliness:				
Year 9 to Year 10	.48	38	17	45
Year 10 to Year 11	.57	99*	01	00
Year 11 to Year 12	.45	99*	01	00
Kandel Depression:				
Year 9 to Year 10	.37	81	11	08
Year 10 to Year 11	.40	30	00	70
Year 11 to Year 12	.35	100	00	00
CBCL Depression/Anxiety:				
Year 9 to Year 10	.66	00	79*	21
Year 10 to Year 11	.72	00	82*	18
Year 11 to Year 12	.69	00	80*	20

Note: *Both mediating parameters statistically significant, two-tailed $p < .05$. The significance of nonshared environmental mediation (E) was not tested given that error variance is included in these estimates and that the null hypothesis that error variance is = 0 is not tenable.

to change in loneliness. Children's self-reported depressive symptoms (Kandel) showed a very similar pattern, with the majority of the variance in change over time accounted for by nonshared environment. Again, only one model (10 to 11 years) showed evidence for a genetic contribution to change over time in depressive symptoms, and there was no evidence that shared environmental variance was operating for change. For parents' ratings of their children's depressive and anxiety symptoms (CBCL), again the results were consistent with the other measures, with nonshared environmental factors accounting for all of the change over time.

Discussion

This study of intraindividual variations in children's and parents' reports of internalizing problems revealed modest to substantial stability from 9 to 12 years of age. Change in internalizing symptoms was most prominent in children's self-reports. By comparison, parents' reports were highly stable, with correlations over time approaching reliability of measurement. Thus, there was evidence for considerable intraindividual change in internalizing problems, although the magnitude of this variation over time depended on the respondent.

The univariate genetic analyses also clearly indicated a difference in the patterns of findings depending on the data source. Individual differences in children's self-reported internalizing problems at any given age included additive genetic variance at most assessments (significant $h2 = .49$ to .64), while there was no evidence for shared environmental variance. In contrast, individual differences in parents' reports of their children's depressive and anxious symptoms consistently included shared environmental sources of variance (significant $C2 = .46$ to .61) and negligible heritable variance. However, consistent across both data sources was the often-substantial nonshared environment component that accounted for two-thirds or more of the variance in internalizing symptoms. These univariate genetic analyses of internalizing problems at a given age do not inform us about underlying gene-environment processes in intraindividual dynamic aspects of change and stability. Therefore, we examined longitudinal bivariate genetic models to test for genetic and environmental sources of variation in stability and change over time. The analysis of stability over time in children's internalizing symptoms revealed a pattern of findings that differed for self-reports and parents' ratings. Intraindividual stability over time in children's self-reported loneliness and depressive symptoms was largely accounted for by genetic sources of variance that were common across the assessments—that is, this stability was genetically mediated. In contrast, nearly all of the stability over time in parents' ratings of their children's depressive and anxious symptoms was accounted for by shared environmental sources of variance that were common across assessments.

The results for children's self-reported internalizing symptoms, but not parent-reported symptoms, are consistent with those reported by O'Connor and colleagues in their analysis of stability and change in adolescent depressive symptoms (O'Connor et al., 1998). In their study, about two-thirds of the stability in depressive symptoms was accounted for by genetic mediation (the remaining stability being accounted for by nonshared environmental mediation). It is noteworthy that O'Connor et al. (1998) used highly reliable cross-source composite measures of depressive symptoms.

The differences between parents' ratings and children's self-reports raise at least two methodological considerations. First, using multiple informants is essential in the study of the intraindividual dynamics of emotional problems in the transition to adolescence. We would have drawn different conclusions had we relied on children's self-reports or parents' perceptions alone. At the same time, combining parents' and children's ratings to yield a cross-informant construct, while often methodologically preferable because of the greater reliability of such measures, can obscure potentially illuminating differences between data sources that may exist. In the case of O'Connor et al. (1998), the pattern of sibling similarity for stability and change in depressive symptoms was similar regardless of data source, therefore making cross-source composites feasible. Second, based on the view that self-reports are better indicators of emotional problems than are parents' reports (Eley et al., 1998), it may be worthwhile to weigh more heavily the findings for children's self-reports. In the CAP data, it was the child-reported symptoms, and not the parents' reports, that replicated the findings of O'Connor et al. (1998). It is noteworthy that all of the sibling similarity in stability for parents' ratings was due to shared environmental variance. This shared environmental variance could be a by-product of shared method variance (same informant, same measure, assessed at the same time for both children), rather than or in addition to actual environmental processes that lead to sibling similarity in patterns of change. For example, the shared environmental mediation may be due to stability in parental attitudes about psychopathology. At the same time, the parent-report measure is probably capturing meaningful within-family variation that is linked to sibling similarity in internalizing problems. Given the contradiction between the findings for these parent-report data and O'Connor et al. (1998), it will be important to continue to examine parent reports and compare them to other sources of data as the children progress through adolescence.

Finally, we turned to sources of variation in intraindividual change in internalizing symptoms over time. With the exception of Asher loneliness ratings from 9 to 10 years, and Kandel depressive symptoms from 10 to 11 years (both of which included significant genetic contributions to change), all of the change in internalizing problems for both self-reports and parents' ratings was attributable to nonshared environmental sources of variance (which includes error variance). Thus, practically all of the intraindividual change in children's internalizing problems (regardless of respondent) was child-specific within families and thus indicative of sibling differences rather than similarities in patterns of change over time. This finding is consistent with O'Connor et al. (1998), where change in depressive symptoms over adolescence was accounted for by nonshared environmental factors.

Nonshared environment is clearly central to change in internalizing symptoms. These data suggest that as children move through the transition into adolescence, environmental influences that impact their feelings of loneliness, sadness, and isolation are most likely to be child-specific within families. This does not rule out the possibility that shared events (like a parental divorce or the loss of a grandparent) in the family have an impact on internalizing symptoms. Instead, it suggests that this impact may not lead to similar internalizing outcomes for siblings in that family. Rather, the siblings may react differently to the same-shared event. Furthermore, there are many experiences that most likely differ for sibling children—most notably, experiences arising out of peer relationships, relationships with nonparental adults such as teachers, and idiosyncratic events like the beginning and ending of romantic relationships. The similar results reported by O'Connor et al. (1998) suggest that nonshared environment continues to play a predominant role in leading to changes in depressive symptoms in adolescence.

Future research should focus on more fine-grained analyses of intraindividual change—for instance, analyses that include measures of biological (e.g., pubertal) and social changes (e.g., school transitions) as covarying processes, rather than relying on chronological age as a general indicator of developmental status. Furthermore, more careful consideration of the similarities and differences in findings across different data sources (parents', teachers', or children's reports) will be required. We have only begun to examine the genetic and environmental processes underlying growth in depressive symptoms in childhood and adolescence; much more research using a variety of quantitative and molecular genetic designs will be needed. Importantly, the gene-environment processes for internalizing problems at any given point in time may be similar or different from the processes that lead to growth or decreases in symptoms over time. For example, certain genes may operate probabilistically on depressive symptoms in childhood, but the action of these genes may be modified by changes in the body arising from the onset of puberty as well as changes in the environment arising from affiliated social changes. Finally, the findings for the static as well as change analyses converge on the finding that nonshared environment is very important in the transition to early adolescence, and may be capturing a host of systematic as well as idiosyncratic experiences that lead to sibling differentiation in internalizing symptoms over time, regardless of the genetic similarity of the siblings.

ACKNOWLEDGMENTS

This research was supported by grants HD-10333 and HD-18426 from the National Institute of Child Health and Human Development and by grant MH-43899 from the National Institute of Mental Health. The authors wish to thank Sally-Ann Rhea, Robin Corley, John Hewitt, John DeFries, and Robert Plomin for their thoughtful comments on earlier drafts of the manuscript.

REFERENCES

Achenbach, T. M. (1991). *Integrative guide for the 1991 CBCL/4–18. YSR, and TRF profiles.* Burlington: University of Vermont.

Asher, S. (1985). "The Loneliness Questionnaire." Personal communication.

Asher, S. R., Hymel, S., & Renshaw, P. D. (1984). Loneliness in children. *Child Development*, *55*, 1456–1464.

Asher, S. R., & Wheeler, V. A. (1985). Children's loneliness: A comparison of rejected and neglected peer status. *Journal of Consulting and Clinical Psychology*, *53*, 500–505.

Austin, M. A., Friedlander, Y., Newman, B., Edwards, K., Mayer-Davis, E. J., & King, M. C. (1997). Genetic influences on changes in body mass index: A longitudinal analysis of women twins. *Obesity Research*, *5*, 326–331.

Cassidy, J., & Asher, S. R. (1992). Loneliness and peer relations in young children. *Child Development*, *63*, 350–365.

Damon, W. (1977). *The social world of the child.* San Francisco: Jossey-Bass.

Deater-Deckard, K., Reiss, D., Hetherington, E. M., & Plomin, R. (1997). Dimensions and disorders of adolescent adjustment: A quantitative genetic analysis of unselected samples and selected extremes. *Journal of Child Psychology and Psychiatry*, *38*, 515–525.

DeFries, J. C., Plomin, R., & Fulker, D. W. (1994). *Nature and nurture during middle childhood.* Cambridge, MA: Blackwell.

Eley, T. C. (1997). Depressive symptoms in children and adolescents: Etiological links between normality and abnormality. *Journal of Child Psychology and Psychiatry*, *38*, 861–866.

Eley, T. C., Deater-Deckard, K., Fombonne, E., Fulker, D. W., & Plomin, R. (1998). An adoption study of depressive symptoms in middle childhood. *Journal of Child Psychology and Psychiatry*, *39*, 337–345.

Elkind, D. (1985). Egocentrism redux. *Developmental Review*, *5*, 218–226.

Francis, D. J., Fletcher, J. M., Stuebing, K. K., Davidson, K. C., & Thompson, N. M. (1991). Analysis of change: Modeling individual growth. *Journal of Consulting and Clinical Psychology*, *59*, 27–37.

Grilo, C. M., & Pogue-Geile, M. F. (1991). The nature of environmental influences on weight and obesity: A behavior genetic analysis. *Psychological Bulletin*, *10*, 520–537.

Harrington, R., Rutter, M., & Fombonne, E. (1996). Developmental pathways in depression: Multiple meanings, antecedents, and endpoints. *Development and Psychopathology*, *8*, 601–616.

Heitmann, B. L., Kaprio, J., Harris, J. R., Rissanen, A., Korkeila, M., & Koskenvuo, M. (1997). Are genetic determinants of weight gain modified by leisure-time physical activity? A prospective study of Finnish twins. *American Journal of Clinical Nutrition*, *66*, 672–678.

Kandel, D., & Davies, J. (1982). Epidemiology of depressive mood in adolescents. *Archives of General Psychiatry*, *39*, 1205–1212.

McGuffin, P., Owen, M. J., O'Donovan, M. C., Thapar, A., & Gottesman, I. I. (1994). *Seminars in psychiatric genetics.* London: Gaskell.

Neale, M. C., & Cardon, L. R. (1992). *Methodology for genetic studies of twins and families.* Dordrecht, Netherlands: Kluwer.

O'Connor, T. G., Neiderhiser, J. M., Reiss, D., Hetherington, E. M., & Plomin, R. (1998). Genetic contributions to continuity, change, and co-occurrence of antisocial and depressive symptoms in adolescence. *Journal of Child Psychology and Psychiatry*, *39*, 323–336.

Parkhurst, J. T., & Asher, S. R. (1992). Peer rejection in middle school: Subgroup differences in behavior, loneliness, and interpersonal concerns. *Developmental Psychology, 28,* 231–241.

Peplau, L. A., & Perlman, D. (1982). *Loneliness: A sourcebook of current theory, research, and therapy.* New York: John Wiley.

Plomin, R., & DeFries, J. C. (1985). *Origins of individual differences in infancy and early childhood: The Colorado Adoption Project.* New York: Academic Press.

Rende, R., Plomin, R., Reiss, D., & Hetherington, E. M. (1993). Genetic and environmental influences on depressive symptomatology in adolescence: Individual differences and extreme scores. *Journal of Child Psychology and Psychiatry, 34,* 1387–1398.

Rubin, K. H., Hymel, S., & Mills, R. S. (1989). Sociability and social withdrawal in childhood: Stability and outcomes. *Journal of Personality, 57,* 237–255.

Stanger, C., Achenbach, T. M., & McConaughy, S. H. (1993). Three-year course of behavioral/emotional problems in a national sample of 4- to 16-year-olds: 3. Predictors of signs of disturbance. *Journal of Consulting and Clinical Psychology, 61,* 839–848.

van den Oord, E. J. C. G., Boomsma, D. I., & Verhulst, F. C. (1994). A study of problem behaviors in 10- to 15-year-old biologically related and unrelated international adoptees. *Behavior Genetics, 24,* 193–205.

Willett, J. B., Ayoub, C. C., & Robinson, D. (1991). Using growth curve modeling to examine systematic differences in growth: An example of change in the functioning of families at risk of maladaptive parenting, child abuse, or neglect. *Journal of Consulting and Clinical Psychology, 59,* 38–47.

SHIRLEY MCGUIRE
JEANIE CLIFFORD

10

Loneliness during the Transition to Early Adolescence

Introduction

This chapter examines children's reports of loneliness while they were making the transition into early adolescence. Loneliness is often conceptualized as an internalizing behavioral problem in developmental psychopathology research. Yet, few researchers have explored genetic and nonshared environmental contributions to loneliness. There are also little data examining links between children's reports of peer isolation and other measures of internalizing problems. Thus, two main research questions were addressed in this chapter: Are individual differences in loneliness heritable during the transition to adolescence? and Are children's perceptions of loneliness linked to parent and teacher reports of internalizing problems over time? In addition, descriptive analyses were conducted examining longitudinal, gender, and adoptive status differences during this important developmental transition. The sample consisted of 590 to 661 Colorado Adoption Project (CAP) children when they were 9, 10, 11, and 12 years old. Eight percent of the CAP children reported feeling really isolated from their peers. While there were no gender differences, children's reports of loneliness decreased across age. Individual differences in loneliness were linked to genetic and nonshared environmental factors. Surprisingly, child reports of loneliness were not related to parent and teacher reports of children's internalizing problems. Children's reports, however, were related to parent and teacher responses to specific questions about loneliness. The nature of loneliness and the importance of biological contributions are discussed.

Loneliness is the awareness of a lack of connectedness with other people coupled with negative emotion (e.g., Ernst & Cacioppo, 1999). Everyone experiences feelings of loneliness now and then. Even preschool children can talk about feelings of isolation in a meaningful way (Cassidy & Asher, 1992), and popular children admit to feeling disconnected from their peer sometimes (Asher, Hymel, & Renshaw, 1984). Some children and adults, however, experience chronic loneliness. These feelings seem to include a poor attributional style in which the present situation (i.e., being friendless) is considered stable, uncontrollable and due to defects within the self (e.g., Boivin & Hymel, 1997; Graham & Juvonen, 1998; Guay, Boivin, & Hodgee, 1999; Laine, 1998; Tur-Kaspa, Weisel, & Segev, 1998). That is, chronically lonely children seem to believe that peer rejection or isolation is their fault and it cannot or will not change. Consequently, extreme loneliness is often grouped with other "internalizing" behavioral problems such as depression, anxiety, and low self-worth (Graham & Juvonen, 1998; Rubin, Chen, McDougall, & Bowker 1995).

Genetic and Environmental Contributions

McGuire and Clifford (2000) were the first to explore the etiology of loneliness from a behavioral genetic perspective. They found significant genetic and nonshared environmental contributions to children's feelings of loneliness at school using a twin-sibling sample and to children's general feelings of loneliness using the CAP sample. Why would loneliness be heritable? Feelings of loneliness appear to be linked to negative emotionality (Ernst & Cacioppo, 1999), a temperament characteristic thought to have biological roots. Studies have documented genetic contributions to the other internalizing problems, such as depressive symptoms and anxious behavior, during childhood (e.g., Eley, Deater-Deckard, Fombonne, Fulker, & Plomin, 1998). In addition, a literature is emerging showing that some areas of self-perception are heritable, including perceptions of scholastic competence, athletic skill, popularity, happiness, and masculinity and femininity (Hur, McGue, & Iacono, 1998; Lykken & Tellegen, 1996; McGue, Hirsch, & Lykken, 1993; McGuire, Manke Saudino, Reiss, Hetherington, & Plomin, 1999; McGuire, Nelderhiser, Reiss, Hetherington, & Plomin, 1994; Mitchell, Baker, & Jacklin, 1989). Thus, loneliness may be heritable due to its connection with other genetically linked traits.

Of course, there is evidence that environmental factors also contribute to individual differences in loneliness. For example, loneliness in children appears to be associated with an insecure attachment during infancy (Berlin, Cassidy, & Belsky, 1995). Not surprisingly, loneliness is consistently linked to peer-group isolation and victimization (e.g., Asher & Wheeler, 1985; Boivin & Hymel, 1997; Cassidy & Asher, 1992; Crick & Ladd, 1993; Renshaw & Brown, 1993). It is likely that these experiences are unique to each child in a family and would result in low, rather than high, sibling similarity (i.e., nonshared environmental influences). This chapter extends McGuire and Clifford's (2000) analyses by examining genetic and environmental links to loneliness across age. The adoption study design of the CAP allowed us to examine the etiology of loneliness from a behavioral genetic perspective. It was hypothesized that loneliness would be significantly heritable and that environmental influences would be nonshared at each age.

Links to Internalizing Problems

Loneliness fits within the internalizing dimension of childhood psychopathology conceptually; however, the link also needs to be examined empirically. This information will be important if loneliness is shown to be heritable—genetic contributions to loneliness may be a result of biological contributions to related problems such as depression or anxiety. The CAP data allowed us to examine associations between parents' and teachers' reports of children's internalizing problems and children's self-reports of loneliness employing a widely used measure in a large, community-based sample. It is important to note that less research in developmental psychopathology has been devoted to measuring and understanding children's internalizing behavioral problems (e.g., loneliness and depression) compared to children's externalizing behavioral problems (e.g., aggression and delinquency). There may be several possible reasons for the research gap. First, externalizing problems may be easier to assess. Externalizing problems are just that—external—and consequently, they are more visible. Aggressive behaviors (e.g., hitting, biting, and fighting) are easier to observe and agree about compared to depressive symptoms (e.g., sulking, withdrawing, and sadness). Second, internalizing symptoms are probably less disruptive to the home and classroom contexts compared to externalizing behaviors. That is not to say that parents or teachers are less concerned about depressed, anxious, or lonely children. Rather, when adults are thinking about and taking care of several children, it may simply be harder to identify withdrawn children. Finally, children's self-perceptions are probably fundamental to the identification of internalizing problems. The best way to know if a child has low self-esteem is to ask the child. This may also be true for loneliness, depression, and anxiety. It is hard, however, to assess children's self-perceptions before middle childhood—that is, before children can talk about themselves in a consistent manner (see Cassidy & Asher, 1992, for an exception in the loneliness area). This means that internalizing problems, such as loneliness, may be more accessible during late childhood and adolescence compared to early childhood. The longitudinal design of the CAP allowed us to examine associations between children's reports of loneliness and parents' and teachers' reports of children's internalizing problems during the preteen and teen years. The second goal of this chapter was to investigate such links when the children were 9, 10, 11, and 12 years old. It was hypothesized that correlations would be low to moderate across reporters; however, the associations would increase over time as children moved into adolescence.

Additional Issues

The CAP design allowed us to explore three additional issues: longitudinal changes, gender differences, and adoptive status differences. Longitudinal change was examined at both the item and mean levels. (The issue of stability is investigated in Deater-Deackard & McGuire, chapter 9, this volume.) Children's answers to individual questions were explored and compared to data published by Asher and colleagues (1984) using a younger sample (506 children from 6 to 9 years old). There is some evidence that mean levels of loneliness are higher during adolescence compared to other age groups (Parkhurst & Asher, 1992; Peplau & Perlman, 1982). Developmental changes in self-concept and abstract cognitive abilities may lead to the emergence of internalizing problems at this time period. In addition, peer acceptance is of great significance during this developmental period (e.g., Harris, 1998). These new abilities and concerns may cause teens to feel the full impact of their circumstances (e.g., peer rejection) mentally and emotionally; they may experience more severe and/or consistent depression, anxiety, and loneliness compared to younger children. Based on the child and adult literatures, it was hypothesized that loneliness would increase as children approached adolescence.

In addition to longitudinal changes over time, gender- and adoptive-status differences were investigated in the CAP sample. Unlike other areas of children's behavior problems, there has been little evidence of gender differences in children's perceptions of loneliness during childhood (Asher et al., 1984; Cassidy & Asher, 1992). The literature on adult loneliness, on the other hand, is rather mixed. Ernst and Cacioppo (1999) point out, however, that women tend to report higher levels of loneliness than men when the questionnaire contains the words *loneliness* and *lonely*. Given that the CAP questionnaire contains these words, it was hypothesized that girls would report more loneliness compared to boys. It was also expected that these differences would be greater for older (i.e., 12 years old) compared to younger (i.e., 9 years old) children. To our knowledge, no one has explored adoptive status differences. Consequently, these analyses were purely exploratory.

In summary, two main issues were addressed in this chapter concerning children's feelings of isolation from their peers. First, links between children's feelings of loneliness and parents' reports of their internalizing problems were explored over time. Second, genetic and environmental contributions to children's loneliness were examined across time. Three descriptive questions were addressed: Do children's reports of loneliness change over time? Are there gender differences? Are there adoptive status differences?

Method

Participants in this analysis included 167 boys and 167 girls in adoptive families and 176 boys and 151 girls in control families. The children completed an adapted version of Asher's Loneliness Questionnaire (Asher, 1985; Asher et al., 1984) when they were 9, 10, 11, and 12 years old. The eight loneliness items were part of a larger scale including self-concept, depression, and family relationship items. Children had to check whether each statement was "Really true," "Sort of true," "In the middle," "A little true," or "Not true" for them. The eight statements were: (1) "I have nobody to talk to," (2) "I feel alone," (3) "I can find a friend when I need one," (4) "I can usually find someone to play with," (5) "I get along with other kids," (6) "I feel left out of things," (7) "There's nobody I can go to when I need help," and (8) "I'm lonely." Internal consistency reliabilities for the total scores were .77 for year 9, .78 for year 10, .80 for year 11, and .86 for year 12 self-reports.

The parents and teachers of the children completed the Child Behavior Checklist (Achenbach, 1991). The CBCL is a reliable and validated measure of children's externalizing and internalizing behavioral problems. Only the internalizing scale was used in this paper, and it contained depression, anxiety, and somatic complains subscales.

Results

Descriptive Data

Degree of Loneliness (Item level)   Children's responses to the eight loneliness questions were examined across time. The majority of the children reported low levels of loneliness, with 48 to 87% of the children stating the loneliness statements were "Not true" for them. Up to 5% of the kids reported feeling truly lonely and up to 8% reported feeling left out of things—evidence of chronic loneliness. Still, these numbers

TABLE 10.1. Means and Standard Deviations for Children's Reports of Loneliness at 9, 10, 11, and 12 Years by Gender and Adoptive Status

| | Adoptive | | Control | |
	Boys ($n = 134$)	Girls ($n = 141$)	Boys ($n = 154$)	Girls ($n = 127$)
Year 9	13.4 (5.3)	14.9 (6.3)	13.4 (5.2)	13.3 (6.0)
Year 10	11.9 (4.6)	12.9 (5.7)	12.3 (5.6)	11.8 (4.9)
Year 11	11.4 (4.8)	11.4 (4.6)	10.9 (4.4)	10.8 (3.9)
Year 12	11.3 (4.6)	11.1 (4.3)	11.9 (5.2)	10.9 (3.7)

are lower than those reported by Asher and his colleagues (1984) in their study of 6- to 9-year-olds and the number of children who reported high levels decreased over time. It could be that very lonely children dropped out of the CAP study, and so, attrition analyses were conducted. Mean differences in children's loneliness at years 9, 10, and 11 were examined comparing children who participated in the study at year 12 to those who did not. These two groups did not differ in their total loneliness scores at year 9, $t(659) = 1.22$, $p = .22$; at year 10, $t(647) = 1.84$, $p = .07$; or at year 11, $t(619) = =0.93$, $p = .35$. In addition, the distribution of scores across the items did not differ across the two groups. The percentage of children who endorsed the "Not True" option (i.e., they were not lonely, were not left out of things, could find friends, etc.) was examined across the eight questions. The two groups were similar with 47 to 92% of the children who did not participate at year 12 and 58 to 86% of those who did participate saying they were not lonely at years 9, 10, and 11. It seems that loneliness decreases while children make the transition into adolescence. It may be that some friendless children learn to make connections with their peers as they pass through middle childhood.

Time, Gender, and Adoptive Status (Mean level)   Table 10.1 shows the means and standard deviations by gender and adoptive status across time. The data were examined using a 2 (gender) × 2 (adoptive status) × 4 (time) mixed analyses of variance. Only children with complete data across the four time points were included ($N = 556$). There were no significant gender effects, adoptive-status effects, or interactions. There was, however, a significant time effect, multivariate $F(3, 1656) = 53.49$, $p < .001$. Follow-up tests showed that children reported significantly more loneliness at year 9 compared to year 10, $F(1, 629) = 41.17$, $p < .001$; year 11, $F(1, 596) = 135.74$, $p < .001$; and year 12, $F(1. 565)$, 86.04, $p < .0001$. Children also reported more loneliness at year 10 compared to year 11, $F(1, 574) = 42.19$, $p < .001$ and year 12, $F(1, 574) =$

TABLE 10.2. Correlations Between Children's Reports of Loneliness and Mothers' and Teachers' Reports of Children's Internalizing Problems Over Time

	Year 9	Year 10	Year 11	Year 12
Year 9	.19* (.14*)	.11* (.18*)	.12* (.18*)	.11* (.06)
Year 10	.17* (.23*)	.21* (.22*)	.11* (.17*)	.10 (.19*)
Year 11	.18* (.07)	.23* (.23*)	.18* (.12*)	.16* (.17*)
Year 12	.13* (.13*)	.15* (.15*)	.15* (.14*)	.12* (.20*)

Note: *$p < .01$. Correlations between mother and child reports are on the left. Correlations between teacher and child reports are on the right and in parentheses. Sample sizes ranged from 478 to 599 for parent report and 368 to 510 for teacher report.

18.28, $p < .001$, even after correcting for the number of comparisons made. The difference was not significant between year 11 and year 12. Contrary to expectations, children's self-reported loneliness decreased over time. It will be important to continue to examine children's perceptions of loneliness as they make their way through adolescence. Is there continuity across the teen years or does loneliness continue to decline over time?

Genetic and Environmental Contributions

The first goal was to examine genetic and environmental contributions to children's reports of loneliness across time. Of the some 600 CAP children, 133 pairs had full data at all four time points. For these pairs, intraclass correlations were calculated for biologically related or full sibling pairs ($N = 69$) compared to biological unrelated or adoptive pairs ($N = 64$). Table 10.2 shows the intraclass correlations for children's reports of loneliness when the CAP children were 9 to 12 years old, after controlling for child's gender. In general, sibling similarity was low, suggesting significant nonshared environmental influences. The intraclass correlations were significant for the full siblings at years 9, 10, and 11, but not at year 12. None of the correlations for the unrelated pairs was significant.

Table 10.2 also contains the parameter estimates for genetic and environment contributions to children's perceptions of loneliness. Model fitting analyses were conducted using variance/covariance matrices. The ACE model was used, which estimates additive genetic, shared environmental, and nonshared environmental contributions to variance in loneliness scores (see chapter 9, this volume, for details about behavioral genetic model fitting). Path estimates were squared to estimate variance due to heritability (h^2), shared environmental influences (c^2), and nonshared environmental influences (e^2) including error. Results

TABLE 10.3. Sibling Intraclass Correlations and Parameter Estimates for Children's Reports of Loneliness by Biological Relatedness of the Pairs

| | Sibling Intraclass Correlations | | Parameter Estimates | | | |
Age	Full Siblings	Unrelated Siblings	h^2	c^2	e^2	$\chi^2 (3)$
9	.20*	.14	.20	.14	.66*	17.7*
10	.29**	.01	.56*	.02	.41*	0.60
11	.32**	.09	.64*	.08	.28*	14.12*
12	.14	.14	.01	.14	.85*	2.58

Note: *$p < .05$. **$p < .01$

showed significant heritability at years 10 and 11 and significant non-shared environmental influences at each time point. There may be interesting environmental changes at 9 and 12 years old that are influencing individual differences in feelings of loneliness during those specific years. McGuire and Clifford (2000), however, reported a heritability of .48 after averaging over all four time points using this CAP sample. This is very close to the heritability of .45 they also reported for the San Diego Sibling Study (SANDSS) using a twin/sibling design and a more current version of Asher's questionnaire. It appears that loneliness, like other internalizing problems during childhood and adolescence, is heritable.

Links to Internalizing Problems

The second goal of this chapter was to examine links between children's reports of loneliness and parents' and teachers' reports of the children's internalizing problems. Table 10.3 shows the correlations between children's reports of loneliness and parents' and teachers' reports of children's internalizing problems. Surprisingly, the correlations were low. The measure of internalizing problems used, however, is a composite of questions concerning depressive symptoms, somatic complaints, and anxiety. Only two questions in the Child Behavior Checklist ask parents and teachers directly about children's loneliness. As a validity check, a second strategy was used to investigate the associations between parent and teacher responses to the two items on the CBC and children's loneliness reports of loneliness. The two questions on the CBC were: "Complains of loneliness" and "Would rather be alone than with others." Children were divided into three loneliness groups (high, medium, and low) based on how their parents or teachers responded to each item using the three-point scale: "very true," "somewhat true," or "not true." To simplify the analyses, only parent and teacher responses at year 9 were used to group the children. Mean difference in children's reports of loneliness (at 9, 10, 11, and 12 years) was the dependent variable and were examined using a 3 (parent or teacher rating group: high, medium, low) × 2 (gender) × 4 (time) mixed ANOVAs.

For parent responses to the first question, "complains of loneliness," there was an overall effect for group, F (2, 507) = 10.17, p < .001. Follow-up analyses showed that this difference was significant at year 9, F (2, 507) = 5.54, p < .01; year 10, F (2, 507) = 6.81, p < .001; year 11, F (2, 507) = 8.16, p < .001; and to a lesser extent year 12, F (2, 507) = 2.81, p < .06. Children whose parents rated "very true" for "complains of loneliness" when their children were 9 years old reported feeling the most loneliness over time compared to the other children. There was a time effect, gender effect, and time-by-gender interaction. Specifically, girls reported more loneliness compared to boys and this disappeared by year 12. This gender effect was not found in the previous analyses using the total sample. These effects did not interact with the group effects.

For teacher responses to the first question, "complains of loneliness," there was also significant effect for group, F (2, 440) = 9.43, p < .001. Follow-up ANOVAs showed significant differences at year 9, F (2, 440) = 5.77, p < .01; year 10, F (2, 440) = 3.96, p < .05; year 11, F (2, 440) = 13.20, p < .001; but not at year 12, F (2, 440) = 1.39, p = .25. Children whose teachers responded that the child complained of loneliness when 9 years old reported feeling lonely at 10 and 11 years, but not 12 years, compared to the others. There was a main effect for time (i.e., loneliness decreased over time), but no other effects or interactions.

For parent responses to the second question, "would rather be alone than with others," there was no significant effect for group, F (2, 506) = 1.09, p = .34. There were the same time, gender, and time-by-gender interaction effects found for the previous analysis using parent reports.

For teacher responses to the second question, "would rather be alone than with others," there was a significant effect for group, F (2, 433) = 6.94, p < .001. There was also an interaction between gender and group, F (6, 862) = 2.11, p < .05. The difference between girls' and boys' reports of loneliness was larger in the high group compared to the medium and low groups. Girls who were identified by their teachers as really wanting to be alone reported significantly more loneliness compared to boys in this group. Given the few gender differences, this finding needs to be replicated. There could be several explanations. Girls who isolate themselves from the group may be more sensitive to being alone—even when it is their preference. Or, girls may be more likely to admit to being lonely than boys under these circumstances. Then again, teachers may be better able to identify lonely girls. A girl wanting to be alone may seem more unusual—and therefore is more noticeable—than a boy wanting to be alone.

In summary, there was little evidence that children's reports of loneliness are linked to other's reports of the children's internalizing problems during the transition to adolescence. Still, there were significant associations between parents' and teachers' responses to specific questions about loneliness and children's reports of loneliness over time, supporting the argument that our version of Asher's Loneliness Questionnaire was valid.

Discussion

Are individual differences in loneliness during childhood heritable across age? The answer appears to be yes when the children were 10 and 11 years old. Not surprisingly, environmental contributions to children's perception of loneliness were nonshared. It is likely that important contributors to individual differences in loneliness are experiences outside the family in peer groups. But, why would loneliness be heritable? Obviously, there is no gene for self-esteem. There are several paths through which biological differences can contribute to our feelings of loneliness. The behavior associated with those evaluations may be genetically influenced. For instance, genetic contributions to loneliness may be due to genetic contributions to emotionality or social interaction styles. McGuire et al. (1994) found that part of the heritability of children's perceptions of academic competence could be explained by genetic contributions to vocabulary and part of the heritability children's perceptions of peer popularity could be explained by genetic contributions to sociability. Biological factors may put us at risk factor for general psychopathology, while environmental factors determine its specific nature (e.g., Kendler, Heath, Martin, & Eaves 1987). This is one of many possible models that need to be tested using multivariate behavioral genetic data.

It would seem that a likely candidate for such multivariate analyses would be other internalizing problems. Surprisingly, parent and teacher reports of children's total numbers of internalizing problems were not highly correlated with children's reports of loneliness. It could be that our measure of loneliness was not valid, but other analyses showed that children's reports were linked to parents' and teachers' responses to specific questions about loneliness. Parent and teacher reports of depressive symptoms may have a different etiology than child reports (see Deater-Deacker & McGuire, chapter 9, this volume). It is also possible, however, that loneliness has been miscast as a member of the internalizing behavioral problems group. This is unlikely, given the links between children's loneliness and serious peer problems (e.g., Asher &

Wheeler, 1985; Boivin & Hymel, 1997; Cassidy & Asher, 1992; Crick & Ladd, 1993; Renshaw & Brown, 1993). Yet another prospect is that loneliness has more in common with self-esteem than it does with depressive, somatic, or anxious behavior. Loneliness may be heritable via biological contributions to the self-evaluation process. Chronic loneliness is thought to involve a maladaptive attribution style. This thinking process may involve cognitive impairments the same way that hostile attribution styles in conduct-disordered children are thought to have an attentional or perceptual component (e.g., Crick & Dodge, 1994). Memory or attentional problems may contribute to a maladaptive view of the world. This is not to say that biological factors work in a deterministic fashion. Rather, psychopathology is the result of biological and environmental risk and protective factors (Rutter & Plomin, 1998). Knowing those factors can help us develop better treatments. Chronic loneliness with an etiology that involves biological factors will require a different intervention from loneliness as a result of experiences alone.

In addition to the main findings, the results of the descriptive analyses were also interesting. How many children experience extreme loneliness during the transition to adolescence? The CAP data show that only about 1 to 8% of the children reported feeling extremely lonely and isolated. They said that it was "really true" that they felt disconnected from their peers and left out of things. While this is lower than the number of children in the Asher et al. (1984) study, it is probably comparable to other areas of psychopathology. There was only one gender effect and no adoptive status differences. Gender differences are consistently found in the developmental psychopathology literature. Boys tend to be higher in externalizing problems and girls tend to be higher in internalizing problems (e.g., Zahn-Waxler, 1993). Given that loneliness is considered an internalizing problem, one would expect that girls would report more loneliness compared to boys. Yet, our data was similar to other studies of children's loneliness that found no gender differences. It will be interesting to follow the CAP children into middle and late adolescence to see if such differences emerge over time.

The findings suggest several interesting avenues for future research. For instance, work is needed to determine if children who report chronic feelings of loneliness over time are a qualitatively different group from those who report temporary feelings of loneliness, or if they are at the end of a continuum. It will be important to understand the correlates and consequences of these differences. In addition, multivariate studies would help us understand the etiology of genetic contributions to loneliness. Environmental-oriented studies of siblings would help us to understand the nature of the unique experiences that contribute to feelings of peer isolation. Given the significance of peer interaction in children's development, it is important that future studies of loneliness include behavioral genetic methods and measures in order to address these issues.

ACKNOWLEDGMENTS

This research was supported by grants HD-10333 and HD-18426 from the National Institute of Child Health and Human Development, and by grant MH-43899 from the National Institute of Mental Health. The authors wish to thank Sally-Ann Rhea and Robin Corley.

REFERENCES

Achenbach, T. M. (1991). *Integrative guide for the 1991 CBCL/4–18, YSR, and TRF profiles.* Burlington: University of Vermont.

Asher, S. R. (1985). *The Loneliness Questionnaire.* Personal Communication.

Asher, S. R., Hymel, S., & Renshaw, P. D. (1984). Loneliness in children. *Child Development, 55,* 1456–1464.

Asher, S. R., & Wheeler, V. A. (1985). Children's loneliness: A comparison of rejected and neglected peer status. *Journal of Consulting and Clinical Psychology, 53,* 500–505.

Berlin, L. J., Cassidy, J., & Belsky, J. (1995). Loneliness in young children and infant-mother attachment: A longitudinal study. *Merrill-Palmer Quarterly, 41,* 91–103.

Boivin, M., & Hymel, S. (1997). Peer experiences and social self-perceptions: A sequential model. *Developmental Psychology, 33,* 135–145.

Cassidy, J., & Asher, S. R. (1992). Loneliness and peer relations in young children. *Child Development, 63,* 350–365.

Crick, N. R., & Dodge, K. A. (1994). A review and reformulation of social information-processing mechanisms in children's social adjustment. *Psychological Bulletin, 115,* 74–101.

Crick, N. R., & Ladd, G. W. (1993). Children's perceptions of their peer experiences: Attributions, loneliness, social anxiety, and social avoidance. *Developmental Psychology, 29,* 244–254.

Deater-Deckard, K., & Plomin, R. (1999). An adoption study of the etiology of teacher and parent reports of externalizing behavior problems in middle childhood. *Child Development, 70,* 144–154.

Eley, T. C., Deater-Deckard, K., Fombonne, E., Fulker, D. W., & Plomin, R. (1998). An adoption study of depressive symptoms in middle childhood. *Journal of Child Psychology and Psychiatry, 39,* 337–345.

Ernst, J. M., & Cacioppo, J. T. (1999). Lonely hearts: Psychological perspective on loneliness. *Applied & Preventive Psychology, 8,* 1–22.

Graham, S., & Juvonen, J (1998). Self blame and peer victimization in middle school: An attributional bias. *Developmental Psychology, 34,* 587–598.

Guay, F., Boivin, M., & Hodges, E. V. E. (1999). Predicting change in achievement: A model of peer experiences and self-system process. *Journal of Educational Psychology, 91,* 105–115.

Harris, J. R. (1998). *The nurture assumption: Why children turn out the way they do.* New York: Free Press.

Hur, Y., McGue, M., & Iacono, W. G. (1998). The structure of self-concept in female preadolescent twins: A behavioral genetic approach. *Journal of Personality and Social Psychology, 74,* 1069–1077.

Kendler, K. S., Heath, A. C., Martin, N. G., & Eaves, L. J. (1987). Symptoms of anxiety and symptoms of depression: Same genes, different environments? *Archives of General Psychology, 44,* 451–457.

Laine, K. (1998). Finnish students' attributions for school based loneliness. *Scandinavian Journal of Educational Research, 42,* 401–413.

Larson, R. W. (1990). The solitary side of life: An examination of the time people spend alone from childhood to old age. *Developmental Review, 10,* 155–183.

Lykken, D., & Tellegen, A. (1996). Happiness as a stochastic phenomenon. *Psychological Science, 7,* 186–189.

McGue, M., Hirsch, B., & Lykken, D. T. (1993). Age and self-perception of ability: A twin study analysis. *Psychology and Aging, 8,* 72–80.

McGuire, S., & Clifford, J. (2000). Genetic and environmental contributions to loneliness in children. *Psychological Science, 11,* 493–497.

McGuire, S., Manke, B., Saudino, K. J., Reiss, D., Hetherington,E. M., & Plomin, R. (1999). Perceived competence and self-worth during adolescence: A longitudinal behavioral genetic study. *Child Development, 70,* 1283–1296.

McGuire S., Neiderhiser, J. M., Reiss, D., Hetherington, E. M., & Plomin, R. (1994). Genetic and environmental influences on perceptions of self-worth and competence in adolescence: A study of twins, full siblings, and step siblings. *Child Development, 65,* 785–799.

Mitchell, J. E., Baker, L. A., & Jacklin, C. N. (1989). Masculinity and femininity in twin children: Genetic and environmental factors. *Child Development, 60,* 1475–1485.

Parker, J. G., & Asher, S. R. (1993). Friendship and friendship quality in middle childhood: Links with peer group acceptance and feelings of loneliness and social dissatisfaction. *Developmental Psychology, 29,* 611–621.

Parker, J. G., & Asher, S. R. (1987). Peer relations and later personal adjustment: Are low-accepted children at risk? *Psychological Bulletin, 102,* 357–389.

Parkhurst, J. T., & Asher, S. R. (1992). Peer rejection in middle school: Subgroup differences in behavior, loneliness, and interpersonal concerns. *Developmental Psychology, 28,* 231–241.

Peplau, L. A., & Perlman, D. (1982). *Loneliness: A sourcebook of current theory, research, and therapy.* New York: John Wiley.

Renshaw, P. D., & Brown, P. J. (1993). Loneliness in middle childhood: Concurrent and longitudinal predictors. *Child Development, 64,* 1271–1284.

Rubin, K. H., Chen, X., McDougall, P., & Bowker, A. (1995). The Waterloo Longitudinal Project: Predicting internalizing and externalizing problems in adolescence. Special issue: Developmental processes in peer relations and psychopathology. *Development and Psychopathology, 7,* 751–764.

Rutter, M., & Plomin, R. (1998). Opportunities for psychiatry from genetic findings. *British Journal of Psychiatry, 171*, 209–219.

Sanderson, J. A., & Siegal, M. (1995). Loneliness and stable friendship in rejected and nonrejected preschoolers. *Journal of Applied Developmental Psychology, 16*, 555–567.

Tur-Kaspa, H., Weisel, A., & Segev, L. (1998). Attributions for feelings of loneliness of students with learning disabilities. *Learning Disabilities Research and Practice, 13*, 89–94.

Zahn-Waxler, C. (1993). Warriors and worriers: Gender and psychopathology. Special issue: Toward a developmental perspective on conduct disorder. *Development and Psychopathology, 5*, 79–89.

JEFFREY R. GAGNE
KIMBERLY J. SAUDINO
STACEY S. CHERNY

Genetic Influences on Temperament in Early Adolescence

A Multimethod Perspective

Introduction

Although twin studies of infant and child temperament consistently yield evidence of moderate genetic influences, adoption studies do not. For example, in the Colorado Adoption Project (CAP), genetically related nonadoptive siblings were no more similar for parent-rated temperament than genetically unrelated adoptive siblings (Plomin, Coon, Carey, DeFries, & Fulker, 1991). However, most twin and adoption research exploring genetic influences on temperament during infancy and early childhood has relied on parent-rating scales, which may be susceptible to contrast effects wherein the differences between siblings may be exaggerated (Saudino, Cherny, & Plomin, 2000). Because contrast effects reduce the similarity of nonadoptive and adoptive siblings, adoption studies that use parent-rating measures may underestimate heritability.

In infancy and early childhood, there are few practical alternatives to parent-rating measures of temperament (behavioral observations are expensive and time-consuming), but the availability of additional raters widens as the child ages. Once the child enters school it is possible to get teacher ratings, which have several advantages. Teachers are familiar with a broader range of children, and therefore their relative ratings may be more accurate. Furthermore, because in most instances different teachers rate the siblings, their ratings should not be prone to contrast biases. In CAP, in contrast to parent ratings across the same age, teacher and tester ratings of temperament at age 7 did show some evidence of genetic influences on emotionality, activity, and sociability (Schmitz, Saudino, Plomin, Fulker, & DeFries, 1996). Therefore, it appears that teacher ratings are an important addition to the exploration of genetic influences on temperament. In the present study, we use teacher ratings to explore genetic and environmental contributions to temperament in early adolescence—a largely unexamined age group with respect to the genetics of personality.

A second source of information regarding temperament that becomes available as children age are the children themselves. As children cognitively mature in late childhood and early adolescence it becomes reasonable to ask them to rate their own behavior. Adoption studies using self-report measures in middle childhood and late adolescence have found evidence of modest genetic influences (Loehlin, Willerman & Horn, 1982, 1985; Scarr, Webber, Weinberg, & Wittig, 1981); however, in CAP, parent-offspring analyses using self-reports of temperament in early adolescence (i.e., from 9 to 16 years) show no genetic effects (Plomin, Corley, Caspi, Fulker, & DeFries, 1998). The average correlation between biological parents and their adopted-away offspring's self-rated temperament was .01. Shared environmental influences were also negligible as indicated by the near zero correlations between adoptive parents and their adopted children. The discrepancy between the findings for self-reports in early and late adolescence (i.e., no genetic influence vs. modest genetic influence) might be due to weaknesses inherent in the parent-offspring design. In CAP, parents and adolescents rated their temperaments on related, but different, scales. Thus, subtle differences between measures could result in lower parent-offspring correlations. Similarly, because different genetic effects could operate on temperament at different ages, with the parent-offspring design, which compares individuals of different ages, genetic effects might go undetected. Both weaknesses are less of a problem with the study of personality in late adolescence because the same measures can be used for parents and offspring, and both parents and offspring have achieved maturity. Therefore, it is possible that the adoptive/nonadoptive sibling design used in the present study might detect genetic influences on self-rated temperament.

There has been much research examining the etiology of individual differences in infant and child temperament and adult personality, but few researchers have examined the factors that influence variability in temperament in early adolescence. In this chapter we use the adoptive/nonadoptive sibling design to explore genetic and environmental contributions to temperament at ages 9, 10, 11, and 12 years measured via parent, teacher, and self-ratings. Based on previous research, our hypothesis was that parent ratings would not be genetically influenced, but that teacher ratings would. Because there was so little previous research conducted with self-ratings in late childhood and early adolescence, we had no specific hypothesis about this measure, but our intuition was that it too would be genetically influenced.

In addition to evaluating genetic influences on each measure, the longitudinal design allows us to examine sources of continuity and change in temperament. Longitudinal studies of child temperament have consistently revealed moderate correlations across age (see reviews by Hubert, Wachs, Peters-Martin, & Gandour, 1982; McDevitt, 1986; Salbach, Morrow, & Wachs, 1991). The obvious implication of this finding is that the behaviors that we define as temperament are consistent, or stable from age to age. On the other hand, correlations that are less than unity indicate that developmental change may be occurring from one age to the next (Clarke & Clarke, 1984). Therefore, there is evidence of both continuity and change in child temperament. Broadly speaking, the general factors that can bring about continuity and change in child temperament are genes and the environment. Previous twin research indicates that there are indeed genetic influences on these individual differences in temperament (see Goldsmith, 1983, and Plomin, 1987, for reviews), and that genetic factors can be a source of both continuity and change in child development due to their dynamic nature (Plomin, 1986; Saudino, Plomin & DeFries, 1996).

One way to investigate the influence of genetic factors in the context of developmental change is to examine the relative contributions of genetic and environmental influences at each age. That is, does heritability—the proportion of phenotypic differences among individuals that can be attributed to genetic influences (Plomin, DeFries, McClearn, & Rutter, 1997)—change across age? Several studies suggest that the heritability of temperament either remains constant or increases with age (Cyphers, Phillips, Fulker, & Mrazek, 1990; Matheny, 1980; McCartney, Harris & Bernieri, 1990; Plomin et al., 1993; Stevenson & Fielding, 1985). A second developmental approach involves genetic influences on continuity and change in development. Genetic influences on continuity suggest that the same genetic factors affect a particular trait (e.g., temperament) across age, while age-to-age genetic change refers to the extent to which genetic factors that affect a trait at one age are similar or differ from those that affect the same trait at another age (Saudino & Cherny, 2001a).

In the present chapter, we investigate changes in heritability across age and examine genetic and environmental sources of continuity and change in temperament during early adolescence in the CAP. To date, most of the research on continuity and change in temperament has

TABLE 11.1. Number of Sibling Pairs at Each Age

	Parent-Rating		Teacher-Rating		Self-Rating	
Age	Adopted	Nonadopted	Adopted	Nonadopted	Adopted	Nonadopted
9	71	91	62	76	88	108
10	68	85	53	69	83	103
11	56	65	45	58	76	90
12	58	57	33	33	67	75

focused on infancy and early childhood and suggests that genetic influences remain stable or increase with age. We wondered if the patterns would hold true for a sample of early adolescents.

Method

Sample

The sibling analyses in this chapter are based on differing numbers of sibling pairs for the three different types of raters at each age (see table 11.1). Sample sizes (i.e., the number of sibling pairs) are successively smaller across age because of subject loss to follow-up, which is frequently the case with longitudinal research. Moreover, there are less teacher-rated data than that for parent or self-ratings due to the fact that not all teachers complied with requests to complete questionnaires.

Temperament Measures

At ages 9, 10, 11, and 12, parent, teacher, and self-reports of temperament were obtained on the Colorado Childhood Temperament Inventory (CCTI; Rowe & Plomin, 1977). The CCTI contains general statements describing the temperament dimensions of emotionality, activity, sociability, and attention—for example, "child cries easily," or "child is very energetic." Parents and teachers were asked to rate each statement on a five-point Likert scale ranging from 1 = "strongly disagree; not at all like the child" to 5 = "strongly agree; a lot like the child." Although the parent and teacher scales differ in the total number of items (i.e., parent 30 items, teacher 35 items), the four subscales examined here differ only with respect to minor wording changes (i.e., each subscale consisted of five items). For self-ratings, each subscale contained four of the five parent rating scale items that were modified so that they were worded in the first person.

Design

The sibling adoption design is based on comparative analyses of the similarity of adoptive and nonadoptive sibling pairs on a particular measure to determine genetic and environmental contributions to the variance of a measure. Genetic influence is implicated when nonadoptive siblings, who on average share 50% of their segregating genes, are more similar to each other on a measure (in this case the CCTI) than adoptive siblings who are not genetically related. Intraclass correlations are typically employed as indices of similarity between the pairs of siblings. Correlations for nonadoptive siblings that exceed those for adoptive siblings imply genetic influence. When correlations for adoptive siblings are greater than zero, shared environmental influences are suggested.

Model-Fitting Analyses

In addition to the correlational analyses, model-fitting analyses were used to estimate genetic and environmental contributions to the CCTI ratings of temperament for the three raters. Maximum-likelihood model-fitting analyses were performed on sibling variance/covariance matrices using the statistical software package Mx (Neale, 1997). To investigate genetic and environmental sources of continuity and change within each rater, we fit a quadrivariate Cholesky model to the longitudinal data. This model allows us to estimate heritability and genetic and environmental contributions to phenotypic stability. Because the sample sizes vary across age, we fit this model to the raw data using a pedigree approach to take advantage of the maximum sample size at each age.

Results

Descriptive Statistics

Tables 11.2–11.4 list the means and standard deviations of the parent, teacher, and self-ratings of temperament, respectively. In general, temperament ratings are moderately consistent from age to age, indicating continuity across development. To evaluate mean differences within raters, repeated measures analyses of variance (ANOVA) with age as a within-subjects variable, and adoptive status and gender as between-subjects variables, were conducted. When significant interactions were found, follow-up Tukey's tests were performed.

Parent Ratings For sociability, there were no main or interaction effects. However, for emotionality, there were significant main effects for adoptive status and gender. Parent ratings of emotionality were

TABLE 11.2. Descriptive Statistics for Parent-Rated Temperament

	Adoptive		Nonadoptive	
	Male M (SD)	Female M (SD)	Male M (SD)	Female M (SD)
Emotionality				
Age 9	15.5 (4.0)	14.4 (4.6)	13.8 (4.1)	13.7 (4.0)
Age 10	14.7 (4.1)	14.3 (4.6)	13.9 (4.2)	13.7 (4.0)
Age 11	15.1 (4.7)	14.4 (4.1)	13.6 (4.2)	13.3 (4.2)
Age 12	15.6 (4.2)	13.5 (3.9)	13.9 (4.3)	12.6 (3.9)
Activity				
Age 9	19.0 (4.6)	19.0 (3.9)	18.6 (3.7)	18.4 (3.8)
Age 10	19.0 (4.3)	18.8 (3.9)	18.1 (4.0)	18.4 (3.6)
Age 11	18.2 (4.7)	18.4 (4.1)	18.0 (4.1)	17.8 (4.1)
Age 12	17.4 (4.6)	18.4 (4.3)	17.1 (3.8)	16.7 (4.0)
Stability				
Age 9	19.9 (4.5)	19.2 (4.3)	18.7 (4.0)	19.1 (3.9)
Age 10	20.0 (4.7)	19.1 (4.5)	18.4 (4.1)	19.0 (3.8)
Age 11	19.9 (4.6)	18.4 (4.7)	18.6 (3.6)	19.7 (3.9)
Age 12	19.3 (4.9)	20.0 (4.8)	18.7 (4.3)	18.4 (4.2)
Attention				
Age 9	18.5 (4.3)	17.8 (3.9)	18.6 (3.5)	19.6 (3.1)
Age 10	18.7 (4.3)	17.3 (4.2)	18.5 (3.6)	19.3 (3.1)
Age 11	17.6 (4.4)	17.9 (4.3)	18.5 (3.7)	19.5 (3.2)
Age 12	18.2 (3.8)	16.6 (4.8)	18.6 (3.5)	20.0 (2.8)

TABLE 11.3. Descriptive Statistics for Teacher-Rated Temperament

	Adoptive		Nonadoptive	
	Male M (SD)	Female M (SD)	Male M (SD)	Female M (SD)
Emotionality				
Age 9	12.0 (5.0)	10.6 (4.4)	10.0 (4.5)	11.0 (4.7)
Age 10	11.0 (4.0)	11.2 (4.6)	10.6 (4.6)	10.2 (4.7)
Age 11	10.3 (4.3)	10.9 (4.8)	9.6 (4.1)	10.3 (4.5)
Age 12	10.3 (4.3)	11.5 (4.1)	8.6 (3.8)	8.6 (3.5)
Activity				
Age 9	18.3 (4.0)	18.0 (4.0)	18.2 (4.1)	18.1 (4.1)
Age 10	18.9 (4.2)	18.7 (4.0)	18.1 (4.2)	18.3 (4.7)
Age 11	18.2 (3.5)	18.3 (4.1)	18.1 (4.1)	17.7 (4.7)
Age 12	17.6 (3.9)	18.1 (3.7)	17.6 (4.1)	18.2 (4.6)
Sociability				
Age 9	14.4 (4.0)	18.2 (4.0)	18.1 (4.0)	18.9 (4.3)
Age 10	18.5 (4.5)	18.7 (4.1)	18.1 (4.1)	18.8 (4.8)
Age 11	18.4 (4.0)	18.5 (3.8)	18.3 (4.1)	18.2 (4.6)
Age 12	18.2 (4.4)	17.1 (3.8)	17.9 (4.5)	19.0 (4.9)
Attention				
Age 9	17.3 (4.2)	17.1 (4.5)	19.0 (4.2)	18.6 (3.9)
Age 10	17.0 (4.4)	17.7 (4.2)	18.4 (4.1)	19.7 (4.1)
Age 11	17.7 (4.1)	17.6 (4.4)	18.4 (4.1)	19.4 (3.6)
Age 12	18.0 (4.3)	16.9 (4.0)	19.5 (3.5)	20.3 (3.0)

TABLE 11.4. Descriptive Statistics for Self-Rated Temperament

	Adoptive		Nonadoptive	
	Male M (SD)	Female M (SD)	Male M (SD)	Female M (SD)
Emotionality				
Age 9	10.0 (4.1)	11.5 (3.9)	10.7 (3.7)	11.0 (3.9)
Age 10	10.0 (4.0)	10.5 (4.3)	9.8 (3.4)	10.2 (3.9)
Age 11	9.3 (3.7)	9.6 (3.9)	9.8 (3.6)	9.6 (3.7)
Age 12	9.5 (3.3)	9.4 (3.7)	9.3 (2.9)	9.6 (3.4)
Activity				
Age 9	15.0 (3.1)	14.9 (3.4)	14.8 (2.9)	14.3 (3.0)
Age 10	15.2 (3.0)	15.4 (3.3)	15.6 (3.1)	15.7 (3.3)
Age 11	16.3 (2.6)	16.0 (3.0)	15.9 (3.1)	15.6 (3.0)
Age 12	15.7 (2.9)	16.3 (2.9)	16.0 (2.7)	15.7 (3.4)
Sociability				
Age 9	15.3 (3.4)	14.4 (3.5)	14.7 (3.3)	14.6 (3.5)
Age 10	15.3 (3.2)	14.7 (3.9)	15.4 (3.1)	15.3 (3.4)
Age 11	15.0 (3.5)	15.0 (3.5)	15.3 (3.0)	15.5 (3.3)
Age 12	15.8 (3.2)	14.9 (3.3)	15.2 (3.1)	15.4 (3.6)
Attention				
Age 9	15.8 (3.2)	14.6 (3.2)	15.0 (3.4)	15.1 (2.8)
Age 10	15.6 (3.4)	15.4 (3.4)	15.4 (2.9)	15.2 (3.0)
Age 11	15.5 (3.5)	15.4 (3.2)	15.0 (2.9)	15.0 (2.7)
Age 12	15.6 (3.0)	15.2 (3.1)	14.6 (2.9)	14.9 (2.9)

significantly higher for adoptive siblings than for nonadoptive siblings ($F = 4.33$, $p = .04$), and males were rated higher than females ($F = 7.38$, $p = .007$). Activity showed a mean effect for age ($F = 11.4$, $p = .0001$), with means decreasing as children get older. Finally, for attention there were significant interactions between gender and adoption status ($F = 8.01$, $p = .005$), and age and adoption status ($F = 2.63$, $p = .05$). On this dimension, nonadoptive females were rated higher than adoptive females, and adoptive siblings were significantly lower at all ages except 10.

Teacher Ratings   For sociability, there was a main effect for gender with females being moderately higher than males at all ages ($F = 4.36$, $p = .04$). There was an age by adoptive status interaction effect ($F = 3.95$, $p = .009$) for emotionality, with adoptive siblings being rated as higher than nonadoptive siblings at ages 9 and 12. This finding is consistent with parent ratings of emotionality. For activity, there was a main effect for gender ($F = 6.41$, $p = .013$). Females had significantly higher activity scores than males, a finding that is contradictory to previous patterns of results that suggest that males are typically higher than females on activity (Eaton & Enns, 1986). However, it is important to note that Eaton and Enns's meta-analytic findings were based primarily on infant and early childhood data. Lastly, for attention, there were significant main effects for adoptive status and age. Nonadoptive siblings had higher ratings than adoptive siblings ($F = 19.09$, $p = .0001$), and attention scores increased slightly with age ($F = 3.01$, $p = .03$).

Self Ratings The reader is reminded that the self-report subscales are based on fewer items than the parent or teacher rating scales (i.e., four vs. five items). Therefore, the means across parent, teacher, and self-ratings are not comparable and the lower means for self-reported temperament in table 11.4 likely reflect the difference in item number. There were significant age effects for sociability, emotionality, and activity, as well as an adoptive status effect for attention. Both sociability ($F = 2.54$, $p = .055$) and activity ($F = 16.03$, $p = .0001$) showed slight increases with age, whereas emotionality decreased slightly across age ($F = 9.93$, $p = .0001$). Attention scores were higher for adoptive siblings than for nonadoptive ($F = 3.90$, $p = .05$). Interestingly, this differs from both teacher and parent ratings, which displayed the opposite pattern, wherein nonadoptive siblings were rated higher than adoptive siblings.

Interrater Agreement

Interrater correlations are presented in table 11.5. Correlations across temperament dimension and age are fairly consistent and reflect modest agreement for all raters (parent-teacher, parent-self, and teacher-self). The modest agreement between parent and teacher ratings is similar to previous findings that suggest that child behavior shows some consistency across rater and situations (Victor, Halverson, & Wampler, 1988); however, it is clear that the measures do not entirely overlap. The lower agreement among parent-self and teacher-self ratings for the temperament dimensions of emotionality and attention are not surprising as these dimensions reflect more internal states, and therefore, it would be less likely that the outside observer would agree with the individual.

Phenotypic Continuity

Age-to-age phenotypic stability correlations are presented in table 11.6. For each temperament dimension and rater type there is considerable continuity across age. The phenotypic stability for self- and teacher ratings are particularly impressive when one considers that self-reports of temperament by children as young as 9 years old show consistency across a three-year time span, and that different teachers rated the temperaments of each child each year. In comparison to the moderate correlations for teacher and self-ratings, correlations for parent ratings are substantially higher. In fact, in some instances, parent ratings were more than double that of teacher or self-ratings. Moreover, stability for

TABLE 11.5. Interrater Agreement Correlations Across Age

	Parent-Teacher	Parent-Self	Teacher-Self
Emotionality			
9 years	.24**	.19**	.14*
10	.09	.17**	.20**
11	.20**	.25**	.07
12	.26**	.27**	.29**
Activity			
9 years	.23**	.21**	.18**
10	.22**	.22**	.26**
11	.18**	.21**	.21**
12	.24**	.43**	.37**
Sociability			
9 years	.32**	.35**	.21**
10	.42**	.34**	.32**
11	.30**	.35**	.21**
12	.32**	.38**	.33**
Attention			
9 years	.26**	.11*	−.04
10	.31**	.17**	.12*
11	.38**	.19**	.16*
12	.38**	.22**	.13

$*p < .05; **p < .01$

TABLE 11.6. Age-to-Age Stability Correlations of Observed Temperament

	Age Interval					
	9–10	9–11	9–12	10–11	10–12	11–12
Emotionality						
Parent	.72	.64	.66	.72	.69	.67
Teacher	.40	.34	.34	.37	.35	.30
Self	.41	.36	.28	.47	.39	.47
Activity						
Parent	.77	.74	.74	.75	.73	.76
Teacher	.40	.30	.24	.40	.38	.36
Self	.32	.38	.26	.44	.30	.39
Sociability						
Parent	.74	.74	.73	.73	.71	.73
Teacher	.45	.37	.34	.43	.34	.36
Self	.53	.52	.40	.54	.51	.51
Attention						
Parent	.71	.68	.67	.69	.68	.69
Teacher	.43	.39	.30	.44	.25	.33
Self	.37	.25	.18	.36	.23	.43

Note: All correlations are significant at $p < .001$

TABLE 11.7. Intraclass Correlations for Adopted and Nonadoptive Siblings

	Parent Rating		Teacher Rating		Self Rating	
	Adopted	Nonadoptive	Adopted	Nonadoptive	Adopted	Nonadoptive
Emotionality						
9 years	0.22[†]	.05	.05	.27*	.02	.09
10	.10	−.14	−.02	.19	.32*	−.09
11	.10	−.08	.10	.18	.06	.05
12	.14	.07	−.29	.25	.16	.03
Activity						
9 years	−.19	−.10	.25*	.20[†]	−.01	.20*
10	−.18	−.15	−.13	.37*	.07	.14
11	−.22	−.30*	−.09	.05	−.15	−.13
12	−.20	−.03	.20	.06	−.11	−.07
Sociability						
9 years	−.16	.10	−.13	.23*	.02	.19*
10	−.12	.19[†]	.05	.32*	.02	.01
11	−.17	.15	−.11	.30*	−.16	.12
12	−.18	−.10	.09	.29[†]	−.12	.05
Attention						
9 years	−.16	−.03	−.04	.23*	−.14	.05
10	−.23[†]	.05	−.02	.24*	.04	.11
11	−.14	.13	.04	.22[†]	−.05	−.05
12	−.09	−.20	−.08	.16	−.20[†]	.10

[†]$p < .10$; *$p < .05$

parent ratings remained high across the three-year interval from ages 9 to 12. It is interesting to note that the obtained age-to-age stabilities for parent ratings approach one-week test-retest reliabilities of the CCTI scales reported by Buss and Plomin (1984). These results are consistent with previous findings that propose that some of this stability in parent ratings can be attributed to the stability of the parent's own perceptions above and beyond actual stability in child behavior (Plomin et al., 1991; Saudino & Cherny, 2001b). In addition, lower continuity for teachers is likely due in part to the fact that different teachers make the ratings each year as compared to parental ratings, which are made by the same parents each year.

Sibling Intraclass Correlations

The intraclass correlations for adoptive and nonadoptive siblings are presented in table 11.7.

Parent Ratings

Across age and temperament dimension, there were no indications of significant genetic influences on the CCTI scales. In fact, for emotionality, the intraclass correlations for adopted siblings exceeded those for nonadoptive sibling by twofold across the four ages—exactly the opposite of what would be expected for a genetic effect. The pattern of the adoptive siblings' correlations exceeding zero does, however, imply shared environmental influences for emotionality. For activity, sociability, and attention, the correlations for nonadoptive siblings did exceed those of the adoptive siblings with the exception of activity at age 11; however, very few of these correlations are significant. The generally low and in many cases negative correlations suggest that parental contrast biases might be operating.

Teacher Ratings   As predicted, correlations for teacher ratings are more consistent with our hypothesis of genetic influences on temperament. With the exception of activity at ages 9 and 12, the intraclass correlations for nonadoptive siblings exceed those of the adoptive siblings. However, only about half of the nonadoptive sibling correlations and only one of the adoptive correlations are significant. Although teacher ratings hint that genetic influences may be operating, there is no clear evidence of developmental trends. That is, the pattern of correlations is inconsistent across age. For example, for sociability and attention, the nonadoptive sibling correlations always exceed the adoptive siblings', but the magnitude of the difference fluctuates from age to age. For activity, the intraclass correlations are similar at age 9 for both nonadoptive and adoptive siblings. Nonadoptive correlations are much higher at age 10 (suggesting genetic influences), nonadoptive are only slightly higher than adoptive at age 11, and at age 12, adopted siblings' correlations exceed the nonadoptive siblings' (opposite of a genetic effect). One plausible explanation for these inconsistencies is the fact that with each age, the number of subjects decreases as a result of subjects lost to follow-up. For all temperament dimensions, at age 9 the sample sizes range from 60–76 for both nonadoptive and adoptive siblings, whereas at age 12, they range from 28–33.

Self-Ratings   Like parent-rated temperament, self-ratings did not suggest significant genetic influences on the CCTI scales. Although there was a trend toward slightly higher correlations for nonadoptive than adoptive siblings, there were few significant correlations for either group. Overall, the pattern of near zero and negative sibling correlations for both sibling groups suggests little sibling resemblance for self-rated temperament.

Variance Components

Because the sibling intraclass correlations suggest that parent and self-ratings did not evince a pattern of genetic influences, longitudinal

TABLE 11.8. Variance Components from Full Longitudinal Model—
Teacher Ratings

	9 Years			10 Years			11 Years			12 Years		
	h^2	c^2	e^2	h^2	c^2	e^2	h^2	c^2	e^2	h^2	c^2	e^2
Emotionality	.49	.06	.45	.37	.02	.61	.19	.15	.67	.39	.02	.60
Activity	.08	.23	.69	.62*	.03	.35	.19	.03	.78	.12	.17	.70
Sociability	.38	.03	.59	.43	.11	.46	.61*	.01	.38	.31	.15	.53
Attention	.42*	.04	.53	.33	.09	.58	.51	.05	.44	.25	.05	.70

Note: h^2 = heritability, c^2 = shared environmental variance, e^2 = nonshared environmental variance
*$p < .05$

model-fitting analyses were conducted only for teacher ratings. Variance components for each temperament dimension at ages 9, 10, 11, and 12 are presented in table 11.8 (heritability estimates as well as estimates of shared and nonshared environmental variance). Consistent with the interpretation of the sibling intraclass correlations, teacher-rated temperament demonstrates moderate heritability across age and dimension. With the exception of activity level, heritability accounts for approximately one-quarter to one-half of the phenotypic variance. The remaining variance in teacher-rated temperament can be accounted for by nonshared environment, which includes measurement error. Overall, there was no substantial pattern of increasing or decreasing heritability with age, suggesting that the heritability of temperament as rated by teachers is relatively consistent from the ages of 9–12. It should be noted, however, that heritabilities were significant only for activity at age 10, sociability at age 11, and attention at age 9. A lack of significant findings can be explained by small sample sizes, and the overall pattern suggests that, with greater power, significant heritability would be found across age and dimension.

Genetic and Environmental Contributions to Phenotypic Continuity

The age-to-age stability correlations of teacher-rated temperament in table 11.6 indicate that there is some overlap between the genetic and/or environmental factors that influence temperament at all ages. With our small sample, our model-fitting analyses were not powerful enough to distinguish between genetic and environmental contributions to age-to-age covariance. However, we could not eliminate all sources of age-to-age covariance from these models without a significant decrement in fit. Nevertheless, as can be seen in table 11.9, age-to-age stability correlations of phenotypic continuity are largely due to genetic factors. Therefore, although we have different teachers rating temperament at different ages, the data hint that there is genetic continuity across teacher ratings at different ages. Indeed, the genetic

TABLE 11.9. Genetic and Environmental Contributions
to the Age-to-Age Stability Correlations—Teachers

Age Interval	Genetic	Shared Environment	Nonshared Environment	Estimated Stability r
Emotionality				
9–10	.30	.03	.07	.40
9–11	.10	.08	.18	.36
9–12	.28	−.01	.06	.33
10–11	.24	.06	.11	.41
10–12	.38	−.02	.01	.37
11–12	.25	−.05	.12	.32
Activity				
9–10	.22	.09	.11	.42
9–11	.12	.08	.09	.29
9–12	.10	.20	.01	.31
10–11	.35	.03	.04	.42
10–12	.28	.08	.07	.43
11–12	.15	.07	.19	.41
Sociability				
9–10	.37	.03	.05	.45
9–11	.40	.02	−.03	.39
9–12	.32	.07	−.04	.35
10–11	.49	.01	−.04	.46
10–12	.28	.10	−.01	.37
11–12	.33	.03	.03	.39
Attention				
9–10	.36	−.04	.10	.42
9–11	.40	.01	−.02	.39
9–12	.27	.04	−.04	.27
10–11	.32	.05	.06	.43
10–12	.21	−.07	.08	.22
11–12	.35	−.03	−.03	.29

Note: Stability estimates taken from the Cholesky modeling vary slightly from those obtained using raw phenotypic data

correlations, indicating the extent to which the same genetic effects operate across age, were substantial ranging from .33 to 1.0, with a median correlation of .92.

Analyses Collapsing Across Age

Aggregation of data increases the reliability of a particular measure by reducing the error variance associated with a single occasion of measurement (Epstein, 1983). Therefore, within each rater (i.e., parent, teacher, self) we averaged scores across age. Although this process makes developmental analyses across age impossible, it was carried out to more accurately investigate the genetic and environmental contri-

TABLE 11.10. Sibling Intraclass Correlations Averaging Temperament Scores Across Age

	Parent Ratings		Teacher Ratings		Self-Ratings	
	Adoptive	Nonadoptive	Adoptive	Nonadoptive	Adoptive	Nonadoptive
Emotionality	.14	.01	−.03	.29*	.20†	−.03
Activity	−.26*	−.20†	−.01	.38*	−.01	.06
Sociability	−.22†	.14	−.05	.39*	−.10	.12
Attention	−.24*	−.02	−.07	.32*	−.12	.08

†$p < .10$; *$p < .05$

butions to temperament for the three raters. To ensure an adequate sample size, average scores were calculated for subjects with data for at least two time periods. Once averaged, the sample included 73 adoptive and 89 nonadoptive pairs; 72 adoptive and 104 nonadoptive pairs; and 62 adoptive and 76 nonadoptive pairs, for parent, teacher, and self-ratings, respectively.

As might be expected from our earlier analyses, the sibling intraclass correlations (see table 11.10) varied depending on the type of rater. In general, parents see little resemblance in the temperaments of their adoptive or nonadoptive children. In fact, the significant negative correlations for adoptive children suggest that parents view the siblings as having opposing temperaments. Similarly, the low adoptive and nonadoptive correlations for self-ratings do not suggest genetic influences. Consistent with our earlier findings, teacher ratings, however, appear to be genetically influenced. When temperament was rated by teachers, nonadoptive siblings demonstrated significant similarity for all dimensions, whereas adoptive siblings showed little or no resemblance. Model-fitting analyses reveal no significant heritabilities for parent and self-ratings; however, teacher ratings of all four CCTI dimensions show substantial genetic variance, with the remaining proportion of variance attributed to the nonshared environment (see table 11.11).

Summary and Discussion

CAP, with its multimethod temperament rating perspective, allows us the unique opportunity to compare parent, teacher, and self-ratings of early adolescent temperament. Most behavioral genetic studies of temperament in infancy and childhood have relied on parent-report measures only. Moreover, genetic influences on temperament in early adolescent samples have seldom been examined within a longitudinal framework (most focus on one time-point). In contrast, the longitudi-

TABLE 11.11. Variance Components Averaging Temperament Scores Across Age

	Parent Ratings			Teacher Ratings			Self-Ratings		
	h^2	c^2	e^2	h^2	c^2	e^2	h^2	c^2	e^2
Emotionality	.00	.07	.93*	.52*	.00	.48*	.00	.11	.89*
Activity	.00	.00	1.0*	.67*	.00	.34*	.14	.00	.86*
Sociability	.36	.00	.64*	.72*	.00	.27*	.29	.00	.71*
Attention	.00	.00	1.0*	.67*	.00	.32*	.19	.00	.81*

Note: h^2 = heritability, c^2 = shared environmental variance, e^2 = nonshared environmental variance; *p < .05

nal nature of CAP allows us to look at temperament ratings across the ages of 9 through 12. Behavioral genetic investigations with a developmental point of view examine genetic and environmental change by focusing on changes in heritability across age and the role of genetic and environmental influences on age-to-age continuity and change. Changes in heritability across age can elucidate time-points wherein transitions occur in development. Genetic influences on age-to-age continuity and change may illuminate the causative processes that factor into developmental changes.

Parent and self-rated temperament did not suggest significant genetic influences on the CCTI scales in the longitudinal sample. However, teacher ratings of temperament did show genetic influences. Model-fitting analyses of teacher-rated temperament demonstrated moderate heritability at all ages, with the remaining variance accounted for by nonshared environmental influences. Although the heritabilities were generally nonsignificant at each age, when scores were averaged across age, all four teacher-rated temperament dimensions displayed significant heritability. Parent and self-ratings, however, continued to show little evidence of a genetic influence. This pattern of results for the teachers as compared to the parents is very similar to those found for hyperactivity/attention related behaviors in twin samples (Simonoff, Pickles, Hervas, Silberg, Rutter, & Eaves, 1998).

It is possible that the lack of genetic influence on parent-rated temperament is due to contrast effects. Previous research with twins suggests that when parents rate the temperaments of their twin siblings, they contrast one twin with the other and magnify behavioral differences (Saudino et al., 2000; Saudino, McGuire, Reiss, Hetherington, & Plomin, 1995). Contrast effects may not be limited to parents' ratings of twins. That is, it is possible that whenever parents rate the temperaments of their children, they evaluate each child in the context of other children that they know well—most likely other children within the family. Thus, just as with twins, the rating of one sibling's temperament is likely to be influenced by the perceived temperament of another sibling. Contrast effects artificially increase differences between MZ and DZ correlations and, therefore, can result in overestimates of genetic influence in twin studies. Contrast effects would have the opposite effect in adoption studies. That is, they would reduce the similarity of nonadoptive and adoptive siblings, and thus, will lead to underestimates of heritability.

The lack of a genetic influence on self-ratings was unexpected, particularly in light of previous adoption study findings that suggest modest genetic influences on self-reports of personality in late adolescence and early adulthood (Loehlin et al., 1982, 1985; Scarr et al., 1981). As with parent ratings, neither adoptive nor nonadoptive siblings displayed any sibling resemblance—arguing against the importance of genetic and shared environmental influences. One possible explanation is that parent perceptions have influenced the adolescents' self-perceptions of temperaments. Thus, if children are told by their parents that they are "the shy one" or the "social one," they may rate themselves accordingly. In this way, contrast effects may be communicated from parent to child. Given the discrepancy between our results and previous adoption studies with older samples, we suspect that the tendency for self-contrast effects may be related to age. That is, we expect that self-reports may yield more evidence for heritability when the children in CAP are older.

The many negative correlations between siblings when temperament is rated by parents or self-reports is consistent with our interpretation of contrast effects. There is no genetic explanation for negative correlations. Moreover, the fact that in CAP, teacher ratings of temperament yield evidence of a genetic influence, but parent ratings and self-ratings of the same dimensions do not, adds evidence to our hypothesis that contrast effects may be at work. The difference in outcomes between parent, teacher, and self-ratings reminds us that relying on a single method of assessment may not provide a complete picture of temperament.

Finally, our longitudinal analyses of teacher-rated temperament suggest there was little evidence of developmental changes in the magnitude of genetic influence across age and that genetic factors contribute to the continuity of temperament across age. This latter finding is particularly interesting when one considers that different raters rated each child at each age. Previous longitudinal twin studies of temperament and personality in infancy, early childhood, and adulthood have also found that genes contribute to the phenotypic continuity across age (McGue, Bacon, & Lykken, 1993; Saudino et al., 1996). These findings offer an interesting perspective on the person-situation debate. The stable, enduring traits that constitute temperament and personality are stable because of genes and not because of the consistency of situational factors across age, as has been proposed by some personality theorists.

REFERENCES

Buss, A. H., & Plomin, R. (1984). *Temperament: Early developing personality traits.* Hillsdale, NJ: Lawrence Erlbaum Associates.

Clarke, A. D. B., & Clarke, A. M. (1984). Constancy and change in the growth of human characteristics. *Journal of Child Psychology and Psychiatry, 25,* 191–210.

Cyphers, L. H., Phillips, K., Fulker, D. W., & Mrazek, D. A. (1990). Twin temperament during the transition from infancy to early childhood. *Journal of the American Academy of Child and Adolescent Psychiatry, 29,* 392–397.

Eaton, W. O., & Enns, L. R. (1986). Sex differences in human motor activity level. *Psychological Bulletin, 100,* 19–28.

Epstein, S. (1983). Aggregation and beyond: Some basic issues on the prediction of behavior. *Journal of Personality, 51,* 360–392.

Goldsmith, H. H. (1983). Genetic influences on personality from infancy to adulthood. *Child Development, 54,* 331–355.

Hubert, N. C., Wachs, T. D., Peters-Martin, P., & Gandour, M. J. (1982). The study of early temperament: Measurement and conceptual issues. *Child Development, 53,* 571–600.

Loehlin, J. C., Willerman, L., & Horn, J. M. (1982). Personality resemblances between unwed mothers and their adopted-away offspring. *Journal of Personality and Social Psychology, 42,* 1089–1099.

Loehlin, J. C., Willerman, L., & Horn, J. M. (1985). Personality resemblances in adoptive families when the children are late-adolescent or adult. *Journal of Personality and Social Psychology, 48,* 376–392.

Matheny, A. P., Jr. (1980). Bayley's Infant Behavior Record: Behavioral components and twin analysis. *Child Development, 51,* 1157–1167.

McCartney, K., Harris, M. J., & Bernieri, F. (1990). Growing up and growing apart: A developmental meta-analysis of twin studies. *Psychological Bulletin, 107,* 226–237.

McDevitt, S. C. (1986). Continuity and discontinuity of temperament in infancy and early childhood: A psychometric perspective. In R. Plomin & J. Dunn (Eds.), *The study of temperament: Changes, continuities and challenges* (pp. 27–38). Hillsdale, NJ: Lawrence Erlbaum Associates.

McGue, M., Bacon, S., & Lykken, D. T. (1993). Personality stability and change in early adulthood: A behavioral genetic analysis. *Developmental Psychology, 29,* 96–109.

Neale, M. (1997). *Mx: Statistical modeling* (4th ed.) Richmond: Department of Psychiatry, Medical College of Virginia.

Plomin, R. (1986). Multivariate analysis and developmental behavioral genetics: Developmental change as well as continuity. *Behavior Genetics, 16,* 25–43.

Plomin, R. (1987). Developmental behavioral genetics and infancy. In J. Osofsky (Ed.), *Handbook of infant development* (2nd ed., pp. 363–417). New York: John Wiley.

Plomin, R., Coon, H., Carey, G., DeFries, J. C., & Fulker, D. (1991). Parent-offspring and sibling adoption analyses of parental ratings of temperament in infancy and early childhood. *Journal of Personality, 59,* 705–732.

Plomin, R., Corley, R., Caspi, A., Fulker, D. W., & DeFries, J. C. (1998). Adoption results for self-reported personality: Evidence for nonadditive genetic effects? *Journal of Personality and Social Psychology, 75,* 211–218.

Plomin, R., DeFries, J. C., McClearn, G. E., & Rutter, M. (1997). *Behavioral genetics.* New York: W. H. Freeman.

Plomin, R., Emde, R., Braungart, J. M., Campos, J., Corley, R., Fulker, D. W., Kagan, J., Reznick, S., Robinson, J., Zahn-Waxler, C., & DeFries, J. C. (1993). Genetic change and continuity from 14 to 20 months: The MacArthur Longitudinal Twin Study. *Child Development, 64,* 1354–1376.

Rowe, D. C., & Plomin, R. (1977). Temperament in early childhood. *Journal of Personality Assessment, 41,* 150–156.

Salbach, E. H., Morrow, J., & Wachs, T. D. (1991). Questionnaire measurement of infant and child temperament: Current status and future diractions. In J. Strelau & A. Angleitner (Eds.), *Explorations in temperament. International perspectives on theory and measurement* (pp. 205–234). New York: Plenum.

Saudino, K. J., & Cherny, S. S. (2001a). Sources of continuity and change in observed temperament. In R. N. Emde & J. K. Hewitt (Eds.), *The transition from infancy to early childhood: Genetic and environmental influences in the MacArthur Longitudinal Twin Study* (pp. 89–110). New York: Oxford University Press.

Saudino, K. J., & Cherny, S. S. (2001b). Parent ratings of temperament in twins. In R. N. Emde & J. K. Hewitt (Eds.), *The transition from infancy to early childhood: Genetic and environmental influences in the MacArthur Longitudinal Twin Study* (pp. 73–88). New York: Oxford University Press.

Saudino, K. J., Cherny, S. S., & Plomin, R. (2000). Parent ratings of temperament in twins: Explaining the "too low" DZ correlations. *Twin Research, 3,* 224–233.

Saudino, K. J., McGuire, S., Reiss, D., Hetherington, E. M., & Plomin, R. (1995). Parent ratings of EAS temperaments in twins, full siblings, half siblings and step siblings. *Journal of Personality and Social Psychology, 68,* 723–733.

Saudino, K. J., Plomin, R., & DeFries, J. C. (1996). Tester-rated temperament at 14, 20, and 24 months: Environmental change and genetic continuity. *British Journal of Developmental Psychology, 14,* 129–144.

Scarr, S., Webber, P. L., Weinberg, R. A., & Wittig, M. A. (1981). Personality resemblance among adolescents and their parents in biologically related and adoptive families. *Journal of Personality and Social Psychology, 40,* 885–898.

Schmitz, S., Saudino, K. J., Plomin, R., Fulker, D. W., & DeFries, J. C. (1996). Genetic and environmental influences on temperament in middle childhood: Analyses of teacher and tester ratings. *Child Development, 67,* 409–422.

Simonoff, E., Pickles, A., Hervas, A., Silberg, J. L., Rutter, M., & Eaves, L. (1998). Genetic influences on childhood hyperactivity: Contrast effects imply parental rating bias, not sibling interaction. *Psychological Medicine*, *28*, 825–837.

Stevenson, J., & Fielding, J. (1985). Ratings of temperament in families of young twins. *British Journal of Developmental Psychology*, *3*, 143–152.

Victor, J. B., Halverson, C. F., & Wampler, K. S. (1988). Family school context: Parent and teacher agreement on child temperament. *Journal of Consulting and Clinical Psychology*, *56*, 573–577.

STEPHANIE SCHMITZ
KIMBERLY J. SAUDINO

12

Links between Temperament and Behavior Problems in Early Adolescence

Introduction

There is growing interest in clinical and developmental psychology in the antecedents of problem behavior in children. One of the possible antecedents of problem behavior is temperament, in particular "difficult temperament" (Thomas & Chess, 1982). In their New York Longitudinal Study (NYLS), Thomas and Chess (1982) showed associations between later adjustment scores and temperamental aspects, as measured with their questionnaires, from age three onward. These measures of temperament have been shown to be genetically influenced (Cyphers, Phillips, Fulker, & Mrazek, 1990). Problem behavior in early and middle childhood was both genetically and environmentally influenced (e.g., Schmitz, Fulker, & Mrazek, 1995).

Rende (1993) tested correlations between different aspects of temperament in infancy and early childhood with problem behavior when the children were seven years old, utilizing data that were collected in the Colorado Adoption Project (CAP). The emotionality aspect of temperament was most consistently related to problem behavior at later ages while there were only sporadic relationships for activity, sociability, and persistence. Research with older children (Stevenson & Gjone, 1997) showed the emotionality aspect of temperament to be most predictive as well. Using data from a longitudinal twin study, Schmitz et al. (1999) reported moderate associations between the emotionality and shyness aspects of temperament and CBCL-measured problem behavior; these associations were mainly genetically mediated.

Bates, Bayles, Bennett, Ridge, & Brown (1991) reported in one of the follow-ups of the Bloomington Longitudinal Project, a longitudinal family study, that caregiver perceptions were the main bases for designations of psychopathology; even though the mother's perception of the child's temperament may not be objective, it was nevertheless valid. They concluded that difficult temperament predicted externalizing and internalizing behaviors in both boys and girls.

Although it appears likely from the literature that temperament predicts later problem behavior, there is little research that explores the genetic and environmental origins of this association. For those aspects of temperament that appear to be consistently correlated with later problem behavior, particularly emotionality, we will assess the extent to which this observed covariation is due to genetic or environmental influences or both.

The current chapter will examine antecedents of problem behavior, as indicated by the Child Behavior Checklist (CBCL/4-18; Achenbach, 1991a) and the Teacher Report Form (TRF; Achenbach, 1991b), with ratings from the temperament domain. Twin, adoption, and genetically noninformative studies show phenotypic associations between temperament rated at prior age points and later aspects of problem behavior, usually in the .20 to .30 range. Twin studies reported genetic and some shared-environmental correlations as underlying these associations. Data from this adoption study will be used to test the replicability of the above results in a nontwin sample, taking advantage of the power of adoption data to test for shared environmental influences.

Methods

Measures

For the temperament domain, the Colorado Childhood Temperament Inventory (CCTI; Buss & Plomin, 1984; Rowe & Plomin, 1977) was completed by the children's parents (1–12 years of age), their teachers (7–12 years of age), the lab tester (7 years of age), and the children themselves (7–12 years of age).

For problem behavior, the Child Behavior Checklist for 4- to 18-year-olds (Achenbach, 1991a) is a screening instrument completed by the children's mother (4–12 years of age) and teachers (TRF; Teacher Report Form, Achenbach, 1991b; 7–12 years of age). The broadband groupings of internalizing and externalizing, as well as the total problems score, will be used as indicators of problem behaviors.

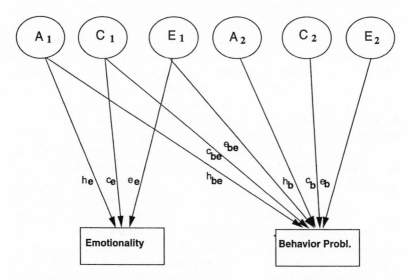

FIGURE 12.1. Bivariate model for genetic and environmental influences on temperament (Emotionality shown here) and problem behavior.

For the current chapter, CBCL/4-18 and TRF ratings at age 12 will be used as the indicator of problem behavior, and we will explore its associations with mother- and teacher-rated temperament ratings during the grade-school years.

Design

To address the question of whether the observed covariation between temperament and problem behavior is due to genetic or environmental influences or both, we can utilize the genetically informativeness of the adoption sample. Adopted siblings resemble each other only due to the environment that they share while nonadopted siblings also share half of their segregating genes, on average, which can act as an additional source of resemblance.

Bivariate model-fitting procedures were used to test for genetic and environmental influences on the covariance between ratings of temperament at each of the four earlier age points, as well as averaged ratings and problem behavior at age 12. This model, shown in figure 12.1, allowed for three factors that are common to both temperament and behavior ratings (one each for genetic, shared, and nonshared environmental influences, shown as paths h_{be}, c_{be}, and e_{be}, respectively), and three factors each that were unique for emotionality and behavior problems, shown as paths h_e, c_e, and e_e for temperament, and h_b, c_b, and e_b for problem behavior, respectively).

Genetic and environmental correlations between the measures indicate the overlap in genetic as well as shared and nonshared environmental influences for the two domains, showing whether the covariation between temperament and problem behavior is due to the same genetic and/or environmental influences. A significant genetic correlation indicates that temperament in earlier years and behaviors at a later time point are influenced by at least some of the same genes. A significant shared environmental correlation would indicate that at least some of the same factors in the siblings' shared environment lead to the association between these variables; this could point to a family rather than an individual intervention approach, or to look at factors in the classroom, if both both children attend the same class. If, however, a nonshared environmental correlation is significant, this would mean that the association between temperament and behavior is due to factors particular to the individual, which could be an illness only one of the siblings suffered or that child's perception of events. A path diagram of this bivariate model is shown in figure 12.1.

Significance of a particular influence was tested by omitting that particular path from the model and refitting the constrained, or reduced, model to the data. Comparing the χ^2 for the reduced model to that for the full model gives a statistical test of the decrement in fit with one degree of freedom. For example, a full bivariate model as described above would estimate nine parameters (three each for genetic, shared, and nonshared environmental effects) and its χ^2 statistic, with its associated p-value, would tell us how well the model fits the data. If we were then to test whether there really is a genetic correlation between temperament and problem behavior, we would omit the genetic correlation from the reduced model and therefore only estimate eight parameters. The χ^2 difference between the full and the reduced model would now be associated with one degree of freedom and statistics books with χ^2 tables give the associated p-value, leading to the conclusion of whether or not the genetic correlation contributes significantly to the model fit. All model fitting was carried out with the structural equation modeling program Mx (Neale, 1997).

Results

Tables 12.1 and 12.2 give the means and standard deviations, separately for boys and girls, by adoption status. Mother and teacher ratings of the Total Problems score and Externalizing were higher for boys than girls (as is to be expected from the different norms) and higher for the adopted children children. For Internalizing there were no sex or group effects. Compared to published norms (Achenbach, 1991a, b), all these means were in the normal range. There were few sex or group differ-

TABLE 12.1. Means and Standard Deviations on Each CBCL/4-18 Broadband Grouping at Age 12 and Comparison to Norm Sample

CBCL Scale (rated by)	Males		
	Adopted	Nonadopted	Norms[a]
Total (mothers)	22.80 ± 17.03	18.69 ± 14.59	23.0 ± 16.7
Total (teachers)	25.71 ± 22.40	21.54 ± 20.94	24.3 ± 26.3
Externalizing (mothers)	9.05 ± 7.59	7.71 ± 6.61	8.9 ± 7.5
Externalizing (teachers)	8.62 ± 10.00	5.98 ± 7.33	7.2 ± 10.3
Internalizing (mothers)	6.12 ± 5.29	5.37 ± 4.77	6.5 ± 5.3
Internalizing (teachers)	4.87 ± 5.20	6.18 ± 7.37	5.3 ± 6.8
	Females		
	Adopted	Nonadopted	Norms[a]
Total (mothers)	20.16 ± 17.76	16.18 ± 13.67	22.7 ± 17.8
Total (teachers)	14.84 ± 15.57	11.97 ± 13.50	15.4 ± 19.5
Externalizing (mothers)	7.45 ± 7.05	5.11 ± 5.05	7.4 ± 6.7
Externalizing (teachers)	4.07 ± 5.57	2.54 ± 4.97	4.3 ± 7.8
Internalizing (mothers)	5.76 ± 5.70	6.11 ± 5.50	7.5 ± 6.6
Internalizing (teachers)	4.92 ± 7.04	5.34 ± 5.74	4.8 ± 6.1

[a]Norms for this age group (Achenbach, 1991a, b)

ences in the Emotionality ratings and none were consistent over the years.

Emotionality was correlated with all aspects of problem behavior at all four age points; no other aspect of temperament was consistently related to problem behavior. Thus, the results reported by Rende (1993) on this sample at a younger age, and by Schmitz et al. (1999) on a twin sample, were replicated with this sample. Table 12.3 shows the correlations between Emotionality ratings at ages 7 through 11 plus their average and the CBCL/TRF ratings at age 12.

Though statistically significant, these correlations are modest in size, indicating that even though temperament at an early age is indicative of mother-rated problem behavior later, the percentage of variance explained is relatively small. This had previously been reported by Rende (1993) and Stevenson and Gjone (1997) for some of the CBCL/4-18 primary scales, and Schmitz et al. (1999) for the broadband groupings. In the twin studies, Emotionality correlated significantly with both Internalizing and Externalizing. These correlations are not just due to similar questions being asked in the CCTI and CBCL/4-18 (so-called content contamination; see also Lengua, West, & Staller, 1998), since there is only a one-item overlap between Internalizing and Emotionality. Correlations between mother and teacher reports are low

TABLE 12.2. Means and Standard Deviations of the CCTI Scale
Emotionality, by Sex and Adoption Status

	Boys		Girls	
Age	Adopted	Nonadopted	Adopted	Nonadopted
Mother Ratings				
Age 7	13.88 ± 3.89	13.95 ± 3.44	14.53 ± 4.26	14.29 ± 3.83
Age 9	14.13 ± 4.23	13.43 ± 4.08	14.33 ± 4.70	14.18 ± 4.26
Age 10	13.73 ± 4.11	13.27 ± 3.96	14.33 ± 4.52	14.21 ± 4.19
Age 11	14.16 ± 4.13	13.31 ± 4.32	14.58 ± 4.54	13.81 ± 3.97
Averaged	14.21 ± 3.75	13.52 ± 3.29	14.48 ± 3.92	14.19 ± 3.60
Teacher Ratings				
Age 7	10.64 ± 4.04	10.80 ± 4.64	10.63 ± 4.34	10.39 ± 4.08
Age 9	11.98 ± 4.75	10.02 ± 4.31	10.28 ± 4.21	11.17 ± 4.74
Age 10	11.43 ± 4.20	10.75 ± 4.87	11.14 ± 4.48	10.86 ± 4.66
Age 11	11.14 ± 4.49	10.21 ± 4.37	10.49 ± 4.54	10.76 ± 4.58
Averaged	11.16 ± 2.99	10.59 ± 3.74	10.55 ± 3.27	10.71 ± 3.46

TABLE 12.3. Correlations between CCTI Emotionality
and CBCL/4-18 at Age 12

	Parents			Teachers		
Age	Total	Ext.	Int.	Total	Ext.	Int.
Parent						
Age 7	.24***	.22***	.24***	.08	−.01	.13*
Age 9	.32***	.29***	.32***	.04	.01	.06
Age 10	.36***	.33***	.35***	.07	.04	.06
Age 11	.36***	.32***	.31***	.14**	.10	.12*
Averaged	.38***	.35***	.37***	.10	.04	.10*
Teacher						
Age 7	.16***	.12**	.12*	.23***	.15**	.18***
Age 9	.25***	.24***	.17***	.29***	.27***	.17***
Age 10	.15**	.12*	.12*	.25***	.27***	.14*
Age 11	.22***	.16***	.17***	.26***	.19***	.19***
Averaged	.27**	.22**	.21**	.36***	.30***	.23***

$***p < .001, **p < .01, *p < .05$

as was to be expected (Achenbach, McConaughy, & Howell, 1987):
the averaged Emotionality ratings correlated .21 and the correlations
between TRF and CBCL/4-18 broadband groupings ranged from .23
(Internalizing) to .35 (Externalizing and Total Problem score).

Table 12.4 gives the sample sizes available for behavior genetic analyses using sibling pairs with table 12.5 showing the within-pair correlations, separately for adopted and nonadopted siblings. The within-pair

TABLE 12.4. Sample Sizes (Pairs)

	Age 7	Age 9	Age 10	Age 11	Age 12
Teachers					
Adopted siblings	60	50	46	37	28
Nonadopted siblings	74	63	58	49	25
Parents					
Adopted siblings	70	60	59	44	47
Nonadopted siblings	85	78	73	54	48

TABLE 12.5. Within-Pair Correlations for CCTI Emotionality and CBCL/4-18

Scales	Adopted Siblings	Nonadopted Siblings
CCTI Emotionality		
Parents		
Age 7	.06	−.21*
Age 9	.20	.03
Age 10	.11	−.15
Age 11	.06	−.12
Averaged	.15	−.14
Teachers		
Age 7	−.11	.16
Age 9	.08	.24
Age 10	−.01	.15
Age 11	.03	.18
Averaged	.02	.34***
CBCL/TRF		
Parents		
Total	.29*	.42**
Externalizing	.26	.37**
Internalizing	.37*	.29*
Teachers		
Total	−.23	.13
Externalizing	−.27	.02
Internalizing	−.16	.12

***$p < .001$, **$p < .01$, *$p < .05$

correlations for Emotionality are generally low, as had previously already been reported for nontwin samples (Plomin, Corley, Caspi, Fulker, & DeFries, 1998; Schmitz, 1994). Differences regarding parameter estimates from twin and from adoption studies still need to be reconciled. Within-pair correlations for mother-rated problem behavior are suggestive of both genetic and shared environmental influences; the adopted siblings resemble each other, which can only be due to

TABLE 12.6. Cross-Correlations for CCTI Emotionality
and CBCL/4-18 at Age 12

Scales	Adopted Siblings			Nonadopted Siblings		
Parents	Total	Ext.	Int.	Total	Ext.	Int.
Age 7	.17	.21	.13	.16	.22*	.05
Age 9	.05	.05	.05	.19*	.19*	.15
Age 10	.10	.08	.11	.19*	.20*	.10
Age 11	.12	.10	.17	.28**	.26**	.21*
Averaged	.12	.12	.11	.22*	.23**	.14
Teachers	Total	Ext.	Int.	Total	Ext.	Int.
Age 7	−.09	−.17	−.03	.16	.18	.08
Age 9	.06	−.02	.10	.01	.08	.02
Age 10	−.04	−.15	.06	.00	−.03	.06
Age 11	−.07	−.12	−.02	.08	−.03	.18
Averaged	.00	−.13	.08	.06	.07	.06

Note: Total stands for Total Problems Score, Ext. for Externalizing, and Int. for Internalizing; **$p < .01$, *$p < .05$

shared environmental influences, and nonadopted siblings resemble each other to a greater degree, which might be due to additional genetic influences.

An initial check on whether the correlations between temperament and problem behavior are due to genetic influences was done by comparing the cross-correlations for adopted and nonadopted siblings—that is, the correlations between the temperament measure in one sibling and the problem-behavior rating in the other. If a genetic etiology mediates these correlations, then the cross-correlations for nonadopted siblings should exceed those for adopted siblings. If only the shared environment mediates these associations, then the cross-correlations should not differ between the two groups. Table 12.6 presents those cross-correlations. The cross-correlations for nonadopted siblings exceed those for adopted siblings for the parental ratings; however, the correlations are relatively small and only significant for the mother-rated Emotionality with Externalizing and Total Problem score for the nonadopted siblings.

To test statistically the significance of the genetic and environmental influence on emotionality and behavioral problems, and the associations between them, models that allowed both common and unique factors were fitted to the data (see figure 12.1). Parent and teacher ratings of emotionality and behavioral problems yield different results with respect to genetic and environmental effects on individual differences in these phenotypes. Parameter estimates for averaged Emotionality and the CBCL/4-18 and TRF broadband groupings are shown in

TABLE 12.7. Parameter Estimates, Averaged Emotionality Data, and CBCL/TRF

Scale	h^2	c^2	e^2	r_G	r_C	r_E
Parents						
Internalizing	.00	.32**	.68	—	21	79
Externalizing	.20	.23	.57	—	26	74
Total Problems	.25	.27*	.48	—	2	98
Emotionality	.00	.10	.90			
Teachers						
Internalizing	.16	.01	.83	67	6	27
Externalizing	.05	.00	.95	33	—	69
Total Problems	.11	.00	.89	60	3	37
Emotionality	.60	.03	.37			

h^2, c^2, e^2 stand for genetic, shared, and nonshared components of variance; r_G, r_C, r_E denote the percentages that genetic and environmental factors contribute to the phenotypic correlation

table 12.7, along with percentages that genetic and environmental influences contribute to the observed correlations.

As can been seen in table 12.7, individual differences in mothers' ratings of Emotionality are almost entirely due to nonshared environmental influences. In contrast, teachers' ratings of Emotionality displayed substantial genetic influence ($h^2 = .60$), accounting for approximately two-thirds of the phenotypic variance, with the remaining variance being due to nonshared environmental influences; however, this heritability estimate is not statistically significant with the small sample sizes for teacher ratings at age 12. The heritabilities for behavioral problems are modest for both mother and teacher ratings, although those for mother ratings tend to be a bit higher. Moreover, for mother ratings, shared environmental factors are equally or more important than genetic influences on behavioral problems.

For Internalizing and Externalizing, the overall correlation with Emotionality ratings at the different ages was statistically significant, but tests of its components (r_G, r_C, and r_E) were not conclusive—that is, each component individually could be left out of the model without a significant change in χ^2 but assuming no correlation led to a significant deterioration in model fit. For the Total Problems score, the overall correlation and the nonshared environmental correlation (r_E) were significant. Despite the lack of significance in our small sample, the overall pattern of results for mother and teacher ratings warrants comment. The zero genetic correlations between mother ratings of Emotionality and behavioral problems were not unexpected, given the zero heritability for mothers' ratings of Emotionality—that is, there cannot be genetic overlap between the two measures when one of them is not genetically influenced. On the other hand, there appears to be

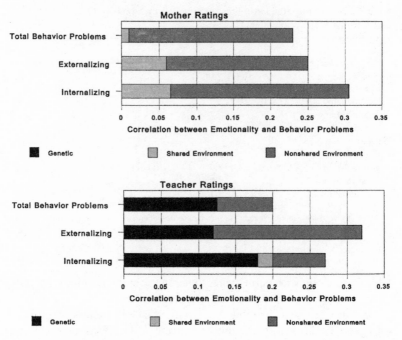

FIGURE 12.2. Top panel: Proportions of correlations between mother ratings of Emotionality and behavioral problems due to shared and nonshared environmental influences.

Bottom panel: Proportions of correlations between teacher ratings of Emotionality and behavioral problems due to genetic, shared and nonshared environmental influences.

substantial genetic overlap between teacher ratings of Emotionality and behavioral problems (genetic correlations ranged from .48 to .72).

Parent and teacher ratings of emotionality and behavioral problems yield different results with respect to genetic and environmental effects on individual differences in these phenotypes. As can be seen in table 12.7 and figure 12.2, the correlations between mothers' ratings of problem behaviors and Emotionality are largely mediated by nonshared environmental factors. In contrast, both genetic and nonshared environmental factors contribute appreciably to the phenotypic correlation between teacher ratings of Emotionality and behavioral problems.

As both mother ratings of behavioral problems and the averaged teacher ratings of Emotionality showed genetic influences and parent CBCL/4-18 ratings correlated significantly with averaged teacher CCTI ratings (see table 12.3, lower left-hand panel), we explored the associations between these measures. As is shown in figure 12.3, the phenotypic correlations between mother-rated behavioral problems and teacher-rated temperament is largely due to shared genetic influences.

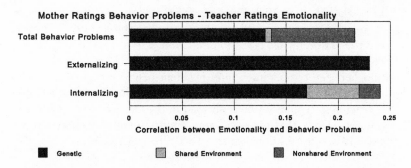

FIGURE 12.3. Proportions of correlations between mother ratings of behavioral problems and teacher ratings of Emotionality due to genetic, shared, and nonshared environmental influences.

Cross-rater genetic correlations ranged from .25 for Total Problems to .73 for Internalizing, indicating a moderate to high overlap in the genes influencing these behaviors. The association between Externalizing and Emotionality is entirely explained by the same set of genetic factors.

Discussion

Even though the correlations between aspects of temperament and problem behavior in this sample were similar to those published on twin data (Gjone & Stevenson, 1997; Schmitz et al., 1999), the results regarding the etiology of such associations could not be replicated.

The main reason for these differences stems from the differences of the estimates of genetic and environmental influences from twin versus adoption studies. Previous studies pointed to the influence of contrasting effects on parental ratings of temperament (Saudino & Cherny, 1997; Saudino et al., 1995). Sibling interaction models were fitted univariately to the temperament and the behavior ratings, but adding the interaction term did not significantly improve the model fit, with the one exception of teacher ratings of Externalizing.

In summary, correlations between problem behavior at age 12 and temperament ratings during grade school are moderate for children exhibiting behavior in the normal range. Previous studies of the covariation of temperament and outcome measures like problem behavior had obtained mixed results. While some studies found childhood temperament to be predictive of later behaviors (Rende, 1993; Stevenson & Gjone, 1997; Thomas & Chess, 1982), other studies did not find these associations (Katz & Gottman, 1993). In the current study, correlations between aspects of the two domains are moderate but statistically significant.

Even though parameter estimates for temperament varied between twin and adoption studies (Plomin et al., 1998; Schmitz, 1994), regardless of the children's age (see Plomin, Coon, Carey, DeFries, & Fulker, 1991, for their low heritability estimates of all CCTI scales), the association between temperament and problem behavior was similar. The role of the common shared environment that Stevenson and Gjone (1997) reported could not be replicated in the current study. Additionally, it was shown that the nonshared environment contributes to the relation between Emotionality and Total Problems score.

The children in this study showed scores in the normal range of behavior; it might be possible that the etiology for extreme problem behavior is different, in that extreme Emotionality might be more predictive of problem behavior later on; however, this question would need to be addressed with a clinical sample.

Moreover, parental perception or rater bias can influence ratings (Hewitt, Silberg, Neale, Eaues, & Erickson, 1992; Schmitz & Fulker, 1993), which can also contribute to observed behavioral continuity (Schmitz & Fulker, 1995). However, since the phenotypic correlations were similar for mother and teacher ratings, and the children usually had a different teacher every year, it is unlikely that the observed associations are due to biased perceptions only. The moderate correlations between mother and teacher ratings is not surprising, given that a review of the literature showed that parent and teacher ratings correlate .27 on average (Achenbach et al., 1987), which might be due to the children behaving differently in the home than in the school setting.

Curiously, the significance of contrast effects reported for younger twins (Saudino & Cherny, 1997; Saudino et al., 1995) could not be replicated. Part of the reason for this could be that the siblings are not rated at the same time. While mothers of twins complete the questionnaires for their children at the same time and thus have more opportunity to compare them, the mothers and teachers complete the questionnaires when the siblings are of the same age, being three years younger on average.

The phenotypic associations between temperament and problem behavior were replicated with the current sample and, though not significant, the genetic correlations underlying these associations are consistent with previous research (Schmitz et al., 1999). The exception to this is when mothers rated both temperament and behavioral problems in their children. This, however, is not surprising, since mother ratings of temperament have been shown to be problematic in behavior genetic studies (Saudino et al., 2000).

The phenotypic correlations between aspects of problem behavior and temperament are moderate, showing that both are not simply assessing the same construct by different methods. At the level of genetic etiology, however, there is substantial overlap between the two for teacher and cross-rater correlations.

This has implications both for the genetic researchers and well as clinicians. For genetic researchers, it is important to know whether they need to be looking for one or several unlinked—that is, uncorrelated sets of genes. Our results suggest that the former is the case in middle and late childhood. For clinicians, our findings of genetic correlations imply a need to find ways to deal with gene expression, gene-environment interactions, or correlations. In addition, covariation between early temperament and later behavior due to nonshared environmental influences would suggest a more individual intervention approach. Thus, researching precursors and their etiologies may help to develop better interventions and preventative strategies for problem behaviors.

REFERENCES

Achenbach, T. M. (1991a). *Manual for the Child Behavior Checklist/4-18 and 1991 Profile*. Burlington: University of Vermont, Department of Psychiatry.

Achenbach, T. M. (1991b). *Manual for the Teacher's Report Form and 1991 Profile*. Burlington: University of Vermont, Department of Psychiatry.

Achenbach, T. M., McConaughy, S. H., & Howell, C. T. (1987). Child/adolescent behavioral and emotional problems: Implications of cross-informant correlations for situational specificity. *Psychological Bulletin, 101*, 213–232.

Bates, J. E., Bayles, K., Bennett, D. S., Ridge, B., & Brown, M. M. (1991). Origins of externalizing behavior problems at eight years of age. In D. J. Pepler & K. H. Rubin (Eds.), *The development and treatment of childhood aggression* (pp. 93–120). Hillsdale, NJ: Lawrence Erlbaum Associates.

Buss, A. H., & Plomin, R. (1984). *Temperament: Early developing personality traits*. Hillsdale, NJ: Lawrence Erlbaum Associates.

Cyphers, L. H., Phillips, K., Fulker, D. W., & Mrazek, D. A. (1990). Twin temperament during the transition from infancy to early childhood. *Journal of the American Academy of Child and Adolescent Psychiatry, 29*, 392–397.

Hewitt, J. K., Silberg, J. L., Neale, M. C., Eaves, L. J., & Erickson, M. (1992). The analysis of parental ratings of children's behavior using LISREL. *Behavior Genetics, 22*, 293–317.

Katz, L. F., & Gottman, J. M. (1993). Patterns of marital conflict predict children's internalizing and externalizing behavior. *Developmental Psychology, 29*, 940–950.

Lengua, L., West, S., & Stadler, I. (1998). Temperament as a predictor of symptomatology in children: Addressing contamination of measures. *Child Development, 69*, 164–181.

Neale, M. C. (1997). *Mx: Statistical modeling* (4th ed.). Richmond: Department of Psychiatry, Medical College of Virginia.

Plomin, R., Coon, H., Carey, G., DeFries, J. C., & Fulker, D. W. (1991). Parent-offspring and sibling adoption analyses of parental ratings of temperament in infancy and childhood. *Journal of Personality, 59*, 705–732.

Plomin, R., Corley, R., Caspi, A., Fulker, D. W., & DeFries, J. C. (1998). Adoption results for self-reported personality: Evidence for nonadditive genetic effects? *Journal of Personality and Social Psychology, 75*(1), 211–218.

Rende, R. D. (1993). Longitudinal relations between temperament traits and behavioral syndromes in middle childhood. *Journal of the American Academy for Child and Adolescent Psychiatry, 32*(2), 287–290.

Rowe, D. C., & Plomin, R. (1977). Temperament in early childhood. *Journal of Personality Assessment, 41*, 150–156.

Saudino, K. J., & Cherny, S. S. (1997). Ratings of temperament in twins: A comparison of alternate models. *Behavior Genetics, 27*(6), 604.

Saudino, K. J., Cherny, S. S., & Plomin, R. (2000). Parent ratings of temperament in twins: Explaining the 'too low' DZ correlations. *Twin Research, 3*, 224–233.

Saudino, K. J., McGuire, S., Reiss, D., Hetherington, E. M., & Plomin, R. (1995). Parent ratings of EAS temperaments in twins, full siblings, half siblings and step siblings. *Journal of Personality and Social Psychology, 68*, 723–733.

Schmitz, S. (1994). Temperament and personality. In J. C. DeFries, R. Plomin, & D. W. Fulker (Eds.), *Nature and nurture during middle childhood* (pp. 120–140). Oxford: Blackwell.

Schmitz, S., & Fulker, D. W. (1993). Behavior ratings in early childhood with the Child Behavior Checklist (CBCL/4-16) (abs). *Behavior Genetics, 23*, 565.

Schmitz, S., & Fulker, D. W. (1995). Continuity due to which factors? An extension to the rater bias model. *Behavior Genetics, 25*, 287(abs).

Schmitz, S., Fulker, D. W., & Mrazek, D. A. (1995). Problem behavior in early and middle childhood: An initial behavior genetic analysis. *Journal of Child Psychology and Psychiatry, 36*, 1443–1458.

Schmitz, S., Fulker, D. W., Plomin, R., Zahn-Waxler, C., Emde, R. N., & DeFries, J. C. (1999). Temperament and problem behavior during early childhood. *International Journal of Behavioral Development, 23*(2), 333–355.

Stevenson, J., & Gjone, H. (Paris, June 1997). *A multivariate behaviour genetic analysis of temperament and behaviour problems in longitudinal study of twins.* (Poster presented at ISRCAP).

Thomas, A., & Chess, S. (1982). Temperament and follow-up to adulthood. In R. Porter & G. M. Collins (Eds.), *Temperamental differences in infants and young children* (pp. 168–175). London: Pittmann.

BETH MANKE
COLEEN CARLSON

13

Using the Social Relations Model to Examine Correlates of Adolescent Humor

The Role of Temperament and Well-Being

Introduction

Humor is often cited as a desirable personality trait and a panacea for all that ails us. It is thought that people who consistently use humor in their everyday interactions are extroverted and outgoing, are more likely to feel better about themselves, experience less depression and loneliness, and are more competent in their social interactions. Despite a plethora of anecdotal evidence supporting these claims, propositions concerning the links between interpersonal humor and personality/temperament and well-being have received relatively little empirical attention, especially as they pertain to children. The purpose of this chapter is to examine the temperament and well-being correlates of humor in adolescent family interactions. Specifically, we set out to investigate whether adolescents' consistent use (actor effects) and elicitation (partner effects) of affiliative and aggressive humor are related to parental reports of children's temperament (i.e., sociability, emotionality, activity, and attention span), internalizing, and externalizing problems; children's reports of depression, loneliness, and self-esteem; and teachers' reports of children's social competence.

This chapter will draw on data from the Colorado Sibling Study (CSS), a subproject of the Colorado Adoption Project (CAP), and represents an extension of a previous investigation of interpersonal humor using the same sample (Manke, Pike, Hobson, & Plomin, 2001). Results from this previous study suggest that children's use of humor in family relationships can be accounted for primarily by characteristics of the "actor" and to a lesser extent the "partner." That is, both affiliative and aggressive humor are explained, in part, by child actor effects or the tendency to use humor similarly with all family members. Results for affiliative and aggressive humor do differ, however, in terms of partner effects. Whereas the use of affiliative humor in family relations can be explained in part by children' s tendencies to elicit similar levels of humor from all family members, partner effects are not significant for aggressive humor. In other words, affiliative humor, such as telling funny stories and acting silly with family members, depends on the general characteristics or dispositions of both persons in the relationship. In contrast, aggressive humor, such as the use of practical jokes and sarcasm, depends on the characteristics of the person engaging in the humorous behavior, not the partner or target of such humor.

Results from the Manke et al. (2001) study further revealed that child actor effects associated with aggressive and affiliative humor may have different origins. That is, genetic influences, but not shared environmental contributions, are important for individual differences in a child's tendency to use similar levels of aggressive humor with all family members. This means that children who are more genetically similar demonstrate more similar patterns of aggressive-humor use with family members. In contrast, child actor effects for affiliative humor are due more to shared environmental influences. This means that growing up in the same family, regardless of genetic similarity, makes siblings' use of affiliative humor similar.

Although this previous study apportioned variance in humor use into actor and partner effects, and further decomposed the variance attributable to actor effects into its genetic and environmental components, it did not examine the specific processes that generate that variance. For example, although we know that genetic factors are important for aggressive humor actor effects, we do not know how these genetic effects are mediated. While these genetic influences may be unique to the use of humor, it is more likely the case that the genetic contributions to aggressive humor are mediated by other genetically influenced traits such as temperament, or internalizing (e.g., depression) and externalizing (e.g., aggression) problems. Thus, the next step in the quest to understand the origins of actor and partner effects associated with humor is to investigate possible correlates of these effects. Once we can identify specific correlates of actor and partner effects, we can incorporate them into multivariate genetic analyses that will allow us to decompose the phenotypic associations into their genetic and environmental components. Ultimately, we will be able to determine whether the genetic and environmental influences detected in the previous study (Manke et al., 2001) are specific to interpersonal humor or mediated by other traits and behaviors.

Unlike the other chapters in this book, the present chapter does not include behavioral genetic analyses, but instead focuses solely on the second step in the process outlined above—identifying specific correlates of actor and partner effects associated with humor. It is our hope that by doing so, we can provide direction for future behavior genetic analyses (i.e., multivariate analyses) of interpersonal humor. It should be noted that despite the absence of genetic analyses in the current chapter, the results presented here are related to those in other chapters in that they are based on a previous behavioral genetic analyses of actor and partner effects using CAP data. What follows is a brief discussion of previous humor research focused on personality and well-being, the constructs we have chosen to investigate as potential correlates of actor and partner effects associated with interpersonal humor.

Humor and Personality

Although no empirical research has investigated the link between children' s temperamental characteristics and the use of interpersonal humor, a few investigations have addressed the association between adult personality and humor. For example, Kuiper & Martin (1993) found that for young adults, "sense of humor" was positively related to the endorsement of items reflecting personality characteristics considered to reflect sociability. In addition, Kuiper & Martin (1998) reported a significant, positive relationship between adults' sociability and each of four different dimensions of humor: sense of humor, coping humor, laughing responsiveness, and liking humor. Furthermore, their results indicated a significant negative relationship between each of these humor factors and depressed personality characteristics. Thus, it would appear that adults who are more sociable and less depressive tend to have more of a "sense of humor," use humor as a coping mechanism, and show greater response to and appreciation of humor.

In a similar vein, Ruch & Kohler (1998) examined the relationship between cheerful and serious personality traits and humor. Adults who were high in trait cheerfulness showed longer lasting exhilaration (smiling and laughing) in response to humor as well as in response to laughing gas. The authors suggest that these results support the notion that there is a temperamental base to humor since, regardless of the stimulus (comic strips versus laughing gas), adults who were high in trait cheerfulness were more likely to laugh than those low on this trait. Based on this, Ruch and Kohler (1998) predicted that cheerfulness would also be associated with one's tendency to use and appreciate humor. Contrary to their predictions, results indicated that seriousness was a better predictor of both humor production and appreciation. That is, more serious adults produced significantly less humor than less serious adults. In addition, seriousness was related to less appreciation of multiple types of humor. Cheerfulness, however, was unrelated to both humor production and appreciation. Thus, while adults with cheerful personalities are more likely to smile and laugh, regardless of the stimulus, it is the degree of seriousness one exhibits that is related to one's tendencies to produce and appreciate humor.

Although the results of Ruch and Kohler's study might suggest that both the appreciation and production of humor are similarly linked with various personality traits, a study by Lanning (1994) suggests differential personality correlates for these dimensions of humor. In his study, a factor analysis of humor appreciation and initiation with items from the Five-Factor Model of Personality (FFP) was conducted. Humor initiation loaded most strongly and positively on the Extroversion dimension. Conversely, humor appreciation did not load on any of the five factors. Interestingly, when all five personality dimensions were partialed out of initiating humor, a significant amount of variance remained. This suggests that personality does not account for all of the variance in humor initiation and that other factors may also play a significant role.

Humor and Well-Being

In recent years that here has been a dramatic increase in the number of workshops, seminars, and popular books that seek to promote greater expression of humor in schools, hospitals, psychotherapy settings, and the workplace. Interest in advocating the use of humor is founded, in part, on the widely held belief that using humor causes or leads to better psychological health. Although evidence of the association between humor and well-being is accumulating, there is little empirical support for the notion that using humor causes better health. Furthermore, few previous studies have focused on the association between humor and well-being among children. In general, adult studies addressing humor and well-being suggest that individuals who report using more humor have higher self-esteem, less discrepancy between their actual and ideal self-concepts, greater stability in their self-concepts over time, less anxiety, and less depression (Kuiper & Martin, 1993). In addition, a study of college students revealed that individuals with a greater sense of humor reported feelings of continued development, were open to new experiences, and had a sense of realizing their potentials (Vaillant & Vaillant, 1992).

A notable exception to the predominant focus on adult humor and well-being is a study of school-age children's humor production and comprehension (Masten, 1986). Results from this study suggest that better humor comprehension and greater humor production are associated with social competence. That is, children who understand and use more humor are viewed by their teachers as being more effectively engaged in the classroom and are more attentive, cooperative, responsive, and productive. In addition, the peers of these children tend to view them as more popular, happy, and as leaders with good ideas for things to do. Interestingly, humor was not related to disruptive aggressive behavior as rated by teachers or peers.

Social Relations Model

In order to examine the association between children's temperamental characteristics and well-being and their use of humor in family interactions, we used an extension of the basic Social Relations Model (SRM: Kenny & La Voie, 1984). The basic SRM involves the study of two-person interactions and states that in measures of dyadic relationships, responses from individuals about their interactions with one another are not independent of the responses obtained from other individuals reporting on the same relationship. In short, the basic SRM states that each family member's use of humor with other family members is a function of four independent components or sources of variance: (1) actor effects—a person's disposition or tendency to behave in a similar way toward all family members, (2) partner effects—a person's tendency to elicit similar interactions from all family members, (3) dyadic relationship effects—factors unique to specific dyadic relationships, and (4) error.

By analyzing the covariation between family members' reports of humor use, several SRM effects can be estimated for both affiliative and aggressive humor. In the present investigation of three-person families (i.e., mother, older sibling, and younger sibling), actor and partner effects are estimated separately for each family member. In addition, dyadic relationship effects are calculated for each of the six relationships: mother → older sibling, mother → younger sibling, older sibling → mother, older sibling → younger sibling, younger sibling → mother, and younger sibling → older sibling.

It should be noted that in order to separate error from dyadic relationship effects, studies must incorporate replications, either over time or across measures (Kashy & Kenny, 1990). Because the current investigation uses data collected at one time-point, and includes one measurement for each dimension of humor (affiliative and distancing), estimates of dyadic relationship effects include measurement error. As a result, these effects are not discussed nor are they examined in reference to the measures of temperament and well-being.

The present study extends the basic SRM by including indices of children's temperament and well-being, thereby permitting the examination of correlations between these measures and significant child actor and partner effects. The advantage of using the SRM to investigate temperament and well-being correlates of adolescent humor lies in the ability to investigate correlates separately for actor and partner effects. That is, we can investigate whether various measures of temperament and well-being are consistently related to a child's general tendency to use or elicit humor from all family members.

Hypotheses

In combination, the above studies provide support for the idea that personality, or temperament, and psychological well-being may be related to children's initiation or use of humor in everyday interactions. Based on previous research, we hypothesize that children's use and elicitation of affiliative humor will be positively related to sociability and negatively related to emotionality, a measure of distress. Although measures of children's activity level and attention span are also included in the present study, hypotheses regarding these dimensions are not made as the links are not readily apparent. In terms of well-being, we propose that affiliative humor will be positively related to general self-esteem and peer competence, and negatively related to internalizing (i.e., depression and loneliness) and externalizing behaviors.

It should be noted that we do not have specific predictions regarding the association between the measures of children's temperament and the use of aggressive humor, as this type of humor has rarely been investigated in previous studies and none of the measures of temperament included in the current study taps hostile or aggressive personality traits. Based on the idea that some forms of humor (i.e., laughing) may facilitate healthy psychological functioning, whereas others types of humor that are used to dominate or manipulate others may lead to negative well-being (Kuiper & Martin, 1998), we propose that children's use of aggressive humor will be negatively related to general self-esteem and peer competence and positively related to internalizing and externalizing behaviors.

It is also difficult to make differential predictions concerning child actor and partner effects for either type of humor, as little research to date has addressed the role of actor and partner in the use of interpersonal humor. Finally, differential predictions concerning older and younger siblings' correlates are not made, as we have no reason to believe that children's temperament or well-being and the use of humor are differentially linked based on birth order or developmental status.

Participants

Participants in the present study were 96 sibling pairs and their mothers taking part in the third phase of the Colorado Sibling Study (CSS: Dunn, Stocker, & Plomin, 1990; Stocker, Dunn, & Plomin, 1989), a longitudinal investigation of children' s familial and extrafamilial relationships during middle childhood and adolescence. As mentioned previously, the CSS is a subproject of CAP. On average, older and younger siblings were 15 and 12 years of age, respectively. Of the 96 families, 56 were nonadoptive (i.e., children living with their biological parents) and 40 were adoptive. Of the 56 sibling pairs living in nonadoptive families, there were 20 boy-boy, 9 boy-girl, 14 girl-boy, and 13 girl-girl sibling pairs. The 40 adoptive families consisted of 5 boy-boy, 15 boy-girl, 15 girl-boy, and 5 girl-girl sibling pairs.

Procedures

Data concerning interpersonal humor were obtained during semi-structured telephone interviews conducted separately with mothers, older siblings, and younger siblings as part of the CSS. Information concerning children's temperament and well-being was drawn from the annual CAP assessments. It is important to note that CSS humor data were collected during one calendar year, whereas data collection for the CAP was based on a standard protocol of assessments administered and organized according to children's chronological age (e.g., 9-year-old CAP data, 10-year-old CAP data, etc.). Because of this, utilizing both data sets simultaneously poses a challenge. One option would be to choose just one CAP year (one age) from which to draw data. Doing this would result in large variations in the time between the CAP and CSS assessments. A second solution, and the one used in the present study, involves drawing data from several CAP data years simultaneously. More specifically, for each child assessed as part of the CSS, the closest CAP year was selected, minimizing the age span between the CSS and CAP assessments for each child. It should be noted that temperament data were drawn from the closest CAP assessment prior to the collection of the CSS humor data, whereas well-being data were drawn from the closet assessment following the collection of the CSS humor data. We recognize that the causal direction between temperament or well-being and humor is an empirical one that we cannot address in the present study, given the lack of longitudinal data on humor. Nevertheless, in order to mirror the ideas expressed in the substantive literature, we decided to treat temperament as a possible antecedent and well-being as a possible outcome of interpersonal humor.

Measures

Interpersonal Humor

Data were collected using a round-robin approach, a design in which data are gathered from each family member about his or her use of humor with each of the other two family members participating in the study. Participants were given the Humor Use in Multiple Ongoing Relationships measure (HUMOR; Manke, 1998), a 12-item measure tapping different humor behaviors. During the interview, participants were asked to indicate how often they engage in each of the humor behaviors with each of the other two family members. Participants' responses were then rated by interviewers on a six-point Likert scale (1 = hardly ever, less than once a month; 6 = very often, several times a day). It is important to note that each of the items included a relationship tag. For example, when older siblings were asked to discuss their use of affiliative humor with their younger siblings they were asked questions such as, "How often do you tell your younger sibling funny stories about things that have happened to you?" and "How often do you laugh at movies, TV, or radio programs that you think are funny when you are with your younger sibling?" When asked to discuss their use of aggressive humor, older siblings were asked questions including, "How often do you play practical jokes on your younger sibling?" and "How often do you playfully insult your younger sibling?" Items were then summed to form total scores representing the participants' use of affiliative and aggressive humor.

Estimates of internal consistency ranged from .65 to .85 for affiliative humor and .58 to .70 for aggressive humor, depending on the relationship in question (i.e., mother-to-older sibling, mother-to-younger sibling, etc.). The use of affiliative and aggressive humor were positively yet moderately correlated (correlations ranged from .30 to .48) across relationships, suggesting that family members who use a great deal of affiliative humor are also likely to use aggressive humor with other family members. All telephone interviews were audiotaped, thereby permitting the calculation of interrater reliability. Estimates of interrater reliability ranged from .81 to .99 for the various humor items and two-week test-retest reliabilities estimated with a subsample of participants ranged from .54 to .86.

Temperament

Information concerning children's temperament was drawn from parental reports gathered during in-person CAP assessments conducted prior (on average four months prior) to the CSS telephone interviews. Parents completed the Colorado Temperament Inventory (CCTI: Rowe & Plomin, 1977), a 30-item measure tapping sociability, emotionality, activity, and attention span. Parents completed the measure of temperament for both older and younger siblings.

Well-Being

A variety of well-being indices were used in the present study. Measures of depression, self-esteem and loneliness were administered as part of the CAP battery assessing children's feelings. Self-esteem was measured using Harter's Self-Perception Profile for Children (1983). Although this measure contains six subscales, only the total score will be used in the present study. Children' s self-reports of depression and loneliness were measured using Kandel's Depressive Mood Inventory (Kandel & Davies, 1982), and Asher's Loneliness Questionnaire (Asher, Hymel, & Renshaw, 1984). Indices of children's internalizing and externalizing problems were also obtained from parents' reports on the Child Behavior Checklist (Achenbach & Edelbrock, 1983). Parents completed this measure for older and younger siblings. Finally, children's social competence was measured using teachers' reports on the Walker-McConnell Scale of Social Competence and School Adjustment (Walker & McConnell, 1988). Although this measure also includes subscales, only the total score representing total social competence was used, as the subscales were highly correlated (intercorellations exceeded .70). Teacher data were available for only 35 of the older and 50 of the younger siblings. Based on this, analyses of teachers' ratings of social competence were conducted with a subsample of all available children.

Results

Analyses were performed in two steps. First, descriptive analyses were performed to test for gender, age, and adoptive status differences in family members' use of affiliative and aggressive humor. Second, a series of extended SRM analyses of adolescent familial humor use examining the temperamental and well-being correlates of significant child actor and partner effects were conducted separately for affiliative and aggressive humor.

TABLE 13.1. Means and (Standard Deviations) for Affiliative and Aggressive Humor by Adoptive Status

	Affiliative Humor		Aggressive Humor	
	Nonadoptive (n = 56)	Adoptive (n = 40)	Nonadoptive (n = 56)	Adoptive (n = 40)
M → OS	16.71 (4.23)	16.35 (4.10)	10.47 (3.30)	11.92 (4.14)
M → YS	16.52 (3.48)	17.20 (3.61)	11.00 (3.74)	11.52 (4.01)
OS → M	14.68 (5.03)	14.60 (5.09)	9.20 (3.01)	9.10 (3.81)
OS → YS	18.19 (5.68)	16.10 (4.91)	14.23 (4.45)	12.48 (5.06)
YS → M	15.14 (5.30)	14.84 (5.75)	9.45 (4.05)	8.16 (2.56)
YS → OS	17.33 (5.08)	15.46 (5.14)	13.58 (4.20)	12.44 (3.99)

Note: M = mother, OS = older sibling; YS = younger sibling

Preliminary Analyses

In order to examine gender and adoptive status differences in affiliative and aggressive humor, a series of 2 (older siblings' gender) × 2 (younger siblings' gender) × 2 (adoptive status) × 2 (type of humor) mixed model ANOVAS using Type of Humor (affiliative and aggressive) as a repeated measure was conducted using the six directed family relationships (e.g., older siblings' use of humor with mothers) as dependent measures. Means by adoptive status are presented in table 13.1. Main effects for Type of Humor were detected for all six directed family relationships, indicating that children and mothers report using more affiliative than aggressive humor in family relationships. No main effects for older or younger siblings' gender were detected, although main effects for adoptive status were revealed in two cases, both involving aggressive humor. First, results for mothers' reports of their use of aggressive humor with older siblings suggested that mothers in adoptive families reported greater use of aggressive humor than did mothers in nonadoptive families (F [1,89] = 5.57, $p < .05$). In contrast, older siblings in adopted families reported using less aggressive humor with younger siblings than did older siblings in nonadoptive families (F [1,89] = 6.06, $p < .05$). Finally, only one higher order interaction involving older siblings' gender, younger siblings' gender, or adoptive status was detected. Specifically, a two-way interaction between younger siblings' gender and adoptive status was revealed for older siblings' use of affiliative humor with mothers, indicating that older siblings with younger brothers in adopted families reported lower levels of affiliative humor with mothers than did older siblings with younger sisters in adopted families $(F [1,89] = 4.15, p < .05)$.

To examine the possible effects of age and within-sibling pair age differences, multiple regression analyses were conducted for the 12 indices of humor use. No main effects for age were detected and in only two cases were sibling age differences a significant predictor of humor use: older siblings' reports of their use of affiliative humor with mothers and mothers' reports of their use of aggressive humor with older siblings. In both cases, a larger age difference between siblings predicted greater humor use.

The size of the present study prohibits the use of gender and age as covariates in the SRM model. To "control" for these effects, all scores were corrected for gender, age, and within-pair sibling age differences prior to conducting the SRM analyses. This was accomplished by using standardized partial residuals from the regression of the humor measures on the gender, age, and within-sibling pair age difference variables (McGue & Bouchard, 1984).

SRM Analyses

The three-person extended SRM is depicted in figure 13.1. The three actor effects, three partner effects, and six dyadic relationship effects are portrayed as solid circles (or latent variables) and are estimated by fitting the SRM to the variance-covariance matrix of the six measures of humor depicted in the six rectangles (manifest variables). In this study, the SRM analyses were performed using LISREL VIII (Jöreskog & Sörbom, 1994). The SRM forces the six dyadic interaction measures of humor to load on the latent factors (the loadings are fixed at 1). This process provides information regarding the amount of variance in the manifest variables accounted for by each component.

Unlike traditional confirmatory factor analysis, which allows factor loadings to vary and thus requires a minimum sample of 100 participants, fewer participants (i.e., 40–50 families) are needed for the estimation of SRM effects (Kashy & Kenny, 1990). Thus, the current study, which is based on data from 96 families (288 individual family members), meets this minimum requirement. The one exception to this is the analyses including teachers' reports of older siblings' social competence, which are based on data drawn from 35 families and as such must be interpreted with caution.

Measured indices of children's temperament and well-being were added to the basic SRM model and are depicted on the left side of figure 13.1 as a dashed rectangle and circle. Correlations between the measures of temperament and well-being and individual level effects of

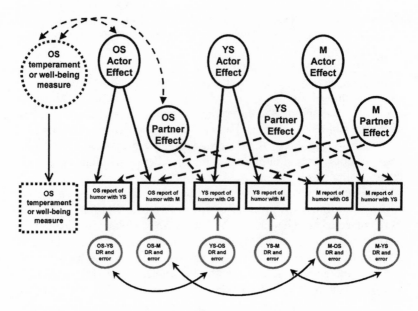

FIGURE 13.1. Path diagram of social relations model (SRM). OS = older sibling, YS = younger sibling, M = mother, and DR = dyadic relationship effect. The six solid rectangles are measured variables of humor use as reported by older sibling, younger siblings, and mothers. The dashed rectangle is a measured variable of children's temperament or well-being. The actor effects are latent variables representing a person's disposition or tendency to use humor in a certain way to all family members. Partner effects are latent variables representing the tendency of family members to elicit similar humor use from all other family members. Dyadic relationship effects are latent variables representing factors unique to each particular dyadic relationship plus error of measurement. The curved, two-way arrows indicate correlations between the variables they connect, and the one-way arrows represent paths, standardized partial regressions of the SRM effect on the measured variable.

actor and partner are depicted as dashed double-headed curved arrows at the top of figure 13.1. Note that this model depicts the correlations between older sibling's temperament/well-being and older siblings' actor and partner effects. An analogous model was developed to examine the correlations between younger siblings' temperament/well-being and younger siblings' actor and partner effects.

Typically, individual reciprocity correlations among individual level effects of actor and partner are also specified in the basic SRM. However, in order to incorporate measures of children's temperament/well-being in the model, and to estimate the correlations between temperament/well-being and actor/partner effects, these individual reciprocity correlations were dropped. Dyadic reciprocity correlations

TABLE 13.2. Variance Estimates from the Social Relations Model (SRM) Analyses

	Affiliative Humor	Aggressive Humor
Actor Effects:		
Mother	6.40*	6.63*
Older Sibling	16.53*	7.71*
Younger Sibling	14.98*	7.25*
Partner Effects:		
Mother	6.58*	1.54
Older Sibling	3.38*	1.04
Younger Stibling	3.33*	2.86
Dyadic Relationship Effects:		
M → OS	7.48*	5.14*
M → YS	2.09	4.17*
OS → M	—	1.21
OS → YS	8.19*	11.32*
YS → M	7.56*	3.65*
YS → OS	6.51*	7.77*
χ^2	$\chi^2 (5) = 2.21$	$\chi^2 (3) = 7.90$
RMSEA	.01	.09

Note: M = mother, OS = older sibling, YS = young sibling. For the fit of the standard model, significance is based on chi-square analysis and the root mean squared error of approximation (RMSEA). Significance of the SRM effects (actor effects, partner effects and dyadic relationship effects) is based on the standard error two-tailed test; $*p < .05$

among the relationship effects are specified and are depicted as double-headed curved arrows at the bottom of figure 13.1.

In principle, estimates of variance are positive. Therefore, after the complete SRM is fit, negative effects are set to zero and the model is retested. Because one cannot correlate unreliable variance estimates, it is necessary that all variance estimates examined as part of the extended SRM be significantly greater than zero in order for the analyses to be valid. That is, in order to examine the correlations between indices of temperament or well-being and child actor and partner effects, variance estimates for both the measure of temperament or well-being and the actor or partner effect in question (e.g., older sibling actor effect) must be significant.

The overall fit of the models is evaluated by a chi-square (goodness-of-fit) test and the root mean squared error of approximation (RMSEA). Both the chi-square test and the RMSEA test the ability of the proposed model to reproduce adequately the observed correlations between the variables and represent the most traditional methods for evaluating model fit.

The results of the basic SRM are the same as those reported in the previous investigation (Manke et al., 2001) and are reproduced here for reference purposes only (see table 13.2). As noted previously, the results

TABLE 13.3. Correlations Between Maternal Reports of Children's Temperament Traits and Children's Actor and Partner Effects

	Older Siblings		Younger Siblings	
	Actor Effect	Partner Effect	Actor Effect	Partner Effect
Affiliative Humor:				
Sociability	−.02	.10	.23^T	.46*
Emotionality	.15	−.28**	−.04	.01
Activity	−.03	−.07	.23^T	.69*
Attention span	.23^T	−.01	−.11	−.06
Aggressive Humor:				
Sociability	.01	—	.05	—
Emotionality	−.11	—	−.15	—
Activity	.06	—	−.10	—
Attention span	.17	—	.03	—

Note: Correlations between indices of children's temperament and older and younger siblings' partner effects for aggressive humor could not be estimated because partner effects for both older and younger siblings were nonsignificant for this type of familial humor. *$p < .05$; **$p < .10$

suggest that mother, older sibling, and younger sibling actor effects are significant for both affiliative and aggressive humor. That is, humor is significantly accounted for by siblings' and mothers' general tendency to use (or not to use) humor consistently with family members. In contrast, partner effects are significant only for affiliative humor, such that the use of aggressive humor in familial interactions is not affected by characteristics of the target or partner.

Because none of the partner effects for aggressive humor were significant, correlations with measures of children's temperament and well-being were not calculated. Correlations with children's temperament traits and well-being were calculated for older and younger sibling actor effects for both affiliative and aggressive humor and older and younger sibling partner effects for affiliative humor. SRM effects involving the mother, such as the mother actor and partner effects, were not correlated with the measures of children's temperament traits or well-being as it seemed unlikely that a mother's consistent use of humor with both of her children or her tendency to elicit similar levels of humor from both siblings would be related to one child's temperament or well-being. The inclusion of data concerning mother-child interactions in the SRM is, nevertheless, necessary for the estimation of child effects that are used in the extended SRM analyses.

Results from the series of extended SRM analyses involving temperament are depicted in table 13.3. Six of the 16 correlations between measures of children's temperament and affiliative humor are significant at the .10 level. The most striking pattern of correlations are those for younger siblings in that both actor and partner effects are positively

TABLE 13.4. Correlations Between Children's Well-Being
and Children's Actor and Partner Effects

	Older Siblings		Younger Siblings	
	Actor Effect	Partner Effect	Actor Effect	Partner Effect
Affiliative Humor:				
Children's Self-Reports				
Depression	−.43*	−.12	−.12	.01
Loneliness	−.38*	−.01	.01	−.11
Self-esteem	.44*	.08	.11	.01
Parental Reports:				
Internalizing problems	−.25**	−.59*	−.08	−.12
Externalizing problems	−.25**	−.35*	.01	.21
Teachers' Reports:				
Social competence	.70*	−.01	.37*	.16
Aggressive Humor:				
Children's Self-reports:				
Depression	−.22	—	.01	—
Loneliness	−.26^T	—	.13	—
Self-esteem	.26^T	—	−.15	—
Parental Reports:				
Internalizing problems	−.01	—	.10	—
Externalizing problems	.09	—	−.02	—
Teachers' Report:				
Social competence	.52*	—	.50*	—

Note: Correlations between indices of children's well-being and older and younger siblings' partner effects for aggressive humor could not be estimated because partner effects for both older and younger siblings were nonsignificant for this type of familial humor. *$p < .05$; **$p < .10$

related to sociability and activity. That is, younger siblings who consistently use, as well as elicit, affiliative humor with family members tend to be more sociable (.23 and .46, respectively) and more active (.23 and .69, respectively). For older siblings, actor effects are positively related to attention span (.22), whereas partner effects are negatively related to emotionality (−.28). These results suggest that older siblings who consistently use more affiliative humor with family members tend to be more task oriented while those who consistently elicit more affiliative humor from family members tend to be less distressed or more emotionally stable. It is interesting to note that for both older and younger siblings, temperamental characteristics were unrelated to actor effects associated with aggressive humor.

Results from the series of extended SRM analyses involving well-being are depicted in table 13.4. Thirteen of the 36 correlations between measures of children's well-being and humor (both affiliative and aggressive) are significant at the .10 level. The most interesting pattern of correlations here involved older siblings' actor effects for affiliative humor. Older siblings who consistently use more affiliative humor with family member report being less depressed (–.43), less lonely (–.38), and having higher self-esteem (.44). Furthermore, they tend to be rated by their parents as having fewer internalizing (–.25) and externalizing (–.25) problems, and are rated by their teachers as being more socially competent (.70). Correlations for older siblings' partner effects associated with affiliative humor were less pervasive; partner effects were negatively related to parental reports of children's internalizing (–.59) and externalizing problems (–.35), suggesting that older siblings who elicit more affiliative humor from family members also exhibit less internalizing and externalizing problems. Contrary to our predictions, the results for younger siblings' actor and partner effects associated with affiliative humor do not parallel those for older siblings, as only the correlation between younger siblings' actor effects and teachers' reports of social competence was significant (.37).

The results for actor effects associated with aggressive humor and well-being are somewhat similar to those for affiliative humor and well-being, in that older siblings who consistently use more aggressive humor with family members report being less lonely (–.26) and having higher self-esteem (.26). In addition, they are rated by their teachers as being more social competent (.52). Contrary to affiliative humor, parental reports of well-being were not related to older siblings' actor effects. As with affiliative humor, the only significant correlation between well-being and younger siblings' actor effects associated with aggressive humor was for teachers' reports of the children's social competence (.50).

Discussion

Discussion of the current findings is broken down into five sections. The first two sections include interpretations of the temperament and well-being correlates, whereas the third and fourth sections contain a general discussion of the differential correlates for actor versus partner effects, and affiliative versus aggressive humor. The final section details future directions in the search for humor correlates.

Humor and Temperament

In general, the extended SRM analyses suggest relatively few links between temperament and children's tendencies to use or elicit humor consistently from family members. Of the few significant correlations revealed, the most consistent links are those related to affiliative humor. Specifically, it appears that younger siblings who consistently use and elicit greater amounts of affiliative humor with/from family members tend to be those who are more sociable and active. In contrast, older siblings who consistently use more affiliative humor tend to have greater attention spans, whereas those who consistently elicit more affiliative humor are less emotional or unstable.

These findings suggest that although temperament variables may be related to children's use of humor, they alone do not adequately serve as the mediating source of either genetic or environmental contributions to affiliative and aggressive humor use. The relatively sparse findings related to children's temperament may be influenced by the measure employed in the current investigation. The Colorado Temperament Inventory (Rowe & Plomin, 1977) tends to tap global or broad temperamental characteristics (e.g., sociability and emotionality). It is possible that the aspects of temperament being measured are not specific enough to capture the nature of affiliative and aggressive humor. Perhaps the inclusion of more specific measures such as cheerfulness, agreeableness, seriousness, self-restraint, sensation seeking, and exuberance would prove more informative in the search for temperament correlates of humor use. Moreover, more "negative" temperament traits such as aggressiveness or hostility might reveal more associations between temperament and children's aggressive humor. If, however, temperament continues to be unrelated to children's use of aggressive humor, we must consider the possibility that the more aggressive forms of humor, such as playing practical jokes and using sarcasm, might be linked to alternative psychological measures, including cognitive ability. Perhaps the ability to plan and execute practical jokes and generate sarcastic remarks requires advanced cognitive skills.

Contrary to our predictions, the results suggest differential associations between temperament and humor for older and younger siblings. Older siblings' use and elicitation of affiliative humor is related to emotionality and attention span, whereas younger siblings' use and elicitation of affiliative humor is related to sociability and activity. Hence, it is possible that the correlates, and perhaps even the nature of humor, changes developmentally. For younger children (average age of 12 years), humor in the context of family relationships seems to be the outgrowth of one's tendency to be outgoing and active. Thus, for younger children, humor may be one of the many ways in which they express affect, interest, and attempt to enhance intimacy. Conversely, for older children (on average 15 years old), humor seems to be more purposeful or deliberate. That is, older children, who are experiencing both the cognitive and physical changes associated with adolescence, may begin to develop a more mature humor style that requires attention or focus and is associated with overall well-being.

An alternative explanation for the differences between older and younger sibling correlates may be the presence of birth-order effects. That is, something unique or different about being the first or second-born child, regardless of current age, may drive the links between humor and temperament. Given the lack of findings documenting consistent birth-order effects for any dimension of social relationships, this explanation would appear less plausible than the presence of a developmental progression. Longitudinal data could be used to address whether the source of these differential correlates is birth order or developmental differences.

Humor and Well-Being

The results concerning well-being and humor are more encouraging: several links were detected again between humor and well-being, however the links are more consistent for affiliative than for aggressive humor and more consistent for older siblings. It would appear that older siblings who consistently use more affiliative humor with family members are more likely to manifest better psychological well-being (e.g., less depression, higher self-esteem, greater social competence, etc.).

Similar to the results for temperament, findings regarding well-being suggest differential correlates for older and younger siblings. That is, older siblings' actor effects for both affiliative and aggressive humor are associated with several indices of well-being. Such was not the case for younger siblings. In short, these results provide further support for the idea that the correlates, and perhaps even the nature of humor, changes developmentally. Perhaps the link between humor and well-being is not solidified until children begin the transition into adolescence. It may be that these older children are capable of using humor more deliberately to cope with daily hassles and life events, thus the links between humor and well-being are more readily apparent.

Future investigations of humor and well-being might consider the idea that levels of life stress need to be examined in order to more fully understand the association between children's use of humor and psychological well-being. That is, it may be the interaction between interpersonal humor and life stress that predicts levels psychological health, rather than the use of humor alone. This line of thinking is supported by the results of several studies with adults suggesting that individuals with a greater sense of humor are less adversely affected by stressful life events, as evidenced by a less dramatic increase in depression and anxiety (Martin & Lefcourt, 1983; Nezu, Nezu, & Blissett, 1988; Vaillant & Vaillant, 1992). The proposed mechanism for the moderating effects of humor is that persons with a good sense of humor assess stressful situations differently from those who appreciate humor less. That is, humorous persons are more likely to appraise stressful events as challenging instead of as a threat (Kuiper, Martin, & Olinger, 1993).

Are There Differential Correlates for Actor and Partner Effects?

The main advantage of using the SRM in the present study is that it makes it possible to investigate whether various measures of temperament and well-being were consistently related to a child's general tendency to use (actor) and/or elicit (partner) humor from all family members. Although not predicted, the results provide limited support for the idea that there are differential correlates for child actor and partner effects for affiliative humor. In regards to temperament, the correlations between sociability and activity and younger sibling's partner effects were significantly greater than those between these same temperament measures and younger siblings' actor effects. Thus, while sociability and activity are related to both younger siblings' actor and partner effects, the links appear stronger for partner effects. If these findings replicate in future studies, it would suggest that temperamental characteristics are more important for the ways in which younger children consistently elicit humor from others. This would imply that temperament may play a more central role in determining whether or not others use humor with an individual rather than the individual's own use of humor.

In contrast, the correlations between various indices of well-being and older siblings' actor effects were significantly greater than those between these same well-being measures and older siblings' partner effects. This might suggest that positive well-being is related to how children use, but not elicit, affiliative humor. One notable exception to this pattern pertains to affiliative humor and internalizing behavior problems. The negative correlation between maternal reports of children's internalizing problems and older siblings' partner effects for affiliative humor was significantly greater than the correlation between internalizing problems and older siblings' actor effects. This suggests that children who exhibit fewer internalizing problems (as reported by mothers not children) have a tendency to elicit more affiliative humor. Perhaps it is more difficult, from the perspective of a parent, to engage in humorous behavior with a child who is evidencing depressive symptoms.

Are There Differential Correlates for Affiliative and Aggressive Humor?

Although most previous research has not made a distinction between the use of aggressive and affiliative humor, results from the present study suggest that there may be important differences between the types of humor. Temperamental characteristics were related to affiliative humor, but not aggressive humor. In addition, it would appear that well-being is more consistently related to affiliative than to aggressive humor. In some respects these differential results are not surprising given that our previous work (Manke et al., 2001) revealed that genetic contributions are important for aggressive humor, whereas shared environmental factors contribute significantly to individual differences in affiliative humor. However, it is surprising that neither temperament nor well-being is consistently correlated with aggressive humor for either older or younger siblings. As was mentioned previously, the reasons for this may lie in our choice of temperament and well-being correlates. That is, different temperament characteristics, such as sensation seeking or hostility, and different indices of well-being, such as problem solving or locus of control, may be better predictors of aggressive humor. Or, it may be that affiliative and aggressive humor are fundamentally different behaviors and have not only different genetic and environmental origins but also distinctly different social functions. Perhaps a more fruitful search for the correlates of aggressive humor lies in the area of cognitive ability.

Future Directions

Future investigations of the correlates of humor use might consider the inclusion of alternative sources of information and the exploration of interpersonal humor in other contexts. Although parents and children themselves are generally thought of as expert reporters of children's characteristics and behavior, the inclusion of teachers and peer reports may be fruitful given that these extrafamilial members have experiences with children in alternative settings (i.e., school, extracurricular activities, etc.). Teachers and peers may have a different perspective not only on children's use of humor, but also on children's temperamental characteristics and well-being. As was evidenced with teachers' reports of children's social competence, information regarding the characteristics of children (e.g., temperament, psychological and physical health, achievement, etc.) collected from extrafamilial sources may serve as the most powerful predictors of children's use of familial humor. Furthermore, we must consider the possibility that the results presented here may be limited to the use of humor in the family context. Perhaps in arenas such as the peer group, where relationships are less likely to be constrained by power differentials, children's tendencies to consistently use or elicit humor may vary greatly from those manifested within the family.

The inclusion of data from multiple raters is beneficial for yet another reason. In studies like the current one, where self-reports are used to estimate actor effects, these effects can be attributed to nothing more than rater bias. Thus, any correlation between actor effects and a self-report measure of well-being or temperament could be interpreted as being due to shared method variance. When child actor effects for humor are correlated with parental, teacher, or peer-rated correlates, however, the proposition that actor effects are merely rater bias is less tenable.

Future studies might also include the examination of dyadic relationship effects. Because the estimates of dyadic relationship effects were confounded with measurement error in the current study, we were unable to examine associations between these effects and children's temperamental traits and well-being. Nevertheless, it is possible that the measures of children's temperament or indices of well-being are more strongly associated with children's tendencies to make unique adjustments to specific family members in terms of humor. For example, although emotionality was unrelated to younger siblings' consistent use or elicitation of humor (i.e., actor and partner effects), it is possible that a child's emotionality is related to factors unique to the parent-child relationship. That is, temperamental correlates of humor may be situation or dyad specific rather than traitlike.

Future investigations would also be enhanced by including longitudinal data on humor use. Such data are necessary if we are to disentangle the causal direction between better psychological health and the use of humor. It is tempting to conclude, based on our correlational data, that the use of humor results in less depression, higher self-esteem, and greater social competence. We can easily envision how a child who uses humor more frequently would feel better about him or herself and be perceived as being more socially competent. However, we must consider the possibility that children who feel better about themselves and are already socially competent feel more confident in their use of humor and thus engage in these behaviors more frequently.

In conclusion, this study suggests that children's psychological well-being, but not temperamental characteristics, is associated with interpersonal humor use. Furthermore, it would appear that the correlates of affiliative and aggressive humor differ, thereby bolstering the idea that these types of humor have distinct social functions. Thus, the next step in the quest to understand the origins of actor and partner effects associated with humor is to continue to investigate possible correlates of these effects. As mentioned previously, perhaps the investigation of cognitive skills and alternative temperamental characteristics will prove more fruitful than the temperament constructs investigated herein. In addition, we can begin to incorporate measures of well-being into multivariate genetic analyses (i.e., a bivariate genetic extension of the SRM) in order to decompose the phenotypic associations found in the present study between well-being and actor effects. Ultimately, it is our hope to determine whether the genetic and environmental contributions detected previously are specific to interpersonal humor, or mediated by other traits and behaviors.

REFERENCES

Achenbach, T. M., & Edelbrock, C. (1983). *Manual for the Child Behavior Checklist and Revised Child Behavior Profile.* Burlington: University of Vermont, Department of Psychology.

Asher, S. R., Hymel, S., & Renshaw, P. D. (1984). Loneliness in children. *Child Development, 55,* 1456–1464.

Dunn, J., Stocker, C., & Plomin, R. (1990). Nonshared experiences within the family: Correlates of behavioral problems in middle childhood. *Development and Psychopathology, 2,* 113–126.

Harter, S. (1983). Developmental perspectives on the self-system. In M. Hetherington (Ed.), P. H. Mussen (Series Ed.), *Handbook of child psychology: Socialization, personality, and social development* (Vol. 4, pp. 275–385). New York: John Wiley.

Jöreskog, K. G., & Sörbom, D. (1994). *LISEREL VIII: User's guide* (5th ed.). Morresville, IN: Scientific Software.

Kandel, D., & Davies, J. (1982). Epidemiology of depressive mood in adolescents. *Archives of General Psychiatry, 39,* 1205–1212.

Kashy, D. A., & Kenny, D. A. (1990). Analysis of family research designs: A model of interdependence. *Communication Research, 17,* 462–482.

Kenny, D. A., & La Voie, L. (1984). The social relations model. In L. Berkowitz (Ed.), *Advances in experimental social psychology* (pp. 141–182). Orlando, FL: Academic Press.

Kuiper, N. A., & Martin, R. A. (1993). Humor and self concept. *Humor, 6,* 251–271.

Kuiper, N. A. & Martin, R. A. (1998). Is sense of humor a positive personality characteristic? In W. Ruch (Ed.), *The sense of humor: Explorations of personality characteristic* (pp. 159–178). New York: Mouton de Gruyter.

Kuiper, N. A., Martin, R. A., & Olinger, L. J. (1993). Coping, humor, stress and cognitive appraisals. *Canadian Journal of Behavioral Science, 25,* 81–96.

Lanning, K. (1994). Dimensionality of observer ratings on the California Adult Q-sort. *Journal of Personality and Social Psychology, 67,* 151–160.

Loehlin, J. C. (1992). *Genes and environment in personality development.* Thousand Oaks, CA: Sage.

Manke, B. (1998). Genetic and environmental contributions to children's interpersonal humor. In W. Ruch (Ed.), *The sense of humor: Explorations of a personality characteristic* (pp. 361–384). New York: Mouton de Gruyter.

Manke, B., Pike, A., Hobson, R., & Plomin, R. (2002). *Humor in adolescent family relations: A Study of individual and dyadic relationship effects.* Manuscript submitted for publication.

Martin, R. A., & Lefcourt, H. M. (1983). Sense of humor as a moderator of the relation between stressors and moods. *Journal of Personality and Social Psychology, 45,* 1313–1324.

Masten, A. S. (1986). Humor and competence in school-aged children. *Child Development, 57,* 461–473.

McGue, M., & Bouchard, T. J. (1984). Adjustment of twin data for the effects of age and sex. *Behavior Genetics 14,* 325–343.

Nezu, A. M., Nezu, C. M., & Blissett, S. E. (1988). Sense of humor as a moderator of the relation between stressful events and psychological distress:

A Prospective analysis. *Journal of Personality and Social Psychology, 54,* 520–525.

Plomin, R., & DeFries, J. C. (1985). *Origins of individual differences in infancy.* New York: Academic Press.

Rowe, D., & Plomin, R. (1977). Temperament in early childhood. *Journal of Personality Assessment, 41,* 150–156.

Ruch, W., & Kohler, G. (1998). A temperament approach to humor. In W. Ruch (Ed.), *The sense of humor: Explorations of a personality characteristic* (pp. 203–230). New York: Mouton de Gruyter.

Stocker, C., Dunn, J., & Plomin, R. (1989). Sibling relationships: Links with child temperament, maternal behavior, and family structure. *Child Development, 60,* 715–727.

Vaillant, G. E., & Vaillant, C. O. (1992). Empirical evidence that defensive styles are independent of environmental influence. In G. E. Vaillant (Ed.), *Ego mechanisms of defense: A guide for clinicians and researchers* (pp. 105–126). Washington, DC: American Psychiatric Press.

Walker, H. M., & McConnell, S. W. (1988). *Walker-McConnell Scale of Social Competence and School Adjustment.* Austin, TX: Pro-Ed.

MICHELLE WARD
LAURA A. BAKER

14

Effects of Prenatal Smoking on 12-Year-Old Children's Cognitive and Behavioral Outcomes

Introduction

A relationship between cigarette smoke and aggression has been revealed in several studies. Findings suggest that people who have been exposed to cigarette smoke exhibit an increase in aggressive behavior. One such study found that nonsmokers who were exposed to second-hand smoke were reliably more aggressive in a laboratory situation than their counterparts who were exposed to clean air (Jones & Bogat, 1978). This study suggests a causal effect of cigarette smoking on observed aggression. These findings are consistent with similar studies on animals, which have also suggested a connection between cigarette smoke and increased motor activity (Richardson & Tizabi, 1994) along with decreased learning efficiency (Bertolini, Bernardi, & Genedani, 1982; Johns, Louis, Becker, & Means, 1982).

Recently, the focus of research in this area has been on the specific effects of prenatal cigarette smoking on future aggression in the children. Several studies have shown evidence of increased antisocial behavior in offspring whose mothers smoked during pregnancy. There is evidence that smoking during pregnancy significantly increases the likelihood of psychiatric symptoms, such as conduct disorder, in male children (Fergusson, Woodward, & Horwood, 1998; Wakschlag et al., 1997; Weissman, Warner, Wickramaratne, & Kandel, 1999). Results for these studies have indicated that mothers who smoked more than half a pack of cigarettes daily were more than four times as likely to have a child that met *DSM-III-R* criteria for conduct disorder. This increased risk was found even when controlling for socioeconomic status (SES), demographic factors, parental psychopathology, pregnancy complications, and other parenting variables. Additionally, prenatal exposure to the constituents of cigarette smoke was associated with attention problems and motor hyperactivity (including diagnosis of attention deficit hyperactivity disorder, ADHD), and impulsivity.

These findings are not limited to problems in childhood. Brennan, Grekin, & Mednick (1999) found similar effects in the adult offspring of mothers who smoked during pregnancy. Maternal smoking significantly predicted violent and nonviolent arrests and life-course persistent offending of males measured at 34 years of age. Similarly, Raesaenen et al. (1999) found that sons of mothers who smoked had a twofold risk of conviction for violent crimes and/or recidivism. These studies on the relationship between prenatal smoking and negative outcome in the offspring have evidenced a dose-response relationship. That is, increasing rates of symptoms were seen as maternal smoking increased.

These studies may indicate a causal effect of prenatal smoking on later aggression. However, it may not be this simplistic. The actual underlying genetic and environmental etiologies of the relationship between smoking and aggression/antisocial behavior are presently unknown. Is this relationship derived solely from prenatal smoking? Perhaps there are additional characteristics of mothers who smoke, such as lower IQ, lack of education, and antisocial tendencies, which impact the cognitive and behavioral outcomes of their children. The nature of these influences could be environmental, genetic, or both. For example, a characteristic of a mother who smokes during pregnancy, such as antisocial predispositions or low IQ, can be transmitted to a child genetically. Cigarette smoking itself may be an indicator of other genetic predispositions, which are passed on to the child. Moreover, there may be additional environmental factors that could explain the relationship between maternal smoking and children's negative outcomes. For example, a mother's smoking behavior during pregnancy may be related to other postnatal factors, such as continued smoking during critical developmental periods of childhood or other negative maternal behaviors that have an environmental impact on children's outcomes.

The effect of prenatal smoking on cognitive and behavioral outcomes in children is the focus of this chapter. These effects will be examined in both adopted and nonadopted (control) children, to evaluate the extent to which smoking during pregnancy may affect general and specific cognitive abilities, personality characteristics, and childhood behavioral problems at 12 years of age. A comparison of effects between adopted children (separated from their birth mothers shortly after birth) and control children (raised by their birth mothers) will be made, along with an investigation of characteristics of mothers who did and did not smoke during pregnancy. We also examined the extent to which maternal characteristics (such as education and cognitive ability) may explain any adverse child outcomes in the smoking groups.

Method

Subjects

The subjects in these analyses are mothers and children who participated in the Colorado Adoption Project (CAP), a longitudinal study investigating adopted and nonadopted ("control") children and their families. In this ongoing project, the children are assessed annually on measures of health, development, cognitive ability, personality, and behavior. Similar measures are assessed in the adoptive and biological parents of the adopted children, as well as the parents of the control children. Parents were assessed either prior to the birth of the child or shortly thereafter, and the adoptive and control parents continue to be tested intermittently throughout the upbringing of the children. Detailed descriptions of the CAP design, sample, and measures have been previously reported (DeFries, Plomin, & Fulker, 1994; Plomin & DeFries, 1985).

The sample for the present study includes a subset of mother-child dyads for which smoking information was available. Dyads were formed using both probands and siblings, since each pregnancy was considered as the unit of analysis (see further discussion at the end of this section below). Analyses are based on a total of 258 adopted children and their biological mothers who gave them up for adoption shortly after birth, and 352 nonadopted ("control") children and their mothers. Cognitive and behavioral measures for the children at age 12 are evaluated, along with birth information (weight and length).

The children in this study include both boys and girls, unlike several previous studies that have focused on boys only. Sex differences were not explored in these analyses, however, due to the small Ns resulting in further breakdown by child's sex after dividing the sample by child's adopted status and mother's smoking behavior.

Procedures

Maternal prenatal smoking was assessed in two ways. One source was birth history information that was gathered from either adoption agencies, for the biological mothers, or through hospitals, for the control mothers. Information about the actual number of cigarettes a woman smoked while pregnant was obtained from the birth history questionnaire, but was available for more control mothers than biological mothers. Additional information was gathered for the control mothers through a mailed questionnaire.

The second source of smoking information was obtained as part of the initial assessment of the biological and control mothers, during a visit to the research laboratory or adoption agency. The mothers completed a three-hour battery of tests and questionnaires either prior to the birth of the child (for most of the biological mothers) or within several months following the birth of the child (for most of the control mothers). In response to a series of questions about their use of alcohol, tobacco, and other drugs, the mothers reported current smoking habits. For those who were tested before the birth of their baby, this information was also evaluated for smoking group classification.

Cognitive and behavioral variables for the mothers were also assessed in the battery of tests described above. For the children, cognitive ability and behavior are investigated throughout childhood and adolescence, though the present study evaluates only measures from a laboratory assessment at 12 years of age.

Measures

Smoking group classification was based primarily on three questions. First, the specific number of cigarettes that a woman smoked daily during her pregnancy was recorded from the birth history questionnaire. The response to this question was open-ended and could be any number, including zero. The second two questions were asked during the laboratory visit, including (1) whether or not the mother was currently a smoker (defined as smoking more than one cigarette every couple of weeks), and (2) how many packs per day she was currently smoking at the time of assessment. This last question was asked for any mother who referred to herself as currently being a smoker. The responses to the number of packs smoked per day could only be whole numbers, and effectively reflected an upper limit on the number of cigarettes smoked. Thus, the response "one pack per day" included mothers who smoked more than one cigarette every two weeks but not more than one pack per day. Similarly, a response of "two packs per day" indicated the mother smoked more than one pack but not more than two packs per day.

The "nonsmoking" group in these analyses included mothers who either smoked zero cigarettes per day during pregnancy (from the birth history questionnaire), and/or responded that she was a nonsmoker at the time of the laboratory visit. The "smoking" group included mothers who smoked at least one cigarette per day (according to the question from the birth history questionnaire), and/or who identified herself as a smoker during a laboratory visit while she was pregnant. These groups are described in greater detail below.

General cognitive ability for the mothers was assessed using the first principal component from a psychometric battery of cognitive ability tests. Four specific cognitive abilities in the mothers were also examined: verbal reasoning, spatial ability, perceptual speed, and memory (see DeFries et al., 1994, for details about the cognitive ability measures).

Behavioral problems were assessed for the mothers through the EASI self-report. This measure assessed emotionality (fear related), emotionality (anger related) activity, sociability, and impulsivity. Information on additional problems was gathered through self-report of the frequencies of several listed items; included for use in this study were depression, hysteria, and sociopathy. Participants were presented with a number of behaviors or symptoms that are related to the problems of interest (i.e., hysteria), and assessments were made based on the number of items that were endorsed for each category. Items used for hysteria and sociopathy categories were modified from the Iowa 500 project (Plomin & DeFries, 1985).

For the children, cognitive ability was assessed from Weschler Intelligence Scale for Children (WISC), as well as from a psychometric battery of specific cognitive abilities. Both general and specific cognitive abilities (verbal ability, spatial ability, perceptual speed, and memory) were examined in the present analyses.

Behavioral outcomes for the children in the present study were based on the parent-rated version of the CBCL (Achenbach & Edelbrock, 1983). This measure consists of 118 items that are rated on a three-point scale. The CBCL yields two second-order factors, internalizing and externalizing problem behavior.

Analyses

Classification of Smoking and Nonsmoking Groups   Those mothers with sufficient information about their smoking behavior during pregnancy were classified into two groups: those who did and did not report smoking during pregnancy. The majority of mothers (80%, or $n = 486$ out of the total sample of 610 mothers) in the present analyses were classified as either smokers or nonsmokers based on one of the questions from the laboratory assessment as described above.

The first group, the nonsmokers, comprised women who called themselves nonsmokers and did not report smoking during her pregnancy. A total of 361 women, or 59% of those for whom sufficient information was available, were classified as nonsmokers. Of these women, 116 were biological mothers and 245 were control mothers, rendering

45% of the biological mothers and 70% of the control mothers nonsmokers.

The second group, the smokers, includes every mother who reported smoking during her pregnancy. These mothers either gave a number (greater than zero) for how many cigarettes she smoked per day during her pregnancy (in the birth history questionnaire), or reported currently smoking one or more packs per day (during the laboratory visit) and were tested while pregnant. It should be noted that for most of the biological mothers, information was not available about the number of cigarettes smoked per day while pregnant. These mothers were classified as smokers based on their responses to the second and third questions described above; they reported currently smoking a pack or more per day and were tested while pregnant. A total of 125 of the women, or 21%, were classified into this smoking category. Of these smokers, 68 were biological mothers and 57 were control mothers—constituting 26% and 16% of the biological and control mothers, respectively.

The other mothers for whom smoking information was available fell into categories not used in the present analyses. Several mothers (14 biological [5%] and 48 control [14%]) reported to be ex-smokers who quit more than one year ago, indicating that they stopped smoking before their pregnancy. Other mothers (32 biological [12%] and 0 control) called themselves smokers but were tested after the birth of their baby; smoking may have begun before, during, or after their pregnancy. Lastly, some mothers (28 biological [11%] and 2 control [.5%]) reported to be ex-smokers and quit smoking less than one year before the birth of their baby. It was not clear whether or not these mothers smoked at all during their pregnancy.

Group Comparisons   We used 2×2 MANOVAs to test for effects of SMOKING (i.e., differences between children whose mothers did and did not smoke during pregnancy), ACGROUP (i.e., differences between adopted and control children), and the interaction between SMOKING and ACGROUP (i.e., the extent to which SMOKING effects may differ between adoptees and controls). In light of the many variables being examined in these analyses, we used the omnibus test approach to help control for Type I errors (see Cliff, 1987). That is, multivariate F values were first examined for significance in a group of variables, followed by inspection of univariate F values for individual variables. Generally, univariate F values were only considered if multivariate F values were significant ($p \leq 0.05$) or marginally significant ($p \leq 0.10$). In the case of marginally significant multivariate F values, the significance of individual variables should be treated with greater caution. The assumption in MANOVA of homogeneity of variance-covariance matrices across groups was tested in each analysis using Box's M. The assumption was met in all instances except for CBCL subscales, which showed considerable skewness for all groups. In this case, a square root transformation of each subscale (externalizing and internalizing problems) resulted in a nonsignificant Box's M. Thus, the transformed CBCL scales were used in MANOVAs for these variables, although raw means and variances are presented in tables 14.1 and 14.2 below.

The SMOKING × ACGROUP interactions were of particular interest in these analyses, as they provide an evaluation of the extent to which SMOKING effects may differ between adopted and control children. Comparisons of SMOKING effects in children raised by their biological mothers to effects in children separated from their mothers at birth allowed for the investigation of the maternal environmental influence beyond the prenatal experience. Effects of SMOKING, ACGROUP, and their interaction were examined separately for birth information, behavioral outcomes, and cognitive outcomes.

Similar MANOVAs were computed for behavioral and cognitive variables in the mothers themselves who did and did not smoke during pregnancy. The extent to which SMOKING effects in child outcomes may have been due to differences in characteristics of mothers who did and not smoke during pregnancy were then examined using analyses of covariance (ANCOVAs), with relevant maternal characteristics (i.e., those that differed for smokers and nonsmokers) as covariates.

Additional analyses were performed comparing the nonsmokers to a subgroup of heavy smokers (10 or more cigarettes per day). Women classified as heavy smokers were either those who were pregnant while reporting that she currently smoked two or more packs per day, or those who reported that during her pregnancy she smoked 10 or more cigarettes per day. The use of 10 cigarettes for the threshold for heavy smoking classification is not arbitrary; previous literature suggests that the behavioral problems associated with prenatal smoke are most pronounced in groups of mothers who smoked 10 or more cigarettes per day (Wakschlag et al., 1997). There are 67 women who fell into this category—24 were biological mothers (17% of the biological mothers), and 43 were control mothers (15% of the control mothers). The specific variable used for classification of heavy smokers was missing for more mothers (particularly biological mothers) than the variables for the broader smoking group. This, plus the exclusion of mothers who smoked lesser amounts, led to smaller Ns for the heavy smoking group. It should be noted that we did not run analyses comparing heavy and light smokers to one another. Although this would be an interesting opportunity to examine dose-response effects, the resulting small Ns in both groups of smokers would obviously compromise power in such analyses. More important, there was considerable heterogeneity among mothers who were not classified as heavy smokers here, and thus we could not formulate a group of "light smokers" with much certainty in

TABLE 14.1. Birth Information, Behavioral Problems, and Cognitive Outcomes for Adopted Children: Means and Standard Deviations by Mother's Smoking Status during Pregnancy

Child Variable	Nonsmokers			Smokers			Heavy Smokers		
	M	SD	N	M	SD	N	M	SD	N
Birth Information									
Birth weight (pounds)	7.191	1.019	112	6.970	.996	65	6.856	1.004	23
Birth length (inches)	19.750	.940	112	19.650	1.010	65	19.530	1.195	23
Behavioral Problems									
CBCL internalizing	2.011	1.194	84	2.229	1.261	51	2.286	1.367	20
CBCL externalizing scale	2.506	1.369	84	2.664	1.537	51	2.754	1.438	20
WISC IQ									
Verbal IQ	107.70	9.78	90	107.20	12.22	57	105.10	13.24	22
Performance IQ	112.89	12.12	90	109.93	10.89	57	109.36	10.98	22
Specific Cognitive Abilities									
Verbal	.0426	.7446	91	-.0095	.8938	55	-.1720	.9793	22
Spatial	.0289	.8487	91	-.1256	.8634	55	-.1690	.7848	22
Speed	-.0094	.8558	91	-.0638	.8673	55	-.1533	.8537	22
Memory	.1045	.7461	91	.0973	.7992	55	-.0306	.8329	22

TABLE 14.2. Birth Information, Behavioral Problems, and Cognitive Outcomes for Control Children: Means and Standard Deviations by Mother's Smoking Status during Pregnancy

	Nonsmokers			Smokers			Heavy Smokers		
	M	SD	N	M	SD	N	M	SD	N
Birth information									
Birth weight (pounds)	7.482	1.063	217	6.809	1.153	54	6.974	1.006	41
Birth length (inches)	20.220	1.072	217	19.620	1.274	54	19.770	1.073	41
Behavioral Problems									
CBCL internalizing	2.055	1.088	201	2.312	1.418	42	2.216	1.479	33
CBCL externalizing scale	2.141	1.235	201	2.695	1.210	42	2.683	1.316	33
WISC IQ									
Verbal IQ	111.38	11.05	226	109.12	10.58	52	109.90	10.19	40
Performance IQ	113.00	11.79	226	110.65	12.94	52	110.85	11.96	40
Specific Cognitive Abilities									
Verbal	.1165	.8654	226	.0026	.8261	52	.0645	.8821	40
Spatial	.1639	.8448	226	-.1679	.8289	52	-.1266	.8147	40
Speed	.1098	.8599	226	-.3445	.8581	52	-.2727	.7714	40
Memory	.0627	.7172	226	-.0718	.6395	52	-.1581	.6291	40

these data. Thus, we restrict our analyses here to the two separate comparisons of: (1) nonsmokers to all smokers, and (2) nonsmokers to only those who could be classified as heavy smokers from the available data.

The present analyses include both probands and siblings in the CAP, and therefore, some mothers are considered more than once across mother-child dyads. Given that smoking information was obtained separately for each pregnancy, however, we retained both siblings and probands in these analyses (i.e., treating one pregnancy as the unit of analysis) in order to maximize the available Ns in comparing children who did and did not experience the constituents of smoke. Additional analyses were run excluding the siblings, so that each mother was considered only once in the sample. The mean differences between smoking and nonsmoking groups remained of comparable direction and magnitude compared to the analyses of the larger sample including siblings. Given the similar patterns of results when including and excluding siblings, we present here the findings from the larger analyses.

Results

Effects of Prenatal Smoking on Child Outcomes

Means and standard deviations for children's birth information, cognitive outcomes, and behavioral outcomes are presented in table 14.1 (for adopted children) table 14.2 (for nonadopted, "control" children), and separately for children of smokers and nonsmokers. Further breakdown for children of the heavy smokers is also provided in tables 14.1 and 14.2.

Separate MANOVAs were computed for each grouping of variables in the children, in order to maximize use of available data—see table 14.3. (Performing one MANOVA on all variables together would reduce the Ns substantially since only cases with complete data for all variables would be included.) Analyses that included all smokers showed significant or marginally significant SMOKING effects for birth info (mult $F = 7.650, df = 1, 443, p < 0.01$), specific cognitive abilities (mult $F = 2.221$, $df = 1, 417, p < 0.07$), and behavioral problems (mult $F = 2.521, df = 1, 373$, $p < 0.08$). More specifically, children of smokers were significantly smaller in both birth weight (univ $F = 15.14, df = 1, 443, p < 0.01$) and length (univ $F = 9.36, df = 1, 443, p < 0.01$), had significantly lower spatial ability (univ $F = 6.20, df = 1, 417, p < 0.01$), lower perceptual speed (univ $F = 7.13, df = 1, 417, p < 0.01$), significantly greater externalizing behavioral problems (univ $F = 4.916, df = 1, 373, p < 0.05$). It is also noteworthy that there was a marginally significant increase in internalizing behavioral problems (univ $F = 2.72, df = 1, 373, p < .10$) among children whose mothers smoked during pregnancy. No general differences in IQ

TABLE 14.3. ACGROUP (Adopted vs. Control) X SMOKING (Nonsmokers vs. All Smokers) MANOVAs for Children's Cognitive and Behavioral Outcomes: Multivariate and Univariate F Statistics and p-values

Variable	SMOKING		ACGROUP		ACGROUP X SMOKING	
	Multivariate F	Univariate F	Multivariate F	Univariate F	Multivariate F	Univariate F
Birth Information	7.65 (.01)		2.49 (.08)		2.53 (.08)	
Weight		15.14 (.01)		.32 (.57)		3.88 (.05)
Length		9.36 (.01)		3.72 (.05)		4.63 (.03)
WISC IQ	1.94 (.14)		2.79 (.06)		.44 (.64)	
Verbal IQ		1.18 (.28)		5.02 (.03)		.53 (.47)
Performance IQ		3.84 (.05)		.10 (.75)		.05 (.83)
Specific Cognitive Abilities	2.22 (.07)		.86 (.49)		.94 (.44)	
Verbal		.74 (.39)		.20 (.66)		.10 (.75)
Space		6.20 (.01)		.23 (.63)		.85 (.36)
Speed		7.13 (.01)		.83 (.36)		3.72 (.05)
Memory		.72 (.40)		1.60 (.21)		.58 (.59)
Behavioral Problems	2.52 (.08)		1.46 (.23)		1.06 (.35)	
CBCL externalizing		4.92 (.03)		1.09 (.30)		1.53 (.22)
CBCL internalizing		2.72 (.10)		.19 (.66)		.02 (.89)

[a]Square root transformation was used in analyses of CBCL because of skewness in both externalizing and internalizing scales

emerged, although there were some marginally significant differences between adoptees and controls overall (mult $F = 2.794$, $df = 1$, 420, $p < 0.06$), with control children scoring significantly higher than adopted children on verbal IQ (univ $F = 5.018$, $df = 1$, 420, $p < 0.03$).

Inspection of the means (tables 14.1 and 14.2) shows the SMOKING effects for birth information, IQ, and behavioral problems appear in both adoptees and controls. Although there is some indication in the means that the differences may be greater in the control children, the ACGROUP × SMOKING interaction was not significant in any analysis, and was only marginally significant for birth information (mult $F = 2.528$, $df = 1$, 443, $p < 0.08$). Thus, while SMOKING effects on birth weight and length may be larger for control children, the behavioral and cognitive differences between children of smokers and nonsmokers are comparable in both adoptees and controls.

Analyses that compared only heavy smokers to nonsmokers showed many similar results for SMOKING effects (see table 14.4). As in analyses of all smokers, there are significant HEAVY SMOKING effects in birth characteristics (mult $F = 4.093$, $df = 2$, 390, $p < 0.02$). However, the marginally significant multivariate F-values for specific cognitive abilities and behavioral problems found in analysis of all smokers disappears in the analysis of heavy smokers. It might be noted with caution that the univariate F-values for HEAVY SMOKING effects on space (univ $F = 3.91$, $df = 1$, 375, $p \le 0.05$) and speed (univ $F = 4.79$, $df = 1$, 375, $p \le 0.05$) abilities are still significant, as is the value for Externalizing behavior problems (univ $F = 3.87$, $df = 1$, 334, $p \le 0.05$). These effects may be attenuated in analyses of heaving smoking due to the considerable reduction in sample size. It may also be the case that some of the smokers may have actually been heavy smokers (>10 cigarettes per day) but could not be classified as such due to the nature of the data available in this study. As noted earlier in this chapter, information was not available for all mothers about the number of cigarettes smoked per day while pregnant. Thus, the subgroup that could be classified as heavy smokers in these analyses may not in actuality be significantly different from those who could not be classified as such. Inspection of the means in tables 14.1 and 14.2, in fact, indicates no appreciable differences in these child outcome variables between heavy smokers and the total group of smokers. Almost certainly it does *not* appear that SMOKING effects are extremely greater in the present analyses of heavy smokers only. Given our inability to formulate a large enough group of "light smokers" with any certainty, however, it is impossible to evaluate any dose-response relationship in this study, as noted earlier.

Behavioral and Cognitive Variables in the Mothers

Means and standard deviations for mother's cognitive and behavioral variables are presented in table 14.5 (Biological Mothers) and table 14.6

TABLE 14.4. ACGROUP (Adopted *vs.* Control) X HEAVY SMOKING (Nonsmokers *vs.* Heavy Smokers) MANOVAs for Children's Cognitive and Behavioral Outcomes: Multivariate and Univariate F Statistics and *p*-values

Variable	HEAVY SMOKING		ACGROUP		ACGROUP X HEAVY SMOKING	
	Multivariate F	Univariate F	Multivariate F	Univariate F	Multivariate F	Univariate F
Birth Information	4.093 (.017)		2.854 (.059)		.324 (.723)	
Weight		8.046 (.005)		1.894 (.170)		.338 (.561)
Length		5.123 (.024)		5.582 (.019)		.649 (.421)
WISC IQ	1.525 (.219)		3.870 (.022)		.092 (.912)	
Verbal IQ		1.607 (.206)		7.127 (.008)		.113 (.737)
Performance IQ		2.693 (.102)		.218 (.641)		.152 (.697)
Specific Cognitive Abilities	1.784 (.131)		.692 (.598)		.392 (.814)	
Verbal		1.137 (.287)		1.584 (.209)		.440 (.508)
Space		3.909 (.049)		.532 (.466)		.148 (.701)
Speed		4.793 (.029)		.006 (.939)		.779 (.378)
Memory		2.823 (.094)		.638 (.425)		.164 (.686)
Behavioral Problems	1.929 (.147)		.835 (.435)		.691 (.502)	
CBCL externalizing		3.869 (.050)		1.179 (.278)		.537 (.464)
CBCL internalizing		1.420 (.234)		.005 (.942)		.097 (.756)

[a]Square root transformation was used in analyses of CBCL because of skewness in both externalizing and internalizing scales

TABLE 14.5. Cognitive and Behavioral Measures for Biological Mothers: Means and Standard Deviations by Smoking Status during Pregnancy

Mother Variables	Nonsmokers			All Smokers		
	M	SD	N	M	SD	N
Cognitive						
General cognitive ability	.29	.99	100	−.08	.88	60
Education level	12.30	1.92	100	11.82	1.70	60
Behavioral						
Impulsivity	12.98	2.79	102	13.29	3.40	62
Emotionality-fear	13.79	3.73	102	13.94	3.44	62
Emotionality-anger	12.20	3.97	102	11.92	3.74	62

TABLE 14.6. Cognitive and Behavioral Measures for Control Mothers: Means and Standard Deviations by Smoking Status during Pregnancy

Mother Variables	Nonsmokers			All Smokers		
	M	SD	N	M	SD	N
Cognitive						
General cognitive ability	.08	.83	144	−.05	.88	38
Education level	15.24	1.96	144	14.58	2.02	38
Behavioral						
Impulsivity	11.97	2.98	146	13.16	2.36	38
Emotionality-fear	12.85	3.45	146	13.18	3.47	38
Emotionality-anger	11.67	3.75	146	12.79	2.85	38

(Control Mothers), separately for smokers and nonsmokers. SMOKING × ACGROUP MANOVAs were also conducted (see table 14.7) on these maternal characteristics, to evaluate the extent to which differences may exist between biological and control mothers themselves who did and did not smoke during their pregnancy. Significant SMOKING group differences were found for the cognitive variables in these mothers (mult $F = 4.25$, $df = 1, 337$, $p < 0.02$). Specifically, mothers who smoked during pregnancy had significantly lower general cognitive ability (univ $F = 5.97$, $df = 1, 337$, $p < 0.02$) and lower educational attainment (univ $F = 5.10$, $df = 1, 337$, $p < 0.03$) compared to nonsmokers. Moreover, these effects appear comparable for biological and control mothers, given the nonsignificant SMOKING × ACGROUP interaction. It is also noteworthy that control mothers had significantly higher education (univ $F = 147.60$, $df = 1, 337$, $p < 0.01$) compared to biological mothers, a finding which has been discussed elsewhere (Plomin & DeFries, 1985).

TABLE 14.7 ACGROUP (Adopted vs. Control) X SMOKING (Nonsmokers vs. All Smokers) MANOVAs for Mother's Cognitive and Behavioral Measures: Multivariate and Univariate F statistics and p-values

Mother Variable	SMOKING		ACGROUP		ACGROUP X SMOKING	
	Multi-Variate F	Uni-variate F	Multi-Variate F	Uni-variate F	Multi-Variate F	Uni-variate F
Cognitive	4.25 (.02)		84.76 (.01)		.89 (.41)	
General cognitive ability		5.10 (.03)		.72 (.40)		1.22 (.27)
Education level		5.97 (.02)		147.60 (.01)		.15 (.70)
Behavioral	1.48 (.22)		2.50 (.06)		1.07 (.36)	
Impulsivity		4.35 (.04)		2.53 (.11)		1.49 (.22)
Emotionality-fear		.31 (.58)		3.89 (.05)		.05 (.82)
Emotionality-anger		.86 (.36)		.14 (.70)		2.36 (.13)

It is particularly important to consider the implications of the SMOKING group differences in maternal characteristics on the child outcome variables being considered in this chapter. Smoking group differences are clearly evident in these analyses, such that children whose mothers smoked during pregnancy appear to exhibit lower birth weight and length, increased behavioral problems, and lower cognitive performance at age 12. However, given the differences in maternal cognitive variables and their known relationship to children's cognitive abilities, it is conceivable that the observed differences in children's cognitive outcome may be due to inherited factors that differ for smoking and nonsmoking mothers. In order to evaluate the extent to which lower education and cognitive ability in the smoking mothers may explain cognitive deficits in their children, an additional set of analyses was performed. Analyses of covariance were performed to test for smoking group differences in the children's outcome variables, after adjusting for group differences in maternal general cognitive ability. ANCOVAs were not computed for behavioral problems or birth information, since none of these variables were significantly correlated with either maternal education or general cognitive ability.

Significant effects of prenatal smoking remained for children's cognitive abilities, even after adjusting for maternal differences in IQ. The multivariate F statistic was significant for children's specific cognitive abilities (mult $F = 1.318$, $df = 1$, 298, $p < 0.02$) and IQ variables (mult $F = 4.175$, $df = 1$, 301, $p < 0.02$). Although mother's cognitive ability is lower for those whom smoked during pregnancy, the effect of prenatal smoking remains significant after controlling for these maternal ability differences.

Summary and Discussion

Using a sample of adopted and nonadopted children and their birth mothers, this study compared the cognitive and behavioral outcomes in children of mothers who smoked during pregnancy and those who did not. The results from the present study indicate that there are birth weight, birth length, and cognitive and behavioral differences between the offspring of the smoking groups. The children of the mothers who smoked during pregnancy were smaller in birth weight and length, and they scored higher on measures of behavioral problems and lower on measures of IQ and cognitive performance at age 12. Analyses performed using a subset of women who could be classified as heavy smokers produced similar results.

It is important to recognize an important difference between adopted and control children in these analyses. In a study using an adoption design, there are additional postnatal environmental influences that remain for the children of the control mothers and not for the adopted-away children of the biological mothers. Though none of the children will escape the genetic influences of his or her mother, the children of the biological mothers have no environmental influence from their biological mothers beyond the womb. The children of the control mothers, on the other hand, continue to be exposed to the characteristics of their birth mother, including her smoking behavior after childbirth. Thus, the children of the control mothers may not have been exposed to cigarette smoke *in utero* alone, but may have been exposed to cigarette smoke from conception until age 12 when they were tested. It would seem quite possible, therefore, that control children might experience more of the problems associated with smoke exposure compared to adopted children whose biological mothers smoked during pregnancy. This was not the case, however, in our analyses—there were no significant interactions between adoptive status and mother's smoking behavior during pregnancy, suggesting the prenatal smoking effects are comparable in adopted and control children. Future studies could explore this further by including smoking information from all mothers (including adoptive) throughout the upbringing of the child.

Analyses investigating cognitive and behavioral characteristics of mothers themselves revealed that mothers who smoked had significantly lower general cognitive ability and educational attainment than mothers who did not smoke. Though this may suggest that inherited factors could be influencing the differences in cognitive ability in the children, the present analyses indicated that the effect of prenatal smoking remains significant, independent of mother's cognitive ability. Clearly, prenatal smoking is associated with adverse outcomes in children, beginning at birth and continuing into adolescence.

REFERENCES

Achenbach, T. M., & Edelbrock, C. (1983). *Manual of the Child Behavior Checklist and the Revised Child Behavior Profile*. Burlington: Department of Psychiatry, University of Vermont.

Bertolini, A., Bernardi, M., & Genedani, S. (1982). Effects of prenatal exposure to cigarette smoke and nicotine on pregnancy, offspring development and avoidance behavior in rats. *Neurobehavioral Toxicology and Teratology, 4*, 545–548.

Brennan, P. A., Grekin, E. R., & Mednick, S. A. (1999). Maternal smoking during pregnancy and adult male criminal outcomes. *Archives of General Psychiatry, 56*, 215–224.

Chang, G., Goetz, M., Wilkins-Haug, L., & Berman, S. (1999). Prenatal alcohol consumption: Self versus collateral report. *Journal of Substance Abuse Treatment, 17*, 85–89.

Cliff, N. (1987). *Analyzing Multivariate Data*. New York: Harcourt Brace, Jovanovich.

DeFries, J. C., Plomin, R., & Fulker, D. W. (1994). *Nature and nurture during middle childhood.* Oxford: Blackwell.

Fergusson, D. M., Woodward, L. J., & Horwood L. J. (1998). Maternal smoking during pregnancy and psychiatric adjustment in late adolescence. *Archives of General Psychiatry, 55,* 721–727.

Johns, J., Louis, T., Becker, R., & Means, L. (1982). Behavioral effects of prenatal exposure to nicotine on guinea pigs. *Neurobehavoral Toxicology and Teratology, 4,* 365–369.

Jones, J. W., & Bogat, G. A. (1978). Air pollution and human aggression. *Psychological Reports, 43,* 721–722.

Plomin, R., & DeFries, J. C. (1985). *Origins of individual differences in infancy.* Orlando, FL: Academic Press.

Raesaenen, P., Hakko, H., Isohanni, M., Hodgins, S., Jarvelin, M., & Tiihonen, J. (1999). Maternal smoking during pregnancy and risk of criminal behavior among adult male offspring in the northern Finland 1966 birth cohort. *American Journal of Psychiatry, 156,* 857–862.

Richardson, S., & Tizabi Y. (1994). Hyperactivity in the offspring of nicotine-treated rats; role of the mesolimbic and nigostriatal dopamine pathways. *Pharmocology Biochemistry Behavior, 47,* 331–337.

Wakschlag, L. S., Lahey, B. L., Loeber, R., Green, S. M., Gordon, R. A., & Leventhal B. L. (1997). Maternal smoking during pregnancy and the risk of conduct disorder in boys. *Archives of General Psychiatry, 54,* 670–676.

Weissman, M., Warner, V., Wickramaratne, P., & Kandel, D. (1999). Maternal smoking during pregnancy and psychopathology in offspring followed to adulthood. *Journal of the American Academy of Child and Adolescent Psychiatry, 38,* 892–899.

A Genetic Analysis of Extremes in Externalizing Behavioral Problems and Negative Family Environments

Introduction

Like most forms of psychopathology, the development of aggressive and delinquent externalizing behavioral problems is thought to be a product of complex gene-environment processes. Behavioral genetics approaches have shown that both environmental and genetic factors work together in causing individual differences in externalizing behavioral problems, as well as individual differences in environments such as parenting behavior and levels of stimulation in the home (Plomin, 1994). These environmental and genetic factors are not independent. Individuals are active participants within their environments, evoking experiences from others and selecting into environments that may further promote or hinder genetically variable characteristics (i.e., gene-environment correlation). At the same time, the expression of the genotype can be conditioned upon the environment, whereby certain genetically variable attributes are heritable only under certain environmental conditions (i.e., gene-environment interaction).

Although these gene-environment processes probably operate throughout the life span, the transition to adolescence is a particularly interesting period of development to be examining gene-environment transactions in developmental psychopathology, in part because of the dramatic increase in children's autonomy that corresponds with this developmental period (Holmbeck, Paikoff, & Brooks-Gunn, 1995). In the case of conduct disorder and other aspects of externalizing behavioral problems, this transition is a time when many of the disordered children (usually boys) select into deviant peer groups and friendships that further promote delinquency (Dishion et al., 1997; Dishion, Spracklen, Andrews, & Patterson, 1996). This increased autonomy is one mechanism contributing to a hypothesized increase in active gene-environment correlation over the course of middle childhood and adolescence (Scarr & McCartney, 1983). As children gain more independence over their daily lives, they have many more opportunities to elicit and select into experiences that are consistent with their own genetically variable characteristics.

Twin and adoption studies are critically important to our understanding of these gene-environment processes underlying the development of psychopathology in childhood and adolescence. The quantitative genetic model provides estimates of variance attributable to genetic factors (heritability), environmental factors that contribute to sibling similarity (shared environment), and environmental factors that contribute to sibling differentiation (nonshared environment). Importantly, these genetic studies are not limited to examinations of child and adolescent "outcomes"—these models can also be used to examine the contribution of child genetically variable attributes on their own environments. Such studies have examined various dimensions of family environments and have found evidence for modest to substantial child genetic contributions, as well as environmental contributions to various aspects of the parenting and home environment (Deater-Deckard, Fulker, & Plomin, 1999). In testing for children's genetic influences on their own environments, this model provides one method of testing for child effects in developmental psychopathology (Lollis & Kuczynski, 1997).

One of the core questions for developmental psychopathology is whether there are genetic and environmental links between normal variation in the unselected population, and extreme variants—what we typically think of as clinically relevant manifestations of disorder. Are the gene-environment processes that are responsible for individual differences in externalizing behaviors the same gene-environment processes that are responsible for clinically defined disorders such as conduct disorder and antisocial personality disorder? Or, are there distinct gene-environment processes for extreme disorders as diagnosed versus subclinical individual differences? Furthermore, given the nonindependence of genetic and environmental factors in development, are the gene-environment processes for normal variations in family environments similar to or different from those responsible for extremely harsh, negative, or abusive family environments?

The first aim of this study was to explore whether the genetic etiology of individual differences in externalizing problems in an "unselected" sample (i.e., variation in the "normal" observed range) is similar to or different from the etiology for children selected as being extreme in aggressive and delinquent behavioral problems during the transition to adolescence. It is possible to analyze separately the individual differences in unselected populations and group differences in selected extreme groups using quantitative genetic analysis, as a means of answering this question.

Individual differences in externalizing behaviors are known to include both genetic and environmental sources of variance (see Deater-Deckard & Plomin, 1999, and Eley, Lichtenstein, & Stevenson, 1999, for recent data and reviews of this large literature). However, we do not know whether, or the extent to which, the etiology of extreme externalizing problems is similar to or different from the etiology of individual differences in the entire unselected population. The genetic and nongenetic antecedents of extreme externalizing behavioral problems may be identical to those of individual differences in the entire range of these behaviors. Alternatively, genetic or environmental influences may differ for extreme groups. An example of this is found in the literature on mental retardation, where there are genetic factors responsible for severe mental retardation that are virtually unrelated to variability in cognitive ability in the entire range found in unselected samples (Rutter, Simonoff, & Plomin, 1996). In one previous study examining selected extreme groups for externalizing behavioral problems (Deater-Deckard, Reiss, Hetherington, & Plomin, 1997), an analysis of individual differences and selected extreme group differences among adolescents showed that the genetic and environmental sources of variance were similar for both the unselected and selected samples. In one other study (Gjone, Stevenson, Sundet, & Eilertson, 1996), differential heritability (i.e., increasing or decreasing heritability across the continuum of scores) for externalizing problems among school-age twins was found to be modest or negligible. These findings suggest that there is not a distinct etiology for more extreme manifestations of behavioral problems, at least within the "normal" or subclinical range that is being assessed in community twin studies like these.

Genetic studies of family environments can also begin to address whether normal variations in family processes play a similar or different role in child and adolescent psychopathology compared to more negative family environments. Scarr (1992) hypothesized that environmental influences are nonlinear in their effects across the continua of environments and developmental outcomes, whereby systematic (e.g., familial) environmental effects are most pronounced at the extremes—for example, in abusive and neglectful home environments. This would suggest that genetic sources of variance in child or adolescent adjustment would be lower, and systematic shared environmental sources of variance higher, in more negative family environments

(Deater-Deckard & Dodge, 1997). In order to examine this possibility, genetic studies of selected negative family environments are also required.

Thus, the second aim of this study was to examine selected extreme groups of children based on several dimensions of the family environment. Selecting more extreme groups assesses the extent to which family processes involved in child and adolescent adjustment problems in more negative family environments are due to child genetic and environmental factors as part of gene-environment correlational processes. As noted above, patterns in the group parameter estimates can be compared to traditional estimates of individual differences based on unselected samples in order to explore similarities and differences between normal variation and extreme variants in family environments.

The goal of this research was to use an adoptive sibling genetic design to estimate the genetic and environmental sources of variance for selected extreme groups of children who were high in aggressive and delinquent behavioral problems, as well as those who were high in various dimensions of negative or aversive family environments. Although the sample size in the present study limited statistical power, these data can serve as an exploratory study of potential patterns of gene-environment processes for selected extreme groups—patterns that can be tested for replication in future research.

An approach referred to as DF regression analysis (from DeFries & Fulker, 1985, 1988) was used to estimate heritability and shared environment for individual differences and group differences. The approach depends on assessing quantitative variation on a disorder-relevant dimension in the siblings of selected probands in a genetically sensitive design. In DF extremes analysis, familial resemblance (in the case of the current study, sibling similarity) is estimated as the extent to which the average score of the siblings of selected, extreme individual children (i.e., probands) on the quantitative measure regresses back to, or is similar to, the unselected population mean, as illustrated in figure 15.1, which shows a standardized scale, with the population mean [μ] = 0. If there is no resemblance for selected probands and their siblings, then the sibling mean will be very similar to the population mean, and the group familiality will be near zero. In contrast, if there is sibling similarity, then the mean for siblings of selected extreme probands will be greater than 0. The group means for adoptive siblings, who are genetically unrelated, and biologically related siblings, who share half of their genes on average, can then be compared to estimate *group heritability* (h_g^2) in the same way that the heritability of individual differences is estimated: by doubling difference between the group mean for biological siblings and adoptive siblings. *Group shared environment* (c_g^2) is also estimated in the same way as shared environment of individual differences: the group mean for adoptive siblings is a direct estimate of

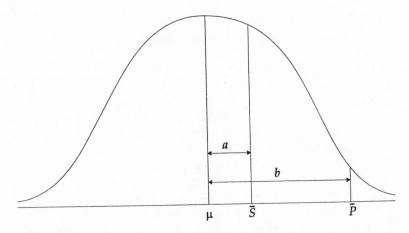

FIGURE 15.1. Estimation of group parameters is based on adoptive and biological sibling group comparisons of extreme, selected proband mean (P) and the mean of proband sibling (S), as compared to the population mean (μ). Group familiality is estimated as a ratio (a/b); it is the proportion of the difference between the proband (P) and population (μ) means that is accounted for by the difference between the sibling (S) and population means. If the adoptive sibling mean of selected probands (S) is no different than the population mean for adoptive siblings, then group-shared environment is negligible. The sibling mean for adoptive siblings (S) is a direct estimate of group-shared environment. Evidence for group heritability is found if the sibling mean for adoptive siblings is lower than the sibling mean for biologically related siblings.

c_g^2, because any sibling similarity in the means of adopted probands and their siblings must be due to common environment.

Although the calculation of these group statistics is similar to the traditional estimates of heritability and shared environment for individual differences, the two types of statistics are very different conceptually. Traditional heritability estimates refer to the contribution of genetic factors to individual differences throughout the range of variation. Group heritability, on the other hand, does not refer to the origins of individual differences at the high end of the distribution; rather, it refers to genetic influence on the average quantitative trait score difference between selected probands and the population mean. The point of this comparison is that the genetic and environmental origins of individual differences in the population can differ from the origins of the average difference between the high extreme and the rest of the population.

Group heritability and shared environment can be estimated using the DF basic regression model. First, scores are standardized and transformed so that they can be expressed as deviations from selected probands' scores:

$$\text{Transformed scores} = (X - X_g)/(P_g - X_g).$$

For both sibling groups (adoptive and biological), the difference between the children's externalizing scores (z-scores) and the overall mean (X_g) for each sibling group was divided by the difference between the proband (selected extreme children) mean (P_g) for that sibling group and the overall mean for each sibling group.

In the DF basic regression model, siblings' (S) scores are predicted from selected probands' scores (P) and degree of genetic relatedness (R), set to 0 for adoptive siblings and 0.5 for biological siblings:

$$S = B_1P + B_2R + A.$$

The group familiality for each sibling group is the average sibling score (S). The regression weight B_2 provides a direct estimate for group heritability (h_g^2). The transformed mean, or group familiality, for adoptive siblings is a function of group shared environment (c_g^2), because any group familiality among unrelated siblings must be due to shared environment factors.

These estimates are based on group means. Indeed, group heritability can be estimated directly from a simple t-test comparing the adoptive and biological probands' siblings' means (DeFries & Fulker, 1988). Because DF extremes analysis is an analysis of means, the parameters that are estimated from this model are not influenced by restricted variance in externalizing behaviors among the selected extreme probands.

Method

Sample

Participants were families with pairs of adoptive (unrelated) and nonadoptive (biologically related) siblings from the Colorado Adoption Project (CAP; see DeFries, Plomin, & Fulker, 1994, for a detailed description of the CAP sample and recruitment procedures). For the analysis of externalizing behavioral problems, participants included 78 pairs of unrelated adoptive siblings (11 same-sex boys, 9 same-sex girls, 58 opposite-sex) and 94 pairs of biologically related siblings (37 same-sex boys, 24 same-sex girls, 33 opposite-sex). Five annual assessments were made when each child was 7, 9, 10, 11, and 12 years old. For the analysis of parent- and child-rated family environments, participants included 95 pairs of unrelated adoptive siblings (13 same-sex boys, 10 same-sex girls, 72 opposite-sex pairs) and 111 pairs of biologically related siblings (40 same-sex boys, 29 same-sex girls, 42 opposite-sex pairs). Three annual assessments were made when each child was 10, 11, and 12 years old. Sibling pairs where one or both children were missing data on all five assessments of teacher-rated externalizing behavioral problems, or the three assessments of parent- or child-rated family environments, were excluded from these analyses.

Measures

Externalizing Behavioral Problems   Teachers rated each child's externalizing behavioral problems by completing Achenbach's (1991) Teacher Report Form (TRF) each year when the child was 7, 9, 10, 11, and 12 years old. The Aggression subscale (25 items, $\alpha = .85$) and Delinquency subscale (9 items, $\alpha = .71$) were analyzed separately. These scores were stable across these five assessments (year-to-year correlations for Aggression from .40 to .61, and for Delinquency from .28 to .44). Therefore, linear composites were computed by averaging each child's Aggression and Delinquency scores across all available (one or more) assessments. These composites were log transformed and standardized for correlational analyses because they were positively skewed.

Family Environment   Parents rated their perceptions of the family environment by completing the 48-item Parent Report (PR; Dibble & Cohen, 1974) when the child was 10, 11, and 12 years old. This questionnaire assesses parents' perceptions of their discipline strategies and feelings of warmth and negativity toward each child. Three factors were derived including warmth (acceptance, child centeredness, sensitivity, positive involvement, shared decision making, $\alpha > .80$), negativity (guilt induction, hostility, withdrawal of affection, $\alpha > .72$), and inconsistency (inconsistent or lax discipline, withdrawal from relationship, $\alpha > .67$). Parents' reports were highly stable (correlations from .67 to .76), so a linear composite was created by averaging across all available (one or more) assessments.

Children's views of the family environment were measured with a brief 20-item version of the Family Environment Scale (FES; Moos & Moos, 1981). Four items from five FES scales were administered: family cohesion, expressiveness, conflict, achievement orientation, and control. Principal components analysis was conducted (the original scales showed poor internal consistency—see Deater-Deckard et al., 1999, for details of this factor analysis). Two scales were derived: a 10-item family positivity scale (high cohesiveness and expressiveness, and low conflict, $\alpha = .78$ to .80) and a 3-item "low achievement orientation" scale (achievement not encouraged, $\alpha = .46$ to .52). Children's reports were stable (correlations from .34 to .60), so the scores were averaged over all available (one or more) assessments to yield linear composites. For the correlational analyses, all composite scores were adjusted for sex mean differences using linear regression, and the standardized residual scores were then analyzed.

Results

For the analysis of extreme groups, those children were selected who showed the highest levels of behavioral problems and the most negative and potentially deleterious family environments. Pairs where both siblings were selected were double-entered, with the p-values of the derived parameters adjusted accordingly. For the analysis of externalizing behavior problems, two groups of children were selected—those with Aggression scores in the highest 15% (one standard deviation above the mean) and those with Delinquency scores in the highest 15%. Sibling pairs where both siblings were selected as affected probands were double-entered. For the analysis of parents' and children's ratings of the family environment, five groups were selected—those with (1) parent-rated negativity in the highest 15%; (2) parent-rated inconsistent discipline in the highest 15%; (3) parent-rated warmth in the lowest 15%; (4) child-rated positivity in the lowest 15%; and (5) child-rated low achievement orientation in the highest 15%.

A summary of the results from previously published findings for individual differences in the unselected sample for teacher-rated child externalizing behaviors (Deater-Deckard & Plomin, 1999) and parent- and child-rated family environments (Deater-Deckard et al., 1999) are also presented. Individual differences heritability (h^2), shared environment (c^2), and nonshared environment (e^2) were estimated using maximum-likelihood model fitting by testing the standard ACE model (Neale & Cardon, 1992).

The following results focus on extreme group and unselected individual differences, with an emphasis on genetic and environmental parameter estimates. Sex and adoptive status mean differences in teacher-rated behavioral problems and parent- and child-rated family environments are described in detail elsewhere (Deater-Deckard & Plomin, 1999; Deater-Deckard et al., 1999). For these analyses, standardized residual scores controlling for child sex were used.

Externalizing Behavioral Problems

The group means (an estimate of sibling similarity in the selected extreme group, with statistical significance determined using a t-test) for the selected extreme groups (+1 SD) of probands and their siblings are shown in the left half of table 15.1. For teacher-rated Aggression, there was no group familiality for adoptive pairs, $r_g = -.06$, but evidence for modest sibling similarity among the biologically related selected probands and siblings, $r_g = .19$, although this familiality coefficient was not significant. More noteworthy is that these group estimates were very similar to the sibling intraclass correlations for the unselected sample, shown in the right half of table 15.1. In the unselected sample,

TABLE 15.1. Teacher-Rated Aggression and Delinquency: Sibling Group Familiality Coefficients (Group Means) for Selected Extreme Groups, and Sibling Intraclass Correlations for Individual Differences in the Unselected Sample (from Deater-Deckard & Plomin, 1999), by Adoptive Status

	Selected +1 SD		Unselected	
	Adoptive	Nonadoptive	Adoptive	Nonadoptive
TRF: Aggression	−.06	.19	−.06	.25*
TRF: Delinquency	.11	.16	.14	.24*

Note: The data for the unselected sample presented in Deater-Deckard & Plomin (1999); *p < .05

TABLE 15.2. Teacher-Rated Aggression and Delinquency: Quantitative Genetic Parameter Estimates for Selected Extreme Groups and Individual Differences in the Unselected Sample (from Deater-Deckard & Plomin, 1999)

	Selected +1 SD			Unselected		
	h_g^2	c_g^2	e^2	h^2	c^2	e^2
Aggression	.48	.00+	.52	.49*	.00⁺	.51
Delinquency	.09	.11	.80	.17	.13*	.70

Note: The data for the unselected sample presented in Deater-Deckard & Plomin (1999); *p < .05; ⁺constrained to = 0, due to a modestly negative adoptive sibling correlation.

adoptive siblings were not correlated for teacher-rated Aggression, $r = -.06$, whereas a modest positive value was found for nonadoptive siblings, $r = .25, p < .05$.

For teacher-rated Delinquency, the sibling group familiality coefficients (left half of table 15.1) for the selected extreme group (+1 SD) are shown. Group familiality was similarly modest and not significant for adoptive, $r_g = .11$, and biological, $r_g = .16$, sibling pairs. Although nonsignificant, these group estimates were generally similar to the sibling intraclass correlations for the unselected sample (right half of table 15.1). Adoptive siblings were modestly albeit not significantly similar for teacher-rated Delinquency, $r = .14$, but biologically related nonadoptive siblings were slightly more similar, $r = .24, p < .05$.

Next, group heritability (h_g^2), group shared environment (c_g^2), and group nonshared environment (e_g^2) were estimated using DF regression analysis (see "Method" above). These are presented in the left side of table 15.2.

For teacher-rated Aggression, group heritability was moderate (.48) but nonsignificant. There was no evidence for group-shared environment, and the remaining group variance was of the nonshared variety

(.52). These estimates were nearly identical to those derived for individual differences in teacher-rated Aggression in the entire unselected sample (right side of table 15.2), which showed evidence for moderate heritability (.49, $p < .05$) and nonshared environment (.51) and no shared environment.

The genetic and environmental parameter estimates for teacher-rated Delinquency are also presented in table 15.2. For the selected extreme group (left half of table), group heritability (.09) and shared environment (.11) were modest and not significant, with group non-shared environment (.80) constituting the remaining variance. Like the findings for Aggression, these group parameter estimates were very similar to those derived for individual differences in the unselected sample (right half of table): modest heritability (.17, nonsignificant) and shared environment (.13, $p < .05$), and substantial nonshared environmental variance (.70). It is noteworthy that such distinct patterns for Aggression and Delinquency were found in both group and individual differences, in light of the fact that the two scales are substantially inter-correlated and are often summed to yield an externalizing syndrome score (Achenbach, 1991).

In summary, there was evidence for genetic variance in teacher-rated Aggression but not Delinquency, and evidence for shared environmental variance in Delinquency but not Aggression. Futhermore, nonshared environmental variance and error accounted for half or more of the variance in Aggression and Delinquency. This pattern of variance estimates was very similar for the selected extreme groups and individual differences in the unselected sample. However, it is important to emphasize that the estimates for the small selected samples were not statistically significant, so that such comparisons should be regarded as tenuous and in need of replication before conclusions can be drawn.

Family Environment

For parents' and children's ratings of the family environment, the sibling group familiality coefficients for the selected extreme groups are shown in the left half of table 15.3. Recall that the scores for parent-rated warmth and child-rated positivity were reversed prior to selection so that the selected groups were lowest in warmth and positivity.

For parent-rated negativity, inconsistency, and warmth, group famil-iality was substantial for both adoptive and nonadopative sibling pairs ($r_g = .70$ to .89, $p < .001$). This pattern suggests the presence of sub-stantial shared environmental variance, and little genetic variance, in these selected extreme groups based on parents' ratings of the family environment. These group coefficients were similar to the sibling intr-aclass correlations found for the unselected sample ($r = .60$ to .84, $p < .001$), shown in the right half of table 15.3.

TABLE 15.3. Parent- and Child-Rated Family Environment: Sibling Group Familiality Coefficients (Group Means) for Selected Extreme Groups, and Sibling Intraclass Correlations for Individual Differences in the Unselected Sample (from Deater-Deckard, Fulker, & Plomin, 1999), by Adoptive Status

	Selected +1 SD		Unselected	
	Adoptive	Nonadoptive	Adoptive	Nonadoptive
PR: Negativity	.74***	.77***	.60***	.78***
PR: Inconsistency	.79***	.81***	.79***	.80***
PR: Warmth[a]	.70***	.89***	.67***	.84***
FES: Family positivity[a]	.29	.43*	.27*	.32*
FES: Low achievement	−.03	.18	.01	.26*

Note: The data for unselected sample presented in Deater-Deckard, Fulker, & Plomin (1999). Scores for each assessment (at 10, 11, and 12 years of age) were averaged to form linear composites. PR = Parent Report (parent-rated), FES = Family Environment Scale (child-rated); *$p < .05$; **$p < .01$; ***$p < .001$; [a]reverse-scored for analysis of extreme groups, so that selected groups were lower in warmth or positivity.

For child-rated positivity and low achievement orientation, a different pattern emerged compared to parents' ratings. The only significant sibling group mean was for nonadoptive sibling positivity ($r = .43$, $p < .05$); by comparison, the adoptive sibling group estimate was more modest and nonsignificant ($r = .29$). The sibling group familiality coefficients for low achievement orientation were nonsignificant. The magnitude of these group means were similar to the sibling intraclass correlations for the unselected sample, as shown in the right half of table 15.3.

Group heritability (h_g^2), group shared environment (c_g^2), and group nonshared environment (e_g^2) estimates were derived for measures of the family environment using the DF basic regression model as described above. These parameter estimates are presented in the left half of table 15.4.

For parents' ratings of the family environment, group heritability was modest to moderate and nonsignificant ($h_g^2 = .05$ to .38), whereas group shared environment was substantial ($c_g^2 = .70$ to .79, $p < .001$). Group nonshared environment was negligible or modest. These group parameter estimates were remarkably similar to the estimates for individual differences in the unselected sample (shown in right half of table 15.4), particularly when one considers the much smaller sample size in the extreme group analyses. For children's reports of the family environment, group heritability was moderate but nonsignificant ($h_g^2 = .27$ to .42), and group shared environment was modest to moderate and nonsignificant ($c_g^2 = .00$ to .29). About half of the group variance in children's perceptions of the family environment was due to nonshared environment and error. In general, this pattern of group genetic and

TABLE 15.4. Parent- and Child-Rated Family Environment:
Quantitative Genetic Parameter Estimates for Selected Extreme
Groups and Individual Differences for Unselected Sample
(from Deater-Deckard, Fulker, & Plomin, 1999)

	Selected +1 SD			Unselected		
	h_g^2	c_g^2	e^2	h^2	c^2	e^2
PR: Negativity	.06	.74***	.20	.38***	.59***	.03
PR: Inconsistency	.05	.79***	.16	.04	.77***	.19
PR: Warmth[b]	.38	.70***	.00	.26***	.67***	.06
FES: Family positivity[b]	.27	.29	.44	.18	.26***	.56
FES: Low achievement	.42	.00+	.58	.52***	.01	.48

Note: The data for unselected sample presented in Deater-Deckard, Fulker, & Plomin (1999).
Scores for each assessment (at 10, 11, and 12 years of age) were averaged to form linear composites.
PR = Parent Report (parent-rated), FES = Family Environment Scale (child-rated); *two-tailed
$p < .05$; **$p < .01$; ***$p < .001$. +parameter constrained = 0.
[a]Group nonshared environment (e_g^2) was estimated as $[1 - (h_g^2 + c_g^2)]$. Because estimates of nonshared
environmental variance (e^2 and e_g^2) include some error variance, it is not appropriate to test the hypothesis that e^2 or $e_g^2 > 0$. Therefore, statistical significance is not reported for the nonshared environment
parameter.
[b]Reverse-scored for analysis of extreme groups, so that selected groups were lower in warmth or
positivity.
[b]Group nonshared environment estimates were computed as $[1 - (h_g^2 + c_g^2)]$; statistical significance
is not known because the standard errors for these estimates are not provided by the DF basic model.

environmental parameter estimates was similar to the pattern of estimates derived for the unselected sample (see right half of table 15.4). Thus, the selected group familiality estimates were similar to the sibling intraclass correlations for the unselected sample, suggesting that the genetic and environmental sources of variance in parents' and children's ratings of the family environment might be similar for the unselected and selected groups. It is important to bear in mind that these selected groups do not represent the most extreme family environments that exist in the population (e.g., abusive and neglectful environments), a topic that we return to in the "Discussion" section that follows.

In summary, for both externalizing behavioral problems and measures of the family environment, overall there were few statistically significant group estimates. However, in examining the effect sizes, the extreme group genetic and environmental parameter estimates for selected samples were very similar to the parameters derived for individual differences in the unselected sample.

Discussion

Quantitative genetic designs are one type of tool for exploring possible links between normal and abnormal behavior, and for delineating the gene-environment processes linking family environments to these problems in adjustment, during the transition to adolescence. The findings from studies of extremes and individual differences will promote our understanding of the full range of externalizing problems in addition to understanding the etiology of the clinical manifestations of these behaviors. Unfortunately, genetic research using clinical samples of children (i.e., those diagnosed with conduct disorder) has operated in parallel but not in conjunction with genetic studies of unselected community samples of children (i.e., children with widely varying degrees of externalizing problems. Applying diagnostic categorization to dimensions of behavioral problems is necessary for many reasons—perhaps most importantly because clinicians must know whom to treat. However, such diagnostic approaches may have little to do with the etiology of the disorder. If the behavior of interest varies continuously in the population as in the case of externalizing behaviors, then dimensional measures that represent the full range of behaviors in the sample can be employed (Deater-Deckard et al., 1997). Selecting more extreme groups of individuals from such samples makes it possible to test whether the etiology of the extremes is similar to or different from the etiology of individual differences in the entire range of behavior.

Longitudinal studies of the development of externalizing behavioral problems suggest that there may be two groups of children to consider. The first is a small group of individuals who show early signs of very high levels of aggression and conduct problems; the second is a much larger group who show elevations in these externalizing behavioral problems in adolescence (Moffit, 1993). These longitudinal studies also have shown that the development of externalizing problems and conduct disorder is intimately linked to problems in parent-child interaction—specifically, the presence of harsh and coercive parenting (Patterson, Reid, & Dishion, 1992). What remains unclear is precisely how these harsh parenting environments combine with genetic liability for aggressive conduct problems in producing individual differences and extreme group differences in conduct disorder symptoms.

The first aim of this study was to examine the group genetic and environmental sources of variance for selected extreme groups of children with the highest levels of aggressive and delinquent behavioral problems in the transition to adolescence. Group genetic variance was moderate but nonsignificant for teacher-rated aggression, and was negligible for delinquency. Similarly, group shared environmental variance was negligible for teacher-ratings of both aggressive and delinquent behavioral problems. Half to nearly all of the group differences variance was accounted for by nonshared environmental variance (which includes error variance). In general, this pattern of group genetic and environmental variance estimates was very similar to the pattern of estimates for the unselected sample. When considering the consistency in these patterns, it is important to remember that the selected extreme groups in this study were not representative of clinically referred children with conduct disorder or related problems. Instead, the selected groups of children are best thought of as "high normal" or "subclinical," but showing elevated levels of problem behaviors.

An exploration of the links between extreme and normal variation in externalizing problems is incomplete without a consideration of children's family environments. There are strong phenotypic associations between children's parenting environments and their aggressive delinquent behavioral problems (Patterson et al., 1992). Far from being independent, active and evocative gene-environment correlational processes have been implicated in the links between these behavioral problems and children's family environments (Ge et al., 1996; O'Connor, Deater-Deckard, Fulker, Rutter, & Plomin, 1998; Pike, McGuire, Hetherington, Priss, & Plomin, 1996).

In the presence of gene-environment correlational processes, the assumption that our measures of family environments assess purely environmental components of a developmental process is unfounded (Plomin & Bergeman, 1991). For example, in the CAP, all of the sibling similarity in the perception of family achievement orientation was accounted for by additive genetic variance during the transition to adolescence (Deater-Deckard et al., 1999). Furthermore, environmental factors are known to create sibling differences (Plomin & Daniels, 1987), as evidenced in the CAP by the often substantial nonshared environmental variance found for externalizing behaviors, as well as measures of the family environment in the transition to early adolescence (Deater-Deckard & Plomin, 1999; Deater-Deckard et al., 1999).

Thus, the second aim of this study was to examine the more extreme negative family environments in the CAP and to estimate group genetic and environmental parameter estimates. The examination of the more negative family environments, through selection of extreme groups of children with negative family environment scores 1 SD above the mean, revealed evidence for significant and substantial group shared environmental variance for parents' reports of parental negativity, inconsistent discipline, and warmth. By comparison, group shared environmental variance was negligible for children's perceptions, but nonshared environmental variance accounted for about half of the variance. There were no statistically significant group genetic parameters for parents or children's reports of the family environment. Generally, the group parameter effect sizes were similar to those found for individual differences in the unselected sample.

Although not statistically significant, it is noteworthy that the shared environmental variance for parent-rated negativity was higher for the extreme group (74%) compared to the unselected sample (59%). This pattern of greater shared environmental variance in measures of family environments is consistent with Scarr's hypothesis that systematic shared environmental factors make an increasingly substantial contribution to the etiology of individual differences in children's behavioral problems for increasingly negative family environments (Deater-Deckard & Dodge, 1997). Given the exploratory nature of the present study, this finding will have to be tested for replication before a conclusion can be made in this regard. Furthermore, Scarr's proposal refers to truly neglectful or abusive environments. Although the selected extreme groups do not represent abusive or neglectful parenting (indeed, this is practically impossible in adoption studies where adoptive parents have been screened for providing loving and enriched environments), the selected groups do represent negative family environments. For instance, a score of 1 SD above the mean on parent-rated negativity corresponds to saying that "I lose my temper when s/he does not do as I ask" about "half the time" (3 on a six-point scale), and a score of 1 SD above the mean on inconsistency corresponds to stating that "I ignore misbehavior" slightly less than "half the time" (2.9 on a six-point scale).

The majority of the group genetic and environmental parameter estimates were not statistically significant. However, the nonsignificant effect sizes for the selected groups were generally similar to the effect sizes found for the unselected sample, suggesting that the genetic and environmental sources of variance may be similar for more extreme negative family environments. A caveat to the findings for the selected groups is that because the selected samples are smaller than the unselected sample used for the analysis of individual differences (h^2, c^2, and e^2), the statistical power is lower for detecting as statistically reliable (at $p < .05$) group parameter estimates (h^2_g, c^2_g, and e^2_g) that are numerically similar to those found for the unselected sample. Nonetheless, reporting the nonsignificant parameters for these extreme groups allows for testing for replication in larger genetically informative samples (Deater-Deckard et al., 1997; Eley, 1997). In spite of the small sample sizes for the selected groups, the DF method provides remarkably robust estimates that are consistent with the individual differences parameter estimates for the unselected sample.

There are several limitations to bear in mind. The findings based on this sample of adoptive sibling pairs may not generalize to samples of twins, or to samples of nonadoptive siblings. Furthermore, the small sample size resulted in low statistical power for examining sex differences and for comparing group and individual differences parameter estimates. Nonetheless, the exploratory analysis of selected extreme groups is an important first step for examining the links between "normal" variation in behavioral problems and family environments, and "abnormal" or pathological levels. In the future, genetically informative studies (that also include measures of the environment) will permit the investigation of this variation as well, so that the most extreme groups (those who are most relevant for research on clinical disorders) can be compared to individual differences in the unselected sample. These selections could be made based on individual differences in the phenotype (i.e., outcome) or the environment (i.e., nongenetic risk factor). More important, selections can be made on one variable (the outcome or environment), and the genetic and environmental mediation of the associations between environments and outcomes can be assessed. Though rarely used, these bivariate extremes analyses (see Deater-Deckard, Pike, & Plomin, 1996) may provide us with more information regarding gene-environment processes in the development of externalizing behavioral problems and troubled family environments in the transition to early adolescence—specifically, whether gene-environment processes linking developmental outcomes and associated environments differ in more extreme, harsh family environments.

Dimensional approaches to assessment in clinical psychology and psychiatry are becoming more accepted because researchers have recognized that there are important individual differences in these disorder-relevant dimensions. Indeed, it is possible that psychopathology merely represents the quantitative extreme of normal variation in personality—for example, depression representing the extreme of neuroticism. If this is the case, then the same genes and environments are likely to be causally linked to both normal and extreme variants. If replicated in future genetic studies, the exploratory findings of this adoption study suggest that this is the case—at least for subclinical extremes—and lead to the prediction that when specific genes are found that are associated with externalizing behavioalr problems, these genes will in fact be associated with not just the clinical extremes of problem behavior but with normal variation throughout the range.

REFERENCES

Achenbach, T. (1991). *Integrative guide for the 1991 CBCL/4–18, YSR, and TRF profiles*. Burlington: University of Vermont Department of Psychiatry.

Deater-Deckard, K., & Dodge, K. A. (1997). Externalizing behavior problems and discipline revisited: Nonlinear effects and variation by culture, context, and gender. *Psychological Inquiry, 8,* 161–175.

Deater-Deckard, K., Fulker, D. W., & Plomin, R. (1999). A genetic study of the family environment in the transition to early adolescence. *Journal of Child Psychology and Psychiatry, 40,* 769–775.

Deater-Deckard, K., Pike, A., & Plomin, R. (1996). Emotional problems in adolescents: A quantitative genetic comparison of unselected individuals vs. selected extreme groups. Paper presented at the biennial meeting of the *International Society for the Study of Behavioral Development*, August 1996, Montreal, Canada.

Deater-Deckard, K., & Plomin, R. (1999). An adoption study of the etiology of teacher and parent reports of externalizing problems in middle childhood. *Child Development, 70,* 144–154.

Deater-Deckard, K., Reiss, D., Hetherington, E. M., & Plomin, R. (1997). Dimensions and disorders of adolescent adjustment: A quantitative genetic analysis of unselected samples and selected extremes. *Journal of Child Psychology and Psychiatry, 38,* 515–525.

DeFries, J. C., & Fulker, D. W. (1985). Multiple regression analysis of twin data. *Behavior Genetics, 15,* 467–473.

DeFries, J. C., & Fulker, D. W. (1988). Etiology of deviant scores versus individual differences. *Acta Geneticaiae Medicae et Gemellologiae: Twin Research, 37,* 205–216.

DeFries, J. C., Plomin, R., & Fulker, D. W. (1994). *Nature and nurture during middle childhood.* Oxford: Blackwell.

Dibble, E., & Cohen, D. J. (1974). Companion instruments for measuring children's competence and parental style. *Archives of General Psychiatry, 30,* 805–815.

Dishion, T. J., Eddy, J. M., Haas, E., Li, F., & Spracklen, K. M. (1997). Friendships and violent behavior during adolescence. *Social Development, 6*(2), 207–225.

Dishion, T. J., Spracklen, K. M., Andrews, D. W., & Patterson, G. R. (1996). Deviancy training in male adolescent friendships. *Behavior Therapy, 27,* 373–390.

Eley, T. C. (1997). Depressive symptoms in children and adolescents: Etiological links between normality and abnormality: A research note. *Journal of Child Psychology & Psychiatry & Allied Disciplines, 38,* 861–865.

Eley, T. C., Lichenstein, P., & Stevenson, J. (1999). Sex differences in the etiology of aggressive and nonaggressive antisocial behavior: Results from two twin studies. *Child Development, 70,* 155–168.

Ge, X., Conger, R. D., Cadoret, R. J., Neiderhiser, J. M., Yates, W., Troughton, E., & Stewart, M. A. (1996). The developmental interface between nature and nurture: A mutual influence model of child antisocial behavior and parent behaviors. *Developmental Psychology, 32,* 574–589.

Gjone, H., Stevenson, J., Sundet, J. M., & Eilertsen, D. E. (1996). Changes in heritability across increasing levels of behavior problems in young twins. *Behavior Genetics, 26,* 419–426.

Holmbeck, G. N., Paikoff, R. L., & Brooks-Gunn, J. (1995). Parenting adolescents. In M. Bornstein (Ed.), *Handbook of parenting, Vol. 1: Children and parenting.* Mahwah, NJ: Lawrence Erlbaum Associates.

Lollis, S., & Kuczynski, L. (1997). Beyond one hand clapping: Seeing bidirectionality in parent-child relations. *Journal of Social and Personal Relationships, 14,* 441–461.

Moffitt, T. E. (1993). Adolescence-limited and life-course-persistent antisocial behavior: A developmental taxonomy. *Psychological Review, 100,* 674–701.

Moos, R. H., & Moos, B. S. (1981). *The Family Environment Scale manual.* Palo Alto, CA: Consulting Psychologists Press.

Neale, M. C., & Cardon, L. R. (1992). *Methodology for genetic studies of twins and families.* Dordrecht, Netherlands: Kluwer Academic Publishers.

O'Connor, T. G., Deater-Deckard, K., Fulker, D. W., Rutter, M., & Plomin, R. (1998). Gene-environment correlations in late childhood and early adolescence. *Developmental Psychology, 34,* 970–981.

Patterson, G. R., Reid, J., & Dishion, T. J. (1992). *Antisocial boys.* Eugene, OR: Castalia.

Pike, A., McGuire, S., Hetherington, E. M., Reiss, D., & Plomin, R. (1996). Family environment and adolescent depressive symptoms and antisocial-behavior—A multivariate genetic analysis. *Developmental Psychology, 32,* 590–603.

Plomin, R. (1994). *The genetics of experience: The interplay between nature and nurture.* Thousand Oaks, CA: Sage.

Plomin, R., & Bergeman, C. S. (1991). The nature of nurture: Genetic influence on "environmental" measures. *Behavioral and Brain Sciences, 14,* 373–427.

Plomin, R., & Daniels, D. (1987). Why are children in the same family so different from each other? *Behavioral & Brain Sciences, 10,* 1–16.

Rutter, M., Simonoff, E., & Plomin, R. (1996). Genetic influences on mild mental retardation: Concepts, findings and research implications. *Journal of Biosocial Science, 28,* 509–526.

Scarr, S. (1992). Developmental theories for the 1990's: Development and individual differences. *Child Development, 63,* 1–19.

Scarr, S., & McCartney, K. (1983). How people make their own environments: A theory of genotype → environment effects. *Child Development, 54,* 424–435.

REBECCA HOBSON
BETH MANKE
SHIRLEY MCGUIRE

16

Maternal Reports of Differential Treatment during Early Adolescence

A Longitudinal Study of Genetic Influences and Links to Child Well-Being

Introduction

Although the main focus of behavioral genetic studies is to disentangle sibling resemblance due to genetic influences from resemblance due to shared environmental factors, the same studies typically demonstrate that two siblings raised in the same family are very different from one another (Dunn & Plomin, 1990; Plomin & Daniels, 1987; Harris, 1998). These sibling differences are due to genetic factors, nonshared environmental influences (i.e., environmental experiences not shared by siblings in the same family), and measurement error. Looking across all traits and behaviors, it appears that genetic differences account for 30 to 50% of these differences between siblings, suggesting that at least half of these differences are due to nonshared environmental influences and error of measurement. The fact that identical twins have demonstrated significant within-pair differences for nearly all behavioral traits further suggests that nonshared environmental influences are important, as identical twins can only be different for environmental reasons (and error of measurement).

Thus, understanding the nature of environmental factors that make two children in the same family different would appear to be a worthwhile pursuit. In fact, much of this research was already under way at the same time behavioral genetic studies were documenting large sibling differences in developmental outcomes. Family researchers have found that siblings differ in social experiences within the family—with their parents and with each other (Brody, Stoneman, & McCoy, 1992; McHale & Pawletko, 1992). Yet, because these studies did not incorporate a genetically sensitive design, it is unclear whether the sibling differences detected were due to environmental or genetic factors.

Probably the most studied aspect of nonshared environment has been maternal differential treatment (MDT), or the extent to which mothers treat two children in the same family differently (see McGuire, in press, for a review). Various studies demonstrate significant and substantial MDT, yet the degree of MDT varies for attention, affection, conflict, and control. Daniels and Plomin (1985) examined the extent to which siblings reported different familial experiences and found nearly half of the siblings reported some degree (a little or much more) of parental differential affection and control. Differential treatment has been shown to occur in other positive domains of parent-child relations such as warmth and attention (McGuire & Roch-Levecq, 2001), but is particularly prevalent in the more negative areas such as conflict (Anderson, Hetherington, Reiss, & Howe, 1994; Brody, Copeland, Sutton, Richardson, & Guyer, 1992; Dunn, Stocker, & Plomin, 1990).

The majority of research on MDT has focused on the extent and consequences of such treatment (e.g., child well-being and sibling relationship quality). Few studies, however, have examined these links longitudinally. In addition, longitudinal associations with adjustment over time have not been examined across early adolescence. Finally, even fewer studies have examined MDT in genetically sensitive designs that control for genetic sources of MDT. The current study explores the stability of MDT, the possible origins (e.g., genetic influences) of such treatment, and the link between MDT and child well-being (overall well-being and competence in extra-familial settings) over three ages during early adolescence.

Stability of MDT over Time

Although separate studies have documented substantial MDT at various child ages (e.g., infancy, middle childhood, late adolescence), most studies do not address whether MDT changes *over time* because the studies are not longitudinal. Examination of stability, however, is important given that the amount and type of differential treatment could vary considerably in infancy, early childhood, and early and late adolescence. MDT may change over time as children reach different ages and developmental milestones. For example, mothers may be more attentive to infants as compared to their school-age siblings (Volling & Elins, 1998).

As older siblings approach adolescence, however, mothers may shift their focus and become more attentive to and controlling of older siblings relative to their younger siblings.

Four studies provide evidence of stability of MDT over time. First, McGuire & Roch-Levecq (2001) addressed stability during infancy. Mothers were interviewed about differential treatment when infants were 14, 20, 24, and 36 months of age. Maternal differential discipline, affection, and attention were found to be low to moderately stable, with correlations ranging from .24 to .33. Stability of MDT has also been addressed during middle childhood. McGuire, Dunn & Plomin (1995) examined the stability of maternal differential discipline, affection, and attention in school-age children (older and younger siblings were 8 and 5 years old, respectively, at time 1 and 11 and 8 years old, respectively at time 2). In this study, several dimensions of MDT were moderately correlated across the two time-points: .60 for ease of discipline, .50 for frequency of discipline, .30 for affection, and .43 for attention.

Conger and Conger (1994) also investigated the stability of parental differential treatment among families with school-age children. Specifically, they examined differential parental hostility and found moderate stability across two time-points, with correlations ranging from .35 to .46. Finally, in a sample of adolescent siblings, Pike, Manke, Reiss, & Plomin (2000) also found parental differential affection and control to be moderately stable, with stability correlations ranging from .19 to .50.

In combination, these studies suggest that MDT is moderately stable during infancy, middle childhood, and adolescence, whether measured using maternal reports (McGuire & Roch-Levecq, 2001; McGuire et al., 1995), child reports (Pike et al., 2000), or observations (Conger & Conger, 1994). Previous studies also indicate that not only are specific domains of differential treatment (i.e., affection, attention) stable over time, but patterns (i.e., favoring the same child across time) of differential treatment are also stable (McGuire et al., 1995).

The current study addresses the stability of maternal reports of differential attention, involvement, and control over three time-points during the transition to adolescence. The transition to early adolescence is an important time to address the stability of MDT owing to such factors as the changing nature of parent-child relations. Early adolescence is often characterized by a shift in a child's focus from family-focused behavior to more peer-oriented activity (Steinberg, 1987). Parent-child relations, as seen by adolescents, may not be as important as they once were, with the focus now on friendships. This may impact the stability of MDT, as adolescents may require less affection, but more attention and control than younger siblings.

Genetic Contributions to Maternal Reports
of Differential Treatment

Is MDT of siblings the result of environmental or genetic factors? Although MDT is typically discussed as a source of nonshared environment free of genetic influence, the mere labeling of this process as environmental does not prevent it from being influenced genetically. While it may seem paradoxical to examine genetic influences on MDT, this research is in line with other behavioral genetic studies that have focused on addressing the genetic and environmental influences on specific family behaviors (Plomin, Reiss, Hetherington, & Howe, 1994; Rowe, 1983). That is, genetic factors can contribute to MDT through genetically influenced traits such as children's personality (e.g., temperament, activity level). For example, a child who is more active may elicit more maternal monitoring than his or her less active sibling. Activity level has been found to be significantly heritable (Plomin, 1990) and thus, genetic contributions to maternal differential treatment may be due to genetic contributions to children's activity levels. That is, mothers could be *reacting* to children's heritable traits. Similarly, a child who is depressed may elicit more maternal attention and affection.

Measures of differential treatment have been treated as behavioral phenotypes in behavioral genetic studies (see Plomin, 1994). Estimates of genetic influence can be based on mean level differences in MDT between groups of siblings who vary in genetic relatedness. These means are typically derived using absolute scores on the Sibling Inventory of Differential Experience (SIDE; Daniels & Plomin, 1985). Absolute scores from the SIDE index the level of differential treatment (same, a bit more, or much more) rather than the direction (older or younger sibling more). Genetic contributions to differential treatment in studies using a twin design would be implied if DZ twins who share only 50% of their genes reported more MDT than MZ twins who are genetically identical. In an adoption design, genetic influence would be evident if as a group, full siblings reported less MDT than unrelated adopted siblings. Unrelated siblings should show the greatest variation in perceptions of MDT because they share no genetic similarities. The current study utilizes an adoption design; therefore, genetic influences would be implied if reports of MDT were greater for unrelated siblings as compared to full siblings.

The extant literature on genetic influences on MDT comes from three studies, all of which have focused on children's reports of parental differential treatment using the SIDE. In the first longitudinal study by Daniels and Plomin (1985), sibling differential experiences were explored with an adoption design. Genetic influences were not found for parental differential treatment as evidenced by nonsignificant mean differences between adoptive and nonadoptive sibling reports of differential treatment (i.e., affection and control). Baker and Daniels (1990) also examined the genetic influence on subjects' reports of parental differential treatment in a sample of adult identical and fraternal twins. In contrast to the findings of Daniels and Plomin (1985), Baker & Daniels did find genetic influences on parental differential treatment (i.e., affection and control), with identical twins reporting significantly less differential treatment than fraternal twins.

The previous two studies addressed genetic influences on parental differential treatment, but their findings were inconsistent: the twin study suggested genetic influences on parental differential treatment, whereas the adoption study found no evidence for genetic influences. Two hypotheses that may explain the difference in results could be that nonadditive genetic influences are operating that would only be seen in identical twins, or that there could be a special identical twin environment operating to make their experiences unusually similar.

In an attempt to clarify previous findings, Pike et al. (2000) examined genetic influences on siblings' reports of parental differential treatment using both a twin and a step-family design at two time ages. In this study, absolute scores from the SIDE were examined for six sibling categories: identical twins, fraternal twins, full (from both intact and remarried families), half, and unrelated siblings. Twin results (i.e., comparisons between identical and fraternal twin means) for maternal differential control support findings from Baker and Daniels' twin study (1990). That is, both studies demonstrated that maternal differential control is genetically influenced. In contrast to the Baker and Daniels study (1990), however, the identical-fraternal twin mean comparison in this study did *not* support the presence of genetic influence on maternal differential affection. That is, the fraternal twin mean for differential maternal affection was *not* significantly greater than the mean for identical twins.

As part of the Pike et al. (2000) study, mean level comparisons across full and unrelated siblings were also conducted in an attempt to replicate the Daniels and Plomin (1985) study. For maternal differential affection and control, mean level comparisons were significant, suggesting that *both* maternal differential control and affection are influenced by genetic factors. These results diverge from those presented by Daniels and Plomin (1985), which indicated that the means for full (nonadoptive) and unrelated (adoptive) sibling pairs were not significantly different from each other, as they relate to maternal differential control and affection. In short, although the Pike et al. (2000) study set out to clarify the results of both a previous twin and adoption study, the results were inconsistent in that the mean comparisons for identical versus fraternal twins and full versus unrelated siblings did not tell the same story, suggesting that much is still to be learned about the genetic and environmental origins of maternal differential treatment.

The previous studies addressed genetic influences on measures of parental differential treatment as reported by *children* and *siblings*. The current study will examine genetic influences on a measure of MDT as reported by *mothers* during their children's early adolescence, in part in an effort to clarify the inconsistent findings that have relied on children's reports of MDT.

Links to Child Well-Being

As mentioned previously, the bulk of research on MDT has focused on the correlates of such treatment for children's well-being (McGuire, in press). Receiving preferential parental treatment (i.e., receiving greater amounts of affection and attention relative to one's sibling) is associated with better adjustment, unless the child recognizes the need for a sibling to receive preferential treatment, such as in families with a chronically ill or disabled child (Bischof & Tingstrom, 1991; McHale & Pawletko, 1992; Wolf, Fisman, Ellison, & Freeman, 1998). Differential treatment has been associated with negative outcomes in the domains of anxiety, depression, and internalizing and externalizing problems for those children receiving less affection and more discipline. For example, McGuire et al. (1995) found that older siblings whose mothers reported disciplining them more often and showed them less affection relative to their younger siblings were rated by both mothers and teachers as displaying more externalizing behaviors. In addition, Conger and Conger (1994) found that differential parental hostility was related to concurrent and subsequent child adjustment (e.g., delinquency). In short, children who receive less preferential parental treatment display lower academic achievement, less autonomy, less social maturity, poorer social relationships, and lower self-esteem than those children who receive similar treatment or preferential treatment from parents (Anderson et al., 1994; Brody et al., 1998, 1992; Deater-Deckard, Pike, Petrill, Cutting, Hughes, & O'Connor, 2001; Kowal & Kramer, 1997; McHale, Updegraff, Jackson-Newsom, Tucker, & Crouter, 2000; Volling & Belsky, 1992; Volling & Elins, 1998). Furthermore, some of the most recent research suggests that parental differential treatment is linked to children's adjustment *apart* from the effects of the level of parenting toward each child separately (Feinberg & Hetherington, 2001). That is, for example, after accounting for the level of parental treatment to older siblings, differential parenting to the siblings (older versus younger) contributes unique variance in older siblings' adjustment.

The current study will examine the relationship between maternal differential attention, involvement, and control and two areas of child well-being: general well-being (i.e., externalizing and internalizing behaviors, overall self-worth) and competence in extra-familial settings (i.e., scholastic and social competence). In an attempt to replicate findings from previous investigations, the current study will examine the associations between MDT and child well-being across three ages during the transition to adolescence. It is expected that preferential treatment (i.e., greater attention and involvement) will be related to higher scores of overall child well-being and competence in extra-familial settings.

Summary

Previous research has examined the stability of MDT, its etiology, and its links to children's adjustment. However, little longitudinal research has been done, especially using genetically sensitive designs during adolescence. Therefore, the goals of the current study were to (1) examine the stability of MDT over time, (2) explore genetic contributions to maternal reports of MDT, and (3) explore the relationship between MDT and child well-being (general well-being and competence in extra-familial settings). Descriptive issues were also addressed such as birth order and gender composition differences in MDT over time.

Method

Subjects

Participants include 109 sibling pairs and their mothers participating in the Colorado Sibling Study (CSS; Dunn et al., 1990; Stocker, Dunn, & Plomin, 1989), a longitudinal investigation of children's familial and extrafamilial relationships during middle childhood and adolescence. CSS represents a subset of siblings from the Colorado Adoption Project (CAP). Although CSS consists of five waves of data collection, data from only the last three waves were used in the present study. MDT data have been reported for wave one and two elsewhere (Dunn et al., 1990; McGuire et al., 1995).

Family members assessed as part of CSS include mothers and two siblings. Siblings were, on average, three years (35.5 months) apart in age (ranging from one to five years' difference). During the third wave of measurement (time 1 in this study), younger siblings were on average 12 years old, ranging from 9 to 15 years of age. Older siblings were on average 15 years old, ranging from 11.5 to 18 years of age. Because children were assessed at two follow-ups (times 2 and 3 in this study), one year apart, they were, on average, one year older at each time point. The 109 families in this investigation consisted of 47 adoptive and 62 nonadoptive (i.e., children living with their biological parents) families. The 47 adoptive families consisted of 5 boy-boy, 19 boy-girl, 18 girl-boy, and 5 girl-girl sibling pairs. The 62 nonadoptive families consisted of 20 boy-boy, 8 boy-girl, 17 girl-boy, and 17 girl-girl sibling pairs.

It should be noted that the sample is characterized by fluid attrition. That is, families participating at time 1 may have dropped out at time 2, but re-entered the study at time 3. Thus, families may have complete data for one, two, or three time-points. Sample sizes are noted on each table. Sixty families have complete data at all three time-points. These families do not differ significantly from those families with data at only one or two time-points in terms of MDT or children's well-being; consequently, we include all data that are available at each time point.

Measures

Mothers' Reports of Differential Treatment

As part of each yearly CSS interview, mothers completed and returned through the mail a questionnaire assessing MDT of their children. Mothers reported on their differential treatment of their two children using a 12-item modified version of the Sibling Inventory of Differential Experience (SIDE; Daniels & Plomin, 1985). This measure asks mothers to rate on a five-point Likert-type scale (1-older sibling much more, 3-siblings the same, 5-younger sibling much more), indicating both the relative direction (younger sibling more, older sibling more, or both children the same) and degree of differential treatment (a little more or much more) mothers exhibit toward their children. Answers given with this format can be rescored to provide an absolute scale that indicates the overall *degree* of MDT (i.e., much more, a bit more, the same). Mothers completed this modified version of the SIDE three times, once each during waves 3, 4, and 5 of CSS.

Items from the modified SIDE were used to create three subscales using the relative scores: *attention* (i.e., need to be particularly sensitive to their feelings, needs more attention, needs more support or leniency), *control* (i.e., have to be strict with, easier to discipline, have to remind one to do chores/things, one is more responsible for misdeeds), and *involvement* (i.e., tend to be prouder of, enjoy doing things with, easy to be affectionate toward). Two items were omitted (particularly interested in things they do, need to punish) because these items did not contribute to internal consistency of the subscales. Internal consistency for the subscales ranged from .48 to .74.

Well-Being

A variety of indices of well-being drawn from the yearly CAP assessments were used in the present study to investigate the hypotheses about the relationship between MDT and child well-being. In order to do this, CSS maternal differential treatment data and CAP well-being data were linked. It is important to note that CSS data were collected in specific calendar years (i.e., 1992, 1993, and 1994), whereas data collection for the CAP was based on a standard protocol of assessments administered and organized according to children's chronological age (e.g., 9-year-old CAP assessment, 10-year-old CAP assessment, etc.). Thus, in order to merge the data, the closest subsequent CAP data were selected for each child assessed at each of the three CSS time-points. This method of merging CAP and CSS data minimizes the time span between the CAP and CSS assessments for each child (See Manke & Carlson, chapter 13, this volume, for a more detailed description of methods used for merging CAP and CSS).

Well-being measures drawn from the CAP assessments include children's self-reports of general self-worth and competence in the areas of scholastic endeavors and conduct. These indices of well-being were assessed using questions based on Harter's (1983) Self-Perception Profile for Children. Children's self-reports of depression and loneliness were measured using Kandel's Depressive Mood Inventory (Kandel & Davies, 1982), and Asher's Loneliness Questionnaire (Asher, Hymel, Renshaw, 1984). These two scales were combined to create an internalizing behavioral scale that parallels the parents' reports. Indices of children's internalizing and externalizing problems were obtained from parents using the Child Behavior Checklist (Achenbach & Edelbrock, 1983). Finally, a measure of children's social competence was obtained from teachers using the Walker-McConnell Scale of Social Competence and School Adjustment (Walker & McConnell, 1988). Three subscales from this measure were used including Leadership, Problem Behavior, and Popularity. It should be noted that on the Problem Behavior subscale, a higher score indicates fewer problem behaviors compared to a lower score. Teacher data were not available for all children at all time-points, and thus, analyses of teachers' ratings of social competence should be interpreted with caution.

Results and Discussion

Descriptive Data

In order to address child gender, child age, and within sibling pair age differences in maternal differential attention, control, and involvement, relative scores from the modified SIDE were used in the following analyses. Three separate 2 (older sibling gender) × 2 (younger sibling gender) × 3 (time) mixed model ANOVAs using older sibling and younger sibling gender as independent variables and the dimensions of

TABLE 16.1. Means (and Standard Deviations) for Maternal Differential Attention, Control, and Involvement[1] by Gender of the Sibling Dyad

Maternal Differential	Boy–Boy (OS / YS)	Girl–Boy (OS / YS)	Girl–Girl (OS / YS)	Boy–Girl (OS / YS)
Attention				
Time 1 ($n = 90$)	3.00 (.67)	2.82 (.78)	3.16 (.74)	3.35 (.83)
Time 2 ($n = 75$)	3.10 (.75)	3.08 (.75)	3.47 (.80)	3.27 (.85)
Time 3 ($n = 86$)	3.27 (.74)	3.05 (.74)	3.39 (.68)	3.43 (.66)
Control				
Time 1 ($n = 91$)	2.75 (.73)	3.24 (.87)	3.13 (.69)	3.21 (1.09)
Time 2 ($n = 76$)	3.01 (.63)	3.25 (.75)	3.23 (.59)	3.13 (.90)
Time 3 ($n = 85$)	2.86 (.69)	3.18 (.80)	3.14 (.88)	3.05 (.77)
Involvement				
Time 1 ($n = 91$)	3.05 (.44)	3.05 (.50)	3.19 (.42)	2.92 (.58)
Time 2 ($n = 76$)	3.10 (.48)	3.08 (.49)	2.90 (.39)	3.00 (.47)
Time 3 ($n = 86$)	3.09 (.50)	3.21 (.73)	3.04 (.44)	2.99 (.39)

Note: [1]Scales represent relative scores from the modified version of the SIDE (Daniels & Plomin, 1985).

OS = older sibling; YS = younger sibling. The total sample consists of 109 sibling pairs, although the number of sibling pairs at each time point varies slightly as is noted above.

MDT as dependent measures with time as a repeated measure were conducted to address gender, sibling gender composition, and time effects. Table 16.1 shows means and standard deviations for MDT by gender composition of the sibling pair for the three time-points. In general, means suggest that younger siblings on average receive more preferential treatment (more attention and involvement, but also more control) than older siblings across most time-points. This is evidenced by the majority of the means being above 3; this indicates that mothers more often responded that younger siblings receive more attention, control, and involvement compared to older siblings. Analyses revealed no significant main effects or interactions involving gender, sibling gender composition, or time for any of the three dimensions of MDT. Regression analyses using older and younger sibling age and within sibling pair age difference separately as criterion variables were performed to determine if a child's age or siblings' age differences accounted for a significant proportion of the variance in reports of maternal differential attention, control, and involvement. No main effects were revealed for either within sibling pair age difference or child age for the subscales of the SIDE, suggesting that the children's age and the age difference between siblings does not explain MDT.

Results of the current study suggest that, on average, mothers report being more attentive, controlling, and involved with younger siblings compared to older siblings. These findings, in part, lend support to previous research that younger siblings are favored on average (McGuire et al., 1995), yet also receive more control from mothers (Brody et al.,

TABLE 16.2. Correlations for Maternal Differential—Attention, Control, and Involvement[1] Over Time

Maternal Differential	Time 1 to 2	Time 2 to 3	Time 3 to 1
Attention	.70*	.71*	.56*
Control	.71*	.65*	.58*
Involvement	.70*	.72*	.54*

Note: [1]Scales represent relative scores from the modified version of the SIDE (Daniels & Plomin, 1985); *p < .05

1992; Stocker et al., 1989). The current study revealed no differences in MDT by child gender, sibling gender composition, child age, or within sibling pair age differences. One possible reason that significant findings were not revealed for child variables such as gender or age may be that the areas addressed by the modified SIDE are quite broad. That is, the questions assessing maternal differential attention, involvement, and control may be too broad to capture differences in the nature of MDT as it would be affected by a child's age or gender and sibling pair gender composition. For example, one question from the modified SIDE addressing control simply asks mothers to address differential treatment on the item, "easier to discipline." The item does not address the context of discipline and how it may differ between siblings. Differences associated with age or gender might be found for more *specific* questions concerning control such as curfew, dating privileges, and expectations for sexual conduct.

The Stability of MDT during Early Adolescence

The stability of maternal differential attention, control, and involvement across three time-points was assessed using the relative scores from the modified SIDE (see table 16.2). Consistent with previous research that has found MDT to be stable in infancy and middle and late adolescence, maternal reports of MDT were found to be moderately stable across all three time-points as evidenced by pervasive and significant stability correlations for the three dimensions of MDT (correlations ranged from .54 to .71). That is, mothers report similar patterns of differential attention, involvement, and control over time. In comparison to previous studies addressing the stability of MDT, correlations from the current study were similar (moderate) or stronger. Correlations with a sample of infants were low to moderate (McGuire & Roch-Levecq, 2001), while those from a sample of children in middle childhood were moderate and somewhat stronger (McGuire et al., 1995). Current findings from maternal reports are in agreement regarding the magnitude of stability with maternal reports of MDT during middle childhood. These findings indicate that MDT is not transitory or idiosyncratic; although some change as well as continuity is indicated.

This instability may be due, in part, to the transitions that are occurring as children enter adolescence. For example, mothers may become more controlling of adolescents who are attempting to become more autonomous, while becoming more responsive to and involved with younger siblings who still view the family as of central importance in their lives. This shift in control and attention would change the stability of MDT over time. Another change or transition that may influence the stability of MDT is a child's entrance into puberty. Adolescents who appear and act more like adults, given the onset of pubertal development, may be given greater responsibility and have higher expectations placed upon them relative to their younger prepubescent siblings. The stability of MDT may change once again as the younger sibling begins puberty. That is, siblings who are discordant for pubertal status may be treated more differently by mothers compared to siblings who are concordant. To date, no study has examined the role that pubertal status plays in MDT during adolescence. Future analyses of pubertal status and concomitant new behaviors such as dating will be examined in CAP that might clarify differences in MDT between adolescents and their younger siblings.

The Genetic Contributions to Maternal Reports of MDT

Estimations of genetic influences are based on mean level differences between groups of siblings who vary in genetic relatedness. To test for genetic influences on maternal reports of maternal differential attention, control, and involvement from the direct measure of MDT, means based on absolute scoring (0 = treat siblings the same/no difference, 1 = treat siblings a little different/moderate difference, 3 = treat siblings very differently/great difference) were compared for adoptive and non-adoptive families. Genetic influences would be implied if the means for adoptive families were greater than means for the nonadoptive families. Three separate 2 (adoptive status) × 3 (time) ANOVAs using adoptive status as the independent variable and maternal differential attention, control, and involvement as dependent variables with time as a repeated measure were conducted to test for mean differences.

Attention and involvement consistently (except for attention at time 16.3) show larger adoptive than nonadoptive sibling means, suggesting genetic influence (see table 16.3). However, the differences are only significant at time 1 for attention [$F(1, 88) = 3.98$, $p < .05$] and at time 3 for involvement [$F(1, 84) = 4.49$, $p < .05$]. Nonetheless, this suggests some genetic influence, but in line with other studies (e.g., Daniels & Plomin, 1985), control shows no evidence of genetic influence.

TABLE 16.3. Means (and Standard Deviations) for Maternal Differential—Attention, Control, and Involvement[1] by Adoptive Status

	Adoptive Siblings		Nonadoptive Siblings	
Attention				
Time 1	.76[a] (.27)	$n = 41$	.63[b] (.34)	$n = 49$
Time 2	.76 (.45)	$n = 31$	.62 (.35)	$n = 44$
Time 3	.59 (.32)	$n = 38$	.60 (.36)	$n = 48$
Control				
Time 1	.74 (.19)	$n = 41$	.68 (.25)	$n = 50$
Time 2	.69 (.37)	$n = 31$	.67 (.29)	$n = 45$
Time 3	.66 (.26)	$n = 37$	.65 (.30)	$n = 48$
Involvement				
Time 1	.43 (.33)	$n = 40$	.33 (.32)	$n = 51$
Time 2	.46 (.29)	$n = 31$	.35 (.34)	$n = 45$
Time 3	.46[a] (.30)	$n = 38$	.31[b] (.32)	$n = 48$

Note: [a,b]Different letters denote significantly different means across adoptive siblings and non-adoptive siblings for that time point. [1]Scales for MDT reflect absolute, rather than relative, differences in sibling treatment (SIDE: Daniels & Plomin, 1985).

Differential attention and involvement may occur, in part, because mothers are responding to children's genetic differences. It may be, for example, that children who are more extroverted and agreeable (traits found to be genetically influenced) elicit more maternal attention and involvement. Although none of the other dimensions of MDT at any of the time points evidenced *significant* genetic influence, the means for adoptive and nonadoptive siblings were in the hypothesized direction (but not for control). That is, for the most part, the means for adoptive siblings were greater than those for nonadoptive siblings, suggesting the possibility of genetic contributions. The lack of significant findings may have been due to the small sample size, ranging from 31 to 41 adopted and 44 to 51 nonadoptive sibling pairs.

Overall Child Well-Being

In order to address the link between MDT and siblings' overall well-being, children's reports of internalizing behavior, self-esteem, and conduct, as well as maternal reports of internalizing and externalizing behaviors, were correlated with each of the MDT subscales at each time-point for older and younger siblings. Relative scores from the modified SIDE for maternal differential treatment were used. As scores from the modified SIDE reflect the relative nature of differential treatment (e.g., 1-older sibling much more, 5-younger sibling much more), the meaning or interpretation of the direction of the correlation will be different for older and younger siblings. For example, a negative correlation between attention and well-being for older siblings would indicate that older siblings whose mothers report being more attentive to younger siblings score lower on measures of well-being. In contrast, a negative correlation between attention and well-being for younger siblings would indicate that younger siblings whose mothers report being more attentive to *them* score lower on measures of well-being.

Table 16.4 reports correlations between MDT and children's well-being. In general, results suggest maternal differential attention and involvement, but not maternal differential control, are associated with general child well-being for both older and younger siblings. Although there are fewer significant correlations at times 1 and 2, results from all three time-points provide a coherent picture of the associations between MDT and general child well-being.

More specifically, older siblings whose mothers are more attentive to their younger siblings report less internalizing behaviors (time 3, $r = -.31$, $p < .05$), less conduct problems (time 2, $r = .28$, $p < .05$), and higher general self worth (time 3, $r = .35$, $p < .05$). Parents also report these older siblings to have less internalizing (time 1, $r = -.24$, $p < .05$; time 2, $r = -.35$, $p < .05$) and externalizing (time 1, $r = -.36$, $p < .05$) behavioral problems. In contrast, younger siblings whose mothers are more attentive to them report lower self worth (time 3, $r = -.23$, $p < .05$). Parents report these younger siblings to have more internalizing (time 3, $r = .25$, $p < .05$) and externalizing (time 3, $r = .30$, $p < .05$) behavioral problems.

Older siblings whose mothers are more involved with younger siblings report more internalizing behavioral problems (time 3, $r = .31$, $p < .05$), lower general self worth (time 3, $r = -.24$, $p < .05$), and feel worse about their competence in the area of conduct (time 3, $r = -.35$, $p < .05$). In addition, parents report these older siblings to have more internalizing (time 2, $r = .32$, $p < .05$; time 3, $r = .34$, $p < .05$) and externalizing (time 1, $r = .44$, $p < .05$; time 3, $r = .46$, $p < .05$) behavioral problems. However, younger siblings whose mothers are more involved with them demonstrate less externalizing behavioral problems (time 1, $r = -.29$, $p < .05$; time 3, $r = -.24$, $p < .05$) as reported by parents.

It is important to note that only one of the 30 correlations involving maternal differential *control* was significant (time 2, $r = .24$, $p < .05$), suggesting that this dimension of MDT has little or no direct relationship to aspects of children's well-being assessed in CAP. Furthermore, our results indicate that maternal differential attention and involvement are heritable, but control is not, suggesting that the correlations between attention and involvement and well-being might involve mothers *reflecting* their children's well-being.

In combination these results suggest different links between maternal differential attention and involvement and well-being. More maternal attention is related to negative well-being (e.g., externalizing behaviors), whereas more maternal involvement is linked with positive

TABLE 16.4. Correlations for Maternal Differential Attention, Control, and Involvement[1] and General Child Well-Being[2]

	Attention			Control			Involvement		
	Time 1	Time 2	Time 3	Time 1	Time 2	Time 3	Time 1	Time 2	Time 3
Child report of internalizing behavior problems									
Older sibling (97, 62, 73)[3]	-.08	-.03	-.31*	-.10	.14	-.08	.08	.04	.31*
Younger sibling (87. 79, 97)	.12	.17	.19	-.15	-.10	-.20	.05	-.03	-.12
Child report of general self-worth									
Older sibling (97, 61, 73)	.13	.17	.35*	.13	-.17	.12	-.18	-.11	-.24*
Younger sibling (87, 78, 93)	.06	-.14	-.23*	.05	-.01	.17	-.09	.01	.11
Child report of competence in the area of conduct									
Older sibling (97, 62, 73)	.16	.28*	.17	.02	-.11	-.01	-.12	-.19	-.35*
Younger sibling (88, 79, 96)	.10	-.04	-.17	-.09	-.13	-.05	-.18	.05	.17
Parent report of internalizing behavioral problems									
Older sibling (74, 59, 45)	-.24*	-.35*	-.15	.23	.11	.06	.23	.32*	.34*
Younger sibling (76, 80, 73)	.22	.13	.25*	-.01	-.06	-.05	-.05	-.01	-.03
Parent report of externalizing behavioral problems									
Older sibling (74, 59, 45)	-.36*	-.08	-.22	.08	.07	-.13	.44*	.26	.46*
Younger sibling (76, 80, 73)	.20	.22	.30*	.04	.24*	-.09	-.29*	-.12	-.24*

Note: [1]Scales represent relative scores from the modified version of the SIDE (Daniels & Plomin, 1985).
[2]Child's report of internalizing behaviors is a composite reflecting depression and loneliness. Child's report of self-esteem reflects the total score of Harter's measure of self-worth, and child's competence in the area of conduct reflects the Harter Conduct subscale (Harter, 1983). Parents' reports of internalizing and externalizing behaviors are derived from the Child Behavior Checklist (Achenbach & Edelbrock, 1983).
[3]Sample sizes for Time 1, 2, and 3, respectively.
*$p < .05$

well-being (e.g., general self-worth). It may be that differential maternal attention is the *result* of children's differential needs. That is, a more difficult child, one who is exhibiting both internalizing and externalizing problem behaviors, may actually elicit more attention (i.e., more leniency or support) from mothers who are concerned. In contrast, differential involvement (e.g., enjoy doing things with) may reflect mutual interests and the engagement in leisure activities. It is possible that a more well-adjusted child (e.g., higher self-worth and fewer internalizing problems) is easier to be around and thus is sought out by parents for engagement in leisure activities. Or, it may be that children who spend more leisure time with parents begin to feel better about themselves. Although no causal conclusions can be drawn from the correlational data, these results suggest that maternal differential attention and involvement are not synonymous—that is, the links between children's well-being and time spent with mothers depends on the underlying reasons for time spent (i.e., because the child's behavior demands it as with attention or because the parent desires to spend time as with involvement).

Competence in Extra-Familial Settings

In order to address the relationship between MDT and child competence in extra-familial settings, maternal differential attention, control, and involvement were correlated with children's reports of scholastic self-esteem, as well as teachers' reports of overall social competence, leadership, problem behavior, and popularity, at each time point for older and younger siblings. It should be noted that a higher score on the teacher reports measure indicates fewer problem behaviors. Relative scores from the modified SIDE were used, and therefore as stated previously, the direction of the correlation has different implications for older versus younger siblings. For example, a negative correlation between attention and competence in extra-familial settings for older siblings would indicate that older siblings whose mothers are more attentive with younger siblings report less competency in extra-familial settings. In contrast, a negative correlation between attention and competence in extra-familial settings for younger siblings would suggest that younger siblings whose mothers are more attentive to *them* report less competency in extra-familial settings.

Although not entirely consistent over times 1, 2, and 3, the results presented in table 16.5 parallel those in table 16.4 in that maternal differential attention and involvement, but not maternal control, are associated with children's competence in extra-familial settings. As with findings for general child well-being, results across the three time-points illustrate a coherent picture of the associations between MDT and child competence in extra-familial settings. Interestingly, maternal

TABLE 16.5. Correlations for Maternal Differential Attention, Control, and Involvement[1] and Children's Competence in Extra-Familial Settings[2]

	Attention			Control			Involvement		
	Time 1	Time 2	Time 3	Time 1	Time 2	Time 3	Time 1	Time 2	Time 3
Child report of scholastic self-esteem									
Older sibling (97, 61, 73)[3]	.17	.35*	.34*	.17	-.08	.06	-.11	-.31*	-.37*
Younger sibling (88, 79, 97)	.01	-.17	-.15	.08	-.18	.11	-.06	.04	.23*
Teacher report of leadership									
Older sibling (53, 39, 24)	.17	.15	.41	.04	-.17	.15	-.19	-.10	-.38
Younger sibling (60, 60, 48)	.03	-.34*	-.20	-.10	-.04	.27	-.02	.22	.05
Teacher report of problem behaviors									
Older sibling (53, 39, 24)	.05	.40*	.45*	.00	-.30	-.07	-.32*	-.42*	-.63*
Younger sibling (61, 60, 48)	.01	-.19	-.30*	-.05	-.35*	.15	-.05	.16	.23
Teacher report of popularity									
Older sibling (53, 39, 24)	.14	.17	.38	-.04	-.21	.33	-.34*	-.15	-.38
Younger sibling (61, 60, 48)	-.06	-.32*	-.23	-.10	-.11	.14	.09	.05	.09

Note: [1]Scales represent relative scores from the modified version of the SIDE (Daniels & Plomin, 1985).
[2]Child's report of scholastic competence reflects the scholastic subscale of the Harter's measure (Harter, 1983). Teachers' reports of leadership, problem behavior, and popularity reflect subscales of the Walker McConnel Scale of Social Competence and School Adjustment (Walker & McConnel, 1988).
[3]Sample sizes for Time 1, 2, and 3, respectively.
*p < .05

differential attention and involvement are more consistently related to older siblings' competence.

More specifically, older siblings whose mothers are more attentive to younger siblings demonstrate higher scholastic self-esteem (time 2, $r = .35$, $p < .05$; time 3, $r = .34$, $p < .05$), and fewer problem behaviors as reported by teachers (time 2, $r = .40$, $p < .05$; time 3, $r = .55$, $p < .05$). In contrast, younger siblings whose mothers are more attentive to them exhibit less leadership ability (time 3, $r = -.34$, $p < .05$), less popularity (time 2, $r = -.32$, $p < .05$), and more problem behaviors (time 3, $r = -.30$, $p < .05$) as reported by teachers.

In contrast, older siblings whose mothers are more involved with younger siblings demonstrate less scholastic self-esteem (time 2, $r = -.31$, $p < .05$; time 3, $r = -.37$, $p < .05$), and more behavioral problems (time 1, $r = -.32$, $p < .05$; time 2, $r = -.42$, $p < .05$; time 3, $r = -.63$, $p < .05$), and less popularity (time 1, $r = -.34$, $p < .05$) as reported by teachers. In contrast, younger siblings whose mothers are more involved with them report higher scholastic self-esteem (time 3, $r = .23$, $p < .05$). It is important to note that, as with the results in Table 16.4, only one of the 30 correlations involving maternal differential control was significant (time 2, $r = -.35$, $p < .05$), again suggesting that this dimension of MDT shows little or no relationship to children's behavior.

In combination, results for maternal differential attention and involvement for both older and younger siblings suggest differential links to children's competence in extra-familial settings. Similar to findings regarding overall child well-being, more relative maternal attention is related to less competence whereas more relative maternal involvement is linked with greater competence in extra-familial settings. It may be no coincidence that there is a similar pattern of results in table 16.4. That is, maternal differential attention is correlated with teachers' ratings of children's adjustment, but control is not, perhaps again due to genetic factors. That is, teachers may be responding to the same characteristics that mothers respond to, characteristics that may be genetically influenced.

Future Directions

The current study is characterized by a number of limitations that point to areas for future research, namely the use of a single reporter, the use of broad dimensions of MDT, and the incorporation of limited time-points. First, this study relied solely on maternal reports. Fathers' reports of differential treatment are not often assessed, although they could provide a unique perspective on their own and mothers' behaviors in regard to children in the same family. Paternal differential treatment may be linked with child outcomes such as self-esteem and sibling relationship quality in ways that differ from MDT (Brody et al., 1992).

Comparing parent and child reports of sibling differential experiences would also be useful. Parents and children may not perceive the nature and extent of MDT similarly. Thus, stability and well-being correlates of MDT may vary by reporter. Associations between maternal differential control and well-being may be more pervasive for child reports, as opposed to maternal reports.

Incorporating multiple reporters of MDT would also allow us to examine the link between MDT and children's well-being with more precision. Previously we suggested that the association between MDT and children's well-being might be genetically mediated. Given our results suggesting genetic influences on MDT (i.e., differential attention and involvement), we feel it is reasonable to propose that the link between MDT and well-being is the result of mothers' responding to children's genetic propensities related to well-being. Because the current study relied on a single report of MDT (i.e., mothers' reports), we could not conduct multivariate analyses to decompose the association between MDT and children's well-being. In order to conduct such analyses, we would need child-specific measures of MDT (e.g., each sibling's report of MDT). In sum, the use of multiple reporters allows for a more comprehensive picture of family functioning.

Another limitation in MDT research and the current study is the general nature of the areas or dimensions of treatment addressed. It may be that the questions measuring MDT are too broad and therefore do not capture subtle day to day MDT. That is, the areas of attention, involvement, and control addressed in the current measure may reflect general parenting strategies. In other words, parents may demonstrate a similar level of control with all children regardless of child characteristics including age, gender, and birth order. These global areas, therefore, may be too broad to capture the associations between MDT and child well-being. Perhaps questions targeting more specific areas of control, for example, might reveal more interesting associations between maternal differential control and child well-being. Maternal control related to dating privileges, curfew, appropriate physical boundaries for outside play, and expectations for sexual conduct may be particularly salient during adolescence when issues of autonomy are relevant.

Future studies should incorporate not only more specific questions addressing MDT but also children's perceptions of fairness and attributions of underlying causes. In a previous study by Kowal and Kramer (1997), children's attributions about fairness in the family were related to perceptions of higher levels of warmth and closeness, less conflict, and less status/power differential in the sibling relationship. These findings, in addition to those of McHale and Pawletko (1992) addressing

MDT in families with a disabled child, point to the importance of addressing not only the occurrence and nature of differential treatment but also the perceived cause and fairness of such treatment. Both children's perceptions of reasons underlying MDT and fairness may have greater impact on such constructs as child well-being and sibling relationship quality than their perceptions of the occurrence of MDT.

Although the current study was unique in that it incorporated data from three time-points over a three-year span during early adolescence, additional time-points would allow for the investigation of more complex time-related questions. For example, one interesting question related to the stability of MDT is whether or not MDT is consistent over *age*? That is, do mothers treat children similar when they are the same age (e.g., 10 years old)? It may be that a specific age elicits a certain amount of attention, involvement, and control in a mother-child relationship. That is, there may be something unique about being 6 years old, 10 years old, or 16 years old. A child beginning school, for example, may elicit from mothers a certain level of attention, involvement, and control that centers on the circumstances of that child. These same levels of maternal attention, involvement, and control are again elicited when the next child reaches school age. The question of consistency of MDT across age could not be addressed in the current study because not enough siblings were the same age over the span of the three time-points.

In sum, the present study suggests that maternal differential treatment during the transition to adolescence is moderately stable, influenced by genetic factors, and associated with children's well-being. Future directions in the area of MDT, and more broadly parental differential treatment, should focus on improving three areas: rater, measurement, and number of time points.

REFERENCES

Achenbach, T. M., & Edelbrock, C. (1983). *Manual for the Child Behavior Checklist and Revised Child Behavior Profile*. Burlington: University of Vermont, Department of Psychology.

Anderson, E., Hetherington, M., Reiss, D., & Howe, G. (1994). Parents' non-shared treatment of siblings and the development of social competence during adolescence. *Journal of Family Psychology, 8*, 303–321.

Asher, S., Hymel, S., & Renshaw, P. (1984). Loneliness in children. *Child Development, 55*, 1456–1464.

Baker, L., & Daniels, D. (1990). Nonshared environmental influences and personality differences in adult twins. *Journal of Personality and Social Psychology, 58*, 103–110.

Bischof, L. G., & Tingstrom, D. H. (1991). Siblings of children with chronic disabilities: Psychological and behavioral characteristics. *Counseling Psychology Quarterly, 4*, 311–321.

Brody, G., Copeland, A., Sutton, L., Richardson, D., & Guyer, M. (1998). Mommy and daddy like you best: Perceived family favourtism in relation to affect, adjustment, and family processes. *Journal of Family Therapy, 20,* 269–291.

Brody, G., Stoneman, Z., & McCoy, K. (1992). Associations of maternal and paternal direct and differential behavior with sibling relationships. *Child Development, 63,* 82–92.

Conger, K., & Conger, R. (1994). Differential parenting and change in sibling differences in delinquency. *Journal of Family Psychology, 8,* 287–302.

Daniels, D., & Plomin, R. (1985). Differential experience of siblings in the same family. *Developmental Psychology, 21,* 747–760.

Deater-Deckard, K, Pike, A., Petrill, S., Cutting, A. L., Hughes, C., & O'Connor, T. G. (2001). Nonshared environment processes in social-emotional development: An observational study of identical twin differences in the preschool period. *Developmental Science, 4,* 1–6.

Dunn, J., & Plomin, R. (1990). *Separate lives: Why siblings are so different.* New York: Basic Books.

Dunn, J., Stocker, C., & Plomin, R. (1990). Nonshared experiences within the family: Correlates of behavior problems in middle childhood. *Development and Psychopathology, 2,* 113–126.

Feinberg, M., & Hetherington, E. M. (2001). Differential parenting as a within-family variable. *Journal of Family Psychology, 15,* 22–37.

Harter, S. (1983). Developmental perspectives on the self-system. In M. Hetherington (Ed.), P. H. Mussen (Series Ed.), *Handbook of child psychology: Socialization, personality, and social development* (Vol. 4, pp. 275–385). New York: John Wiley.

Harris, J. H. (1998). *The Nurture assumption: Why children turn out the way they do.* New York: Touchstone.

Kandel, D., & Davies, J. (1982). Epidemiology of depressive mood in adolescents. *Archives of General Psychiatry, 39,* 1205–1212.

Kowal, A., & Kramer, L. (1997). Children's understanding of parental differential treatment. *Child Development, 68,* 113–126.

McGuire, S. (in press). Nonshared environment research: What is it and where is it going? *Marriage and Family Review.*

McGuire, S., Dunn, J., & Plomin, R. (1995). Maternal differential treatment of siblings and children's behavioral problems: A longitudinal study. *Development and Psychopathology, 7,* 515–528.

McGuire, S., McHale, S., & Updegraff, K. (1996). Children's perceptions of the sibling relationship in middle childhood: Connections within and between family relationships. *Personal Relationships, 3,* 229–239.

McGuire, S., & Roch-Levecq, A. (2001). Mothers' perceptions of differential treatment of infant twins. In R. N. Emde & J. K. Hewitt (Eds.), *Infancy to early childhood: Genetic and environmental influences on developmental change* (pp. 247–256). New York: Oxford University Press.

McHale, S. M., & Pawletko, T. (1992). Differential treatment of siblings in two family contexts. *Child Development, 63,* 68–81.

McHale, S. M., Updegraff, K. A., Jackson-Newsom, J., Tucker, C. J., & Crouter, A. C. (2000). When does parents' differential treatment have negative implications for siblings? *Social Development, 9,* 149–172.

Pike, A., Manke, B., Reiss, D., & Plomin, R. (2000). A genetic analysis of differential experiences of adolescent siblings across three years. *Social Development, 9,* 96–114.

Plomin, R. (1990). *Nature and nurture.* Pacific Grove, CA: Brooks/Cole Publishing Company.

Plomin, R., & Daniels, D. (1987). Why are children in the same family so different from each other? *The Behavioral and Brain Sciences, 10,* 1–16.

Plomin, R., Reiss, D., Hetherington, M., & Howe, G. (1994). Nature and nurture: Genetic contributions to measures of the family environment. *Developmental Psychology, 30,* 32–43.

Rowe, D. (1983). A biometric analysis of perceptions of family environment: A study of twin and singleton sibling kinships. *Child Development, 54,* 416–423.

Steinberg, L. (1987). Impact of puberty on family relations: Effects of pubertal status and pubertal timing. *Developmental Psychology, 23,* 451–460.

Stocker, C., Dunn, J., & Plomin, R. (1989). Sibling relationships: Links with child temperament, maternal behavior, and family structure. *Child Development, 60,* 715–727.

Volling, B., & Belsky, J. (1992). The contributions of mother-child and father-child relationships to the quality of sibling interaction. *Child Development, 63,* 1209–1222.

Volling, B., & Elins, J. (1998). Family relationships and children's emotional adjustment as correlates of maternal and paternal differential treatment: A replication with toddlers and preschool siblings. *Child Development, 69,* 1640–1656.

Walker, H. M., & McConnell, S. W. (1988). *Walker-McConnell Scale of Social Competence and School Adjustment.* Austin, TX: Pro-Ed.

Wolf, L., Fisman, S., Ellison, D., & Freeman, T. (1998). Effect of sibling perception of differential parental treatment in sibling dyads with one disabled child. *Journal of the American Academy of Child and Adolescent Psychiatry, 37,* 1317–1325.

KIMBERLY J. SAUDINO
JEFFREY R. GAGNE
MADELINE BECKER

17

Genetic Influences on Life Events in Middle Childhood and Early Adolescence

Introduction

One of the most interesting findings to emerge from recent behavioral genetic research that genetic factors contribute substantially to many measures designed to assess the environments of individuals (See Plomin, 1994, for a review). When treated as a phenotype in genetic analyses, ostensible "environmental measures" such as ratings of the family environment, peer groups, social support, and divorce often show as much genetic influence as measures of personality (Plomin & Bergeman, 1991). Self-reports of life events are one widely used environmental measure that has consistently displayed genetic influence in late adolescent and adult samples (e.g., Billig, Hershberger, Iacono, & McGue, 1996; Kendler, Neale, Kessler, Heath, & Eaves, 1993; Moster, 1990, cited in McGue, Bouchard, Lykken, & Finkel, 1991; Plomin, Lichtenstein, Pedersen, McClearn, & Nesselroade, 1990).

The finding of genetic influences on life events suggests that, to some extent, experiences and life events do not happen capriciously to the individual. In fact, life events appear to be heritable to the extent to which an individual has some influence on the events themselves. For example, life events researchers often make a distinction between "independent" and "nonindependent" events. *Independent* events are those events that are independent of the individual's actions (e.g., death of spouse, illness). The individual has no control over independent life events, and consequently, these events are sometimes referred to as "uncontrollable" life events. *Nonindependent* events refer to life events that are potentially dependent on the individual's behavior (e.g., loss of job, getting married). Because the individual can play an active role in determining or shaping nonindependent life events, they are sometimes referred to as "controllable" life events. Across a variety of samples, genetic analyses of life events measures in late adolescence and adulthood converge in finding significant genetic influence on nonindependent (controllable) events, but not for independent (uncontrollable) events (Billig et al., 1996; Kendler et al., 1993; Moster, 1990, in McGue et al., 1991; Plomin et al., 1990).

The present study extends this research by exploring genetic influences on life events in middle childhood and early adolescence. There are many ways in which the life events of our younger sample may differ from those of adults. For example, the types of life events experienced and the degree to which individuals have control over their environment likely change as individuals enter adulthood. Children and young adolescents are dependent on their parents (or other adults) to meet their basic needs (i.e., food, shelter). Parents typically set rules, which children are expected to follow, and make decisions with regard to major life changes (e.g., it is the parent who decides to move residences). Thus, many events that would be nonindependent or controllable events for adults are independent or uncontrollable for children and young adolescents (e.g., change in financial status, divorce). Therefore, we were curious to see if life events in middle childhood and early adolescence would also show genetic influences.

We also wondered if the magnitude of genetic influence on life events might change across age. As children mature and enter adolescence they becomes more independent of their parents, and this increased autonomy may result in individuals playing a more active role in the life events that they experience. Indeed, in their theory of niche picking, Scarr and McCartney (1983) proposed that as children move from late childhood to early adolescence, there is an increase in the active seeking or creation of environments that are correlated with their genetically influenced characteristics. If there are developmental increases in the extent to which individuals have influence over their experiences, we can then expect to find increased genetic influence on life events from middle childhood to early adolescence.

Measures and Procedures

In the Colorado Adoption Project (CAP), there are two measures of life events and two types of informants. These are summarized in table 17.1.

TABLE 17.1. Summary of Life Events Measures Used in CAP

Life Events Measure	Age of Child				
	7	9	10	11	12
Social Readjustment Rating Scale—Parent					
SRRQ total	✓	✓	✓		
SRRQ parent stress	✓	✓	✓		
Social Readjustment Rating Scale—Child					
SRRQ child stress	✓	✓	✓		
Life Events Scale for Adolescents—Child					
LESA total		✓	✓	✓	✓
LESA stress		✓	✓	✓	✓

Note: TOTAL scores reflect the total number of events that had occurred within 12 months prior to testing. STRESS scores reflect the total number of events weighted by the parent or child's rating of how upsetting the event was.

When the children were ages 7, 9, and 10, both parents and children completed the Social Readjustment Rating Questionnaire for elementary school children (SRRQ; Coddington, 1972). The SRRQ is a 33-item scale that is based on the Social Readjustment Rating Scale (Holmes & Rahe, 1967), modified for use with children (Coddington, 1972). The SRRQ is designed to assess major life events that occur in the lives of children (e.g., beginning school, death of parent or grandparent, divorce of parents, failure of grade in school). Parents were asked to complete the questionnaire by indicating if the event had occurred in the child's life during the past year. For those events that did occur, parents were asked to indicate on a four-point scale (0 = not at all, 3 = very much) how stressful the event was for the child. For the children, the SRRQ was administered using a semistructured interview format. The interviewer, blind to the parents' ratings of the events, asked the child to discuss the life events that the parent indicated had occurred in the past year. The child was asked, "How did [the event] make you feel? How upset were you?" and asked to rate the degree of upsettingness on the same four-point scale that was used by the parents. There were three SRRQ scores computed at each age: (1) the total number of events as indicated by the parent [SRRQ total]; (2) the overall stressfulness rating as indicated by parents [SRRQ parent stress]; and (3) the overall stressfulness rating as indicated by the child [SRRQ child stress]. The two stressfulness scores reflect the total number of events weighted by the parent or child's rating of how upsetting the event was.

At 9, 10, 11, and 12 years of age, children were interviewed using a second life events measure, the Life Events Scale for Adolescents (LESA; Brooks-Gunn & Petersen, 1984). This measure differs from the

SRRQ in that it does not assess *major* life events such as a death in the family or parent's divorce; rather, the LESA assesses more minor events and daily hassles that are typically encountered during adolescence (e.g., dating, friendships, fights with siblings, quarrels with parents). Children were orally presented with a list of 54 items and asked whether or not the event had occurred within the past 12 months. For each event that occurred, the child was then asked, "How did you feel about it?" and asked to rate the upsettingness of each event on a seven-point scale (1 = very upset, 7 = very pleased). As with the SRRQ, a score indicating the total number of events experienced within the previous 12 months [LESA total] and a score reflecting the total number of events weighted by the child's rating of how upsetting the event was [LESA stress] were derived.

We used total events scores and total stress scores for both measures because neither the SRRQ nor the LESA has separate scales for independent and nonindependent events. Nonetheless, 25 of the 33 items (76%) on the SRRQ reflect major life events that are largely *independent* of the child or adolescent (i.e., events that happened to other individuals and/or over which the child had no control). For example, seven items refer to the health of other family members, five items assess changes in the parent's marital relationship, and six items assess parent's finances, work, or incarceration. Thus, the SRRQ tends to measure life events that are more specific to the family than to the child or adolescent in question. The LESA has a very different focus. Only 12 of the 54 items (22%) on the LESA reflect events that were independent of the adolescent. The large majority of items (78%) on the LESA assess events in which the adolescent potentially played an active role (i.e., *nonindependent* events). For example, 11 items assess events related to the adolescent's dating or social relationships, 14 items assess quarrels or changes in the adolescent's relationships with parents and siblings, and 6 items refer to the adolescent's behavior at school. Thus, although we did not specifically examine independent and nonindependent events, our two measures roughly map onto to these two categories of events (i.e., SRRQ = independent; LESA = nonindependent).

Design

The sibling adoption design involves comparing the similarity of adoptive and nonadoptive sibling pairs to assess genetic and environmental contributions to the phenotypic variance of a measure. Genetic variance is implied when nonadoptive siblings who share approximately 50% of their segregating genes are more similar for a measure than adoptive siblings who are not genetically related. Intraclass correlations typically serve as indices of sibling similarity. Correlations for non-adoptive siblings that exceed those for adoptive siblings suggest genetic

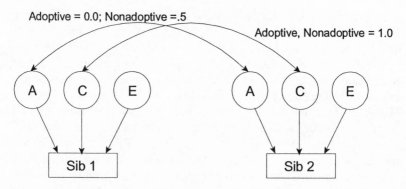

FIGURE 17.1. Univariate genetic model.

influence. Shared environmental influences are suggested when the correlation for adoptive siblings are greater than zero.

Model-Fitting Analyses

In addition to correlational analyses, model-fitting analyses were used to estimate genetic and environmental contributions to the CAP measures of life events. Maximum-likelihood model-fitting analyses were performed on sibling variance/covariance matrices using Mx (Neale, 1997). The full univariate model, depicted as a path diagram in figure 17.1, uses sibling covariances to decompose the phenotypic variance of a measure into genetic and environmental components. The rectangles represent the phenotypic values of each sibling. The circles represent latent genetic and environmental factors. The curved double-headed arrows indicate correlations between the variables they connect, whereas the single-headed arrows represent standardized partial regressions of the measured variable on the latent variable.

The A factors refer to additive genetic influences—the sum of the average effect of all genes that influence a trait. Based on the degree of genetic relatedness, the A factors correlate 0 and .5 for adoptive and nonadoptive siblings, respectively. Shared environment (C) refers to the influence of shared rearing environments on sibling resemblance. Because all siblings were reared in the same family, both adoptive and nonadoptive correlate 1.0. Finally, the E factors reflect nonshared environmental influences and measurement error. Nonshared environmental influences are those environmental factors that are unique to each member of a sibling pair and that make siblings different from each other (e.g., illnesses or accidents, and measurement error) and therefore these are depicted in the path diagram as residual arrows for each sibling representing the remaining variance not explained by the genetics or shared environment.

Results

Descriptive Statistics

Table 17.2 lists the means and standard deviations of the SRRQ total number of life events, SRRQ parent ratings of stressfulness, SRRQ child ratings of stressfulness, the total number of life events on the LESA, and the stressfulness ratings for the LESA. Overall, the mean number of life events endorsed on the SRRQ scales was low. This is consistent with previous adolescent and adult research using similar measures (e.g., Cohen, Burt, & Bjorck, 1987; DuBois, Felner, Brand, Adan, & Eaves, 1992; Kendler et al., 1993). In contrast to the SRRQ, the number of items endorsed on the LESA is substantially higher. The difference between the SRRQ and the LESA likely reflects the fact that the SRRQ assesses major life events whereas the LESA taps minor events and daily hassles that are more typical in adolescent life.

To evaluate mean differences, repeated measures analyses of variance (ANOVA) with age as a within-subjects variable, and adoptive status and gender as between-subjects variables, were conducted for each measure. When significant interactions were found, follow-up Tukey's tests were performed. An alpha level of .05 was used for all statistical tests.

For all three SRRQ scales, males and females did not significantly differ; however, there were significant main effects for age and adoptive status, SRRQ total: F_{Age} (2,746) = 22.03, $p < .01$; F_{Astat} (1,373) = 8.60, $p < .01$; SRRQ parent stress: F_{Age} (2,542) = 3.04, $p < .05$; F_{Astat} (1,271) = 7.80, $p < .01$; SRRQ child stress: F_{Age} (2,528) = 7.06, $p < .01$; F_{Astat} (1,264) = 7.22, $p < .01$. The total number of events reported decreased with age; however, there was no clear linear age trend for parent or child ratings of stressfulness. The means for adoptive families were significantly lower than those for nonadoptive families on all three measures, perhaps indicating the adoption agency's desire to place children in optimally functioning families. It should be noted, however, that although significant, the mean differences between adoptive and non-adoptive families were slight (i.e., the effect sizes were small ranging from .27 to .34).

Both the total number of events and the stressfulness ratings on the LESA displayed significant main effects for age, LESA total: F_{Age} (3,1026) = 6.65, $p < .001$; LESA stress: F_{Age} (3,993) = 7.37, $p < .001$. For both measures, scores were highest at age 12. There were significant gender by adoptive status interactions for both measures, LESA total:

TABLE 17.2. Means and Standard Deviations for CAP Life Events Measures at Ages 7, 9, 10, 11, and 12 Years of Age

Measure	7 Years			9 Years			10 Years			11 Years			12 Years		
	M	SD	N	M	SD	N	M	SD	N	M	SD	N	M	SD	N
SRRQ total	2.82	1.74	(394)	2.37	1.80	(419)	2.13	1.74	(403)						
SRRQ parent stress	4.13	3.35	(389)	4.70	3.42	(356)	4.62	3.60	(332)						
SRRQ child stress	3.50	3.30	(387)	4.35	3.24	(353)	4.10	3.22	(326)						
LESA total				11.50	5.11	(419)	11.18	5.03	(403)	10.93	5.09	(380)	12.05	5.99	(355)
LESA stress				46.28	17.50	(407)	45.58	17.92	(398)	46.49	19.77	(375)	48.83	22.58	(351)

Note: SRRQ = Social Readjustment Rating Questionnaire. LESA = Life Events Scale for Adolescents

TABLE 17.3. Intercorrelations Between the Social Readjustment Questionnaire and the Life Events Scale for Adolescents at 9 and 10 Years of Age

Social Readjustment Rating Questionnaire	Life Events Scale for Adolescents			
	TOTAL	n	STRESS	n
Parent				
9 years	.22**	(419)	.10*	(356)
10 years	.16**	(403)	.13*	(331)
Child				
9 years			.09	(353)
10 years			.08	(326)

*p < .05; **p < .01

$F_{\text{GenderXAstat}}$ (3,1026) = 6.65, $p < .001$; LESA stress: $F_{\text{GenderXAstat}}$ (3,993) = 7.37, $p < .001$. However, follow-up Tukey's tests revealed no significant differences between gender/adoptive status groupings.

Phenotypic Correlations

Parent-Child Agreement The question of parent-child agreement for reports of life events has rarely been examined. In CAP, both parent and child rated the stressfulness of events on the SRRQ and, therefore, provided us with a unique opportunity to evaluate the convergence across raters. The correlations between parent and child ratings were high, $r_7 = .69, p < .0001$; $r_9 = .73, p < .0001$; $r_{10} = .75, p < .0001$; however, these correlations are influenced by the number of events reported. Partial correlations adjusting for the total number of events were $r_7 = .36, p < .0001$; $r_9 = .37, p < .0001$; $r_{10} = .41, p < .0001$. Thus, there is moderate agreement between parent and children ratings of the stressfulness of life events. In terms of mean differences between parent and self-ratings, paired t-tests indicated that at all ages, parents' ratings of the stressfulness of events were significantly higher than those reported by their children, age 7: $t = 4.51, p < .0001$; age 9: $t = 2.42, p < .05$; age 10: $t = 3.62, p < .001$.

Cross-Scale Agreement Table 17.3 presents the intercorrelations between the SRRQ and LESA measures at 9 and 10 years of age. Overall, the correlations across measures were modest both for total number of life events experienced and for the stressfulness ratings of events. In fact, the correlations for child ratings of stressfulness across the two measures were not significantly different from zero. Thus, it would appear that the two measures tap different aspects of adolescent life events.

TABLE 17.4. Age-to-Age Stability for Life Events Measures in CAP

	SRRQ			LESA	
Age Interval	Total	Parent Stress	Child Stress	Total	Stress
7–9 years	.20**	.13*	.03		
7–10	.23**	.19**	.15*		
9–10	.34**	.37**	.25**	.47**	.44**
9–11				.39**	.34**
9–12				.48**	.34**
10–11				.54**	.48**
10–12				.51**	.43**
11–12				.52**	.50**

Note: SRRQ = Social Readjustment Rating Questionnaire. LESA = Life Events Scale for Adolescents;

$*p < .05; **p < .01$

Age-to-Age Stability   With the exception of the correlation for child reports of stress on the SRRQ across ages 7 and 9, age-to-age correlations for all measures were significant, indicating that life events show some stability across age (see table 17.4). This is particularly true for the LESA measures, which show moderate age-to-age correlations as compared to the more modest correlations for the SRRQ measures. In general, children who experience a high number of events at one age tend to experience a high number of events at other ages. Similarly, the degree of stress associated with life events at one age is significantly related to stress at other ages. The difference in stability between the two life events scales might, once again, reflect the types of life events items assessed. For example, on the SRRQ, major life events such as divorce or parental death are unlikely to be repeated across age; however, quarrels with parents or friends (items from the LESA) could be expected to occur frequently.

Sibling Intraclass Correlations

Intraclass correlations for adoptive and nonadoptive siblings are presented in table 17.5. As can be seen, the results vary across measure and age.

Social Readjustment Rating Questionnaire   In general, there is little evidence of genetic influences on the SRRQ scales. In fact, at age 7, genetically unrelated adoptive siblings were *more* similar than nonadoptive siblings on all three SRRQ measures. For both parent and child ratings of the stressfulness of experienced events, the correlations for adoptive siblings, although modest, were significant whereas the correlations for nonadoptive siblings were nonsignificant and close to zero. This pattern

TABLE 17.5. Sibling Intraclass Correlations and Estimates
of Variance Components from the Univariate Model

	Adoptive		Nonadoptive		Variance Components[a]		
Age and Measure	r	n (pairs)	r	n (pairs)	h^2	c^2	e^2
SRRQ							
Age 7							
Total	.13	(86)	.08	(103)	.00	**.14**	**.86**
Parent stress	.27*	(84)	.01	(101)	.00	**.13**	**.87**
Child stress	.24*	(84)	.01	(108)	.00	**.11**	**.89**
Age 9							
Total	.18†	(95)	.24*	(111)	.03	**.23**	**.74**
Parent stress	.14	(67)	.21†	(80)	.03	**.18**	**.79**
Child stress	.01	(66)	.13	(78)	.21	.02	**.77**
Age 10							
Total	.10	(86)	.14	(104)	.06	**.14**	**.80**
Parent stress	.11	(57)	.31*	(73)	.30	**.16**	**.54**
Child stress	.19	(52)	.14	(72)	.00	**.17**	**.83**
LESA							
Age 9							
Total	−.06	(95)	.17†	(111)	**.39**	.00	**.61**
Stress	.10	(88)	.11	(107)	.03	.09	**.88**
Age 10							
Total	.14	(86)	.26*	(104)	**.29**	**.14**	**.57**
Stress	.04	(83)	.22*	(103)	**.38**	.05	**.57**
Age 11							
Total	.08	(76)	.28*	(93)	**.43**	.07	**.50**
Stress	−.04	(76)	.16	(90)	**.35**	.00	**.65**
Age 12							
Total	−.23*	(69)	.26*	(76)	**.41**	.00	**.59**
Stress	−.21†	(67)	.28*	(75)	**.44**	.00	**.56**

Note: SRRQ = Social Readjustment Rating Questionnaire. LESA = Life Events Scale for Adolescents. h^2 =
genetic (additive) variance, c^2 = shared environmental variance, e^2 = nonshared environmental variance.
[a]Values that are in bold indicate parameters that were retained in the best-fitting model.
†$p < .10$; *$p < .05$

suggests that there might be some special shared environmental influence on life events that is operating in adoptive families at this age. At
ages 9 and 10 years, nonadoptive siblings are only *slightly* more similar
than adoptive siblings for the total number of events experienced on
the SRRQ. Correlations that are similar in magnitude for both adoptive and nonadoptive siblings suggest shared environmental influences;
however, it should be noted that at age 10 neither the adoptive nor the
nonadoptive correlations were significant. There are hints that parent
ratings of how stressful the experienced events were for the child may
be genetically influenced at ages 9 and 10. The correlations for nonadoptive siblings on the SRRQ parent stress scales were significant or

approached significance, but those for adoptive siblings did not. Child ratings of the stressfulness of life events do not, however, appear to be genetically influenced. Neither the nonadoptive nor the adoptive correlations were significant. Moreover, there was no consistent pattern of nonadoptive siblings being more similar than adoptive siblings.

Life Events Scale for Adolescents   Sibling correlations for the Life Events Scale for Adolescents (LESA) are more consistent with the hypothesis of a genetic influence on life events (table 17.5). With the exception of stress scores at 9 years, correlations for nonadoptive siblings exceed those for adoptive siblings, suggesting genetic influences on self-reports of life events in early adolescence. Moreover, nonadoptive siblings generally show a pattern of significant or near significant similarity for self-reports of life events on this measure, whereas adoptive siblings show little resemblance. In fact, the pattern of negative correlations at age 12 suggests that adoptive siblings show substantial *differences* in their reports of life events in early adolescence.

Model-Fitting Analyses

Variance components (i.e., estimates of heritability, shared and non-shared environmental variance) for each life events measure are presented in table 17.5. These model-fitting estimates are consistent with our interpretation of the sibling intraclass correlations. At all ages, the SRRQ measures of life events did not display significant genetic influence (i.e., the genetic parameter was not retained in the best-fitting model). Individual differences in the SRRQ measures tend to be due largely to nonshared environmental influences (which includes measurement error), although there are some modest shared environmental influences. In contrast to the SRRQ, both the total number of life events experienced and the ratings of stressfulness on the LESA appear to be genetically influenced in early adolescence. With two exceptions, ratings of the stressfulness of experienced events at age 9 and the total number of life events at age 10, heritability estimates for life events on the LESA were significant, substantial, and similar in magnitude across age. Moreover, there was little evidence of shared environmental influences on the LESA measures. The environmental influences that are important to life events as measured by the LESA are of the nonshared variety.

Discussion

Why do the SRRQ and the LESA yield different results with regard to genetic and environmental influences on life events in middle childhood and early adolescence? At first we were puzzled by the discrepant outcomes across measures; however, our findings made sense once we considered the types of items that each scale includes. Given that the SRRQ is largely a measure of independent events and the LESA a measure of nonindependent events, our findings are consistent with previous behavioral genetic studies of life events in late adolescence and adulthood, which find significant genetic influences on controllable (nonindependent) life events, but not for uncontrollable (independent) events. The finding of shared environmental influences on the SRRQ also fits with previous research. In their study of genetic influences on life events in late adolescence, Billig et al. (1996) differentiated between independent life events that occur within the family and those that occur outside the family. Shared environmental influences accounted for 81% of the variance in independent events that were familial, whereas nonshared environmental influences accounted for most (79%) of the variance in nonfamilial independent events. As indicated above, the majority of the items on the SRRQ refer to events that are familial (and independent); thus, our finding of shared environmental influence on this measure is in agreement with Billig et al. (1996). Our results do, however, differ in that the magnitude of shared environmental influence is substantially lower than that reported by Billig et al. (1996). A possible reason for this may be due to the fact that Billig et al. (1996) used a lifetime measure of life events in twins, whereas in CAP we assessed only those events occurring in a 12-month period prior to a specific age. Twins experience familial life events at the same age; however, this is not the case with siblings. In CAP, the average age difference between siblings is approximately three years. Therefore, although both children within a family were assessed at the same chronological age (i.e., 7, 9, 10, 11, and 12), the life events measures for each child at the same age assess different epochs in the family. That is, the 7-year assessment for Sib 1 assesses the family environment at a different time than the 7-year assessment for Sib 2. Although both siblings experience the same familial events (e.g., parents divorce, the family moves, etc.), they do so at different ages, and thus they cannot contribute to shared environment. Consequently, these familial events will be missed when comparing siblings at the same chronological age. It is likely that if we had used a lifetime measure of life events we would have found more shared environmental influence.

The finding that nonindependent (controllable) life events in early adolescence are genetically influenced raises the question of how measures of the environment such as life events come to show genetic effects. Environments and experiences have no DNA, so how can they be affected by genetic factors? The answer to this apparent paradox is that the environment is not independent of the individual—individuals play an active role in creating their own environments. Genetic variance for measures of the environment arise as a result of gene-environment (GE) correlation. That is, the environment covaries with genetically influenced characteristics of the individual (Plomin, DeFries, & Loehlin, 1977). If measures of life events reflect genetically influenced characteristics of the individual, we can expect to find genetic influence on such measures.

The next step is to identify those genetically influenced characteristics that are responsible for genetic variance on life measures in adolescence. Personality is a good candidate. Researchers examining the mechanisms through which individuals are exposed to certain situations have found that individuals tend to choose situations or activities that reflect their personality (e.g., Diener, Larsen, & Emmons, 1984; Emmons, Diener, & Larsen, 1985; Magnus, Diener, Fujita, & Pavot, 1993). Therefore, genetically influenced personality traits could affect how people select, construct, or perceive their environments. Multivariate genetic analyses have been used to explore the extent to which genetic effects on life events measures in late adolescence and adulthood overlap with genetic effects on personality. Such studies suggest that personality traits can account for a portion of genetic effects on life events measures. For example, Billig et al. (1996) found that there was significant genetic covariance between a measure of nonindependent life events and the personality dimension of constraint in their late adolescent sample. Perhaps more surprising are the results from the Swedish Adoption/Twin Study of Aging that found that genetic influences on life events were *entirely* mediated by neuroticism, extraversion, and openness to experience (Saudino, Pedersen, Lichtenstein, McClearn, Plomin, 1997). That is, these three core dimensions of personality accounted for *all* of the genetic variance on life events in a sample of older adult females.

Can personality explain the genetic influence on life events in our young adolescent sample? If personality does mediate genetic influences on life events, two conditions must hold. First, the measure of personality must show a genetic influence. Second, personality and life events must be related (i.e., there must be a significant phenotypic correlation between the two variables). These conditions were not easily met in our CAP sample. Neither parent nor self-ratings of temperament displayed significant heritability at these ages (see Gagne, Saudino, & Cherny, chapter 11, this volume). This finding is contrary to twin studies that find substantial genetic influence on temperament and is likely a result of contrast effects—that is, the tendency to overestimate differences between siblings (see Saudino, 1997, for a discussion of contrast effects). Nonetheless, a measure that does not show genetic influence cannot explain genetic variance in another measure. Teacher ratings of temperament did show some evidence of genetic influence; however, the associations between temperament and life events were modest (ranging from .00 to −.18) and the pattern was inconsistent across age. Thus, as assessed in CAP, temperament does not appear to mediate genetic influences on life events. However, our measures of temperament at these ages were not ideal—parent and self-ratings appear prone to contrast biases and teacher ratings are based on the teacher's knowledge of the child in a specific, structured environment. It is possible that other measures of personality might produce different results.

The finding of genetic influences on life events in early adolescence is intriguing because it implies that the environment may not be independent of the child. This notion contrasts to the traditional view of the environment as an exogenous force that acts upon the individual. Our results suggest that characteristics of the child might covary with the child's environment. Identifying the specific characteristics at work remains a goal of future research. In the meantime, our results suggest that in some instances the environment can *reflect* as well as *affect* characteristics of young adolescents.

REFERENCES

Billig, J. P., Hershberger, S. L., Iacono, W. G., & McGue, M. (1996). Life events and personality in late adolescence: Genetic and environmental relations. *Behavior Genetics, 26*, 543–554.

Brooks-Gunn, J., & Petersen, A. C. (1984). Problems in studying and defining pubertal events. *Journal of Youth and Adolescence, 13*, 181–196.

Coddington, R. D. (1972). The significance of life events as etiologic factors in the diseases of children. *Journal of Psychosomatic Research, 16*, 7–18.

Cohen, L. H., Burt, C. E., & Bjorck, J. P. (1987). Life stress and adjustment: Effects of life events experienced by young adolescents and their parents. *Developmental Psychology, 23*, 583–592.

Diener, E., Larsen, R. J., & Emmons, R. A. (1984). Person X Situation interactions: Choice of situations and congruence response models. *Journal of Personality and Social Psychology, 47*, 580–592.

DuBois, D. L., Felner, R. D., Brand, S., Adan, A. M., & Evans, E. G. (1992). A prospective study of life stress, social support, and adaptation in early adolescence. *Child Development, 63*, 542–557.

Emmons, R. A., Diener, E., & Larsen, R. J. (1985). Choice of situations and congruence models of interactionism. *Personality and Individual Differences, 6*, 693–702.

Holmes, T. H., & Rahe, R. H. (1967). The social readjustment rating scale. *Journal of Psychosomatic Research, 11*, 213–218.

Kendler, K. S., Neale, M., Kessler, R., Heath, A., & Eaves. L. (1993). A twin study of recent life events and difficulties. *Archives of General Psychiatry, 50*, 789–796.

Magnus, K., Diener, E., Fujita, F., & Pavot, W. (1993). Extraversion and neuroticism as predictors of objective life events: A longitudinal analysis. *Journal of Personality and Social Psychology, 65*, 1046–1053.

McGue, M., Bouchard, T. J., Lykken, D. T., & Finkel, D. (1991). On genes, environment, and experience. *Brain and Behavioral Sciences, 14,* 400–401.

Neale, M. (1997). *Mx: Statistical modeling* (4th ed.). Richmond: Department of Psychiatry, Medical College of Virginia.

Plomin, R. (1994). *Genetics and experience: The developmental interplay between nature and nurture.* Newbury Park, CA: Sage.

Plomin, R., & Bergeman, C. S. (1991). The nature of nurture: Genetic influence on "environmental" measures. *Behavior and Brain Sciences, 14,* 373–427.

Plomin, R., DeFries, J. C., & Loehlin, J. C. (1977). Genotype-environment interaction and correlation in the analysis of human behavior. *Psychological Bulletin, 84,* 309–322.

Plomin, R., Lichtenstein, P., Pedersen, N. L., McClearn, G. E., & Nesselroade, J. R. (1990). Genetic influence on life events during the last half of the life span. *Psychology and Aging, 5,* 25–30.

Saudino, K. J. (1997). Moving beyond the heritability question: New directions in behavioral genetic studies of personality. *Current Directions in Psychological Science, 6,* 86–90.

Saudino, K. J., Pedersen, N. L., Lichtenstein, P., McClearn, G. E., & Plomin, R. (1997). Can personality explain genetic influences on life events? *Journal of Personality and Social Psychology, 72,* 196–206.

Scarr, S., & McCartney, K. (1983). How people make their own environments: A theory of genotype-environment effects. *Child Development, 54,* 424–435.

ERICA L. SPOTTS
JENAE M. NEIDERHISER

The Developmental Trajectory of Genotype-Environment Correlation in Early Adolescence

Introduction

Evidence for the presence of genotype-environment (GE) correlations has been accumulating in recent years (for example, Ge et al., 1996; Pike, McGuire, Reiss, Hetherington, & Plomin, 1996; O'Connor, Deater-Deckard, Fulker, Rutter, & Plomin, 1998). While there is not much question as to the existence of GE correlations, there are still questions about patterns of correlations that might change over time and what role GE correlations have in development. This chapter is designed to begin the exploration of these questions.

A GE correlation is simply a correlation between a genotype and the environment to which the genotype is exposed. There are three ways that a genotype and an environment can be correlated: passively, evocatively, and actively (Plomin, DeFries, & Loehlin, 1977; Scarr & McCartney, 1983). Passive GE correlation refers to the transmission of both genes and environment from parents to their children. For example, parents who are very warm and outgoing will not only pass on those genetically influenced tendencies but may also provide an environment for their child that fosters warm, outgoing behavior in the child. Evocative GE correlation refers to environments that are evoked by the child's genotype. To again use the example of warm and outgoing behavior, a child who exhibits these characteristics will tend to elicit pleasant responses from others. Finally, active GE correlation occurs when children actively select environments that are correlated with their genotype. A warm and outgoing child will most likely seek out other children who are also warm, rather than children who tend toward more negative behaviors. Although evocative and active GE correlations are distinct conceptually, they are difficult to disentangle empirically. Therefore, for the remainder of this report, nonpassive GE correlation will be used to refer to any GE correlations that may be either evocative or active in nature. It is also important to note that GE correlations may be either positive or negative (Plomin et al., 1977). Positive GE correlations exist when the behavior or trait shared by the parent and child are similar. For example, a parent might have a quick temper, the genes for which are shared by the child, who also has a quick temper; additionally, the child will be exposed to the environment created by the parent's quick temper and might learn to handle situations with anger. Negative GE correlations are indicated when the trait or behavior in the parent inspires dissimilar behavior in the child. For example, a parent might have a quick temper, and might share those genes with the child. However, the parent's response to those genetic influences in the child may be to tolerate no angry outbursts from the child.

Scarr (Scarr, 1992; Scarr & McCartney, 1983) has delineated a developmental theory incorporating passive, evocative, and active GE correlations. This theory predicts that passive GE correlation would be more apparent in infancy than in later childhood and adolescence. While infants are able to elicit some environmental responses, for the most part they must make do with the environment provided by their caregivers. As children develop cognitively and physically throughout childhood and adolescence, they are increasingly able to act upon the environment and seek out environments that suit them. At this point, evocative and active GE correlations would predominate over passive GE correlations.

The Colorado Adoption Project (CAP), a longitudinal adoption study that has followed adopted and matched nonadopted control children from ages 1 to 12, provides a unique opportunity to test Scarr and McCartney's theory. The CAP has already reported findings from infancy to middle childhood that provide some support for the decreasing importance of passive GE correlation and the increasing importance of nonpassive GE correlation with increasing child age (e.g., Hershberger, 1994; O'Connor et al., 1998; Plomin & DeFries, 1985; Plomin, DeFries, & Fulker, 1988). Between the ages of 1 and 4, there was some evidence of decreasing passive GE correlations for cognitive abilities and behavior problems, but not for personality characteristics (Plomin & DeFries, 1985; Plomin et al., 1988). It must be noted that the evidence for passive effects was not overwhelming. There was also evidence during early childhood for nonpassive GE correlations in the realms of temperament and personality. During middle childhood, there was evidence for passive GE correlations, particularly of the negative sort (Hershberger, 1994). Negative passive GE correlations were found primarily for externalizing and behavior problems. Findings of passive GE correlations in middle childhood run counter to Scarr's theory. The same report found evidence for an increase in evocative GE correlations, as one might expect, during middle childhood (Hershberger, 1994). O'Connor and colleagues (1998) found evidence for evocative GE correlations for externalizing behaviors in early adolescence. Now that the CAP sample has reached early adolescence it is possible to examine GE correlations during this important developmental period. The present analysis will integrate previous reports of GE correlation with new explorations using the most recent CAP data to provide a larger picture of GE correlation from infancy to early adolescence.

Method

The data used in these analyses were obtained from the 1-, 3-, 5-, 7-, 9-, 10-, 11-, and 12-year assessments of subjects participating in the CAP (DeFries, Plomin, & Fulker, 1994; Plomin & DeFries, 1985; Plomin et al., 1988). The CAP employs a full adoption design by including birth (biological) parents who have relinquished their child for adoption at birth, adoptive parents who are rearing a child they adopted at birth, and nonadoptive control families (biological parents who are rearing their own children). The adoptive and nonadoptive families have been followed longitudinally from the child's age of 12 months, with extensive assessments of the children and global assessments of the child's family environment.

Sample

The sample sizes for each family type at each time are presented in table 18.1. Overall the sample size for the environmental measures ranged from 187 to 216. For birth mothers, the sample size ranged from 216 to 238, and from 228 to 240 for nonadoptive control mothers. Sample size ranges are reported because there is some variation from measure to measure owing to missing data. Only birth mothers are included in the present report because the number of participating birth fathers was relatively small.

Only children adopted at, or soon after, birth were included in the CAP (number of days to placement ranged from 0 to 156). Biological parents and adoptive families were recruited through two Denver

TABLE 18.1. Sample Sizes for Participating Nonadoptive Control Families and Adoptive Families

	Nonadoptive Control Families	Adoptive Families
Year 1	238–240[a]	259
Year 3	217–218	247
Year 5	196–197	215
Year 7	207	231
Year 9	203–204	218
Year 10	222–224	216
Year 11	224	221–224
Year 12	208–210	197–198

Note: [a]Sample size varies for different measures

adoption agencies. Biological parents were asked to participate if it was considered likely that they would relinquish their child for adoption. Biological parents were tested either prior to or soon after delivery of the child. Adoptive parents were asked to participate after the adoption of their child, but were not assessed until their adopted child reached 12 months of age. Nonadoptive control families were recruited through local hospitals. The adoptive families and nonadoptive control families were matched to ensure comparability between these two family types. Families who expressed interest in participating in the projects were selected if they could be matched to an adoptive family based on sex of the infant, number of children in the family, age of the father (within five years), occupation of the father, and total years of father's education (within two years). Biological, adoptive, and nonadoptive parents all completed a three-hour test battery administered in small groups. See Plomin and DeFries (1985) for more details about the sample.

The majority of families participating in CAP are Caucasian (90% of biological parents and 95% of adoptive and nonadoptive parents). The adoptive and nonadoptive parents are representative of the Denver metropolitan area in terms of socioeconomic status in 1978, shortly after the CAP began. Biological parents are younger than adoptive and nonadoptive parents (average age of biological parents = 20, of adoptive parents = 33, nonadoptive = 31 years). Although the years of education and occupational status of biological parents are also lower than adoptive and nonadoptive parents, the status of the grandparents is comparable (i.e., the parents of the biological parents and the parents of the adoptive and nonadoptive parents). This suggests that the discrepancies in socioeconomic indicators of adoptive and nonadoptive parents vs. biological parents may be due to the difference in their ages.

Measures

Methods of data collection for years 1 through 9 can be found elsewhere (DeFries et al., 1994; Plomin & DeFries, 1985; Plomin et al., 1988). The 10- and 11-year data collection was conducted using telephone testing and interviews. Prior to the telephone interview, parents were contacted to schedule the interview, to explain the testing procedure, and to confirm that the test packet had been received. Two telephone interviews were conducted when the children were approximately 10 and 11 years old. During the first call, a short interview was conducted with a parent (mother 90% of the time) and a longer interview with the child. During the second call a cognitive test battery was administered. Parents were also asked to complete questionnaires at that time. At age 12 the children were tested in the laboratory. The laboratory session lasted for five hours, with two breaks.

Two sets of measures were examined in these analyses: birth and adoptive parent characteristics and the adopted child's environment.

Parent Characteristics. Cognitive abilities, personality, interests, talents, common behavioral problems, commonly used drugs, and other miscellaneous information was collected for the biological, adoptive, and nonadoptive parents. Only cognitive abilities, personality and talents were employed in the present report.

Cognitive Abilities. Thirteen tests from the Hawaii Family Study of Cognition (DeFries et al., 1974) were included. The scores were adjusted for age, age squared, and gender separately for each group of parents. The first unrotated principal component is used as an index of general cognitive ability and is the only measure used in the present report.

Personality. Personality was assessed using two self-report measures. The three-second order factors of extraversion, neuroticism, and independence and tough poise from form A of Cattell's Sixteen Personality Factor Questionnaire (16PF; Cattell, Eber, & Tatsuoka, 1970) were employed in the present report. In addition, the four subscales of emotionality: fear; emotionality: anger; activity; and sociability from the Extraversion, Activity, and Sociability Inventory (EASI; Buss & Plomin, 1975) were examined.

Interests and Talents. Parents' self-reported interest and talent in the arts (music and art), group sports, individual sports, and mechanical and domestic arts were assessed. Because the correlation between interests and talents is very high, only talents were examined in the present report.

Behavioral Problems. Questions about headaches, speech problems, depression, hysteria, sociopathy, phobias, sleep problems, compulsive behavior, motion sickness, and menstrual problems were asked of all parents. Assessments of depression, hysteria, and sociopathy were used for these analyses. The items for these scales were modified from the Iowa 500 Project (Tsuang, Crowe, Winokur, & Clancy, 1977).

Child Environment Measures

Two measures of the home environment were included in CAP: the Home Observation for Measurement of the Environment (HOME; Caldwell & Bradley, 1978) and the Family Environment Scale (FES; Moos & Moos, 1981) questionnaire. Only the FES was administered at all assessments, therefore only the FES will be examined in the present report.

The FES is a 90-item self-report questionnaire completed by mothers. For the CAP, the format was altered from true-false to a five-point rating scale. The FES consists of 10 scales: cohesion, expressiveness, conflict, independence, achievement orientation, intellectual-cultural orientation, active-recreational orientation, moral-religious emphasis, organization, and control. A modified version of the FES was also completed by the children at 9, 10, 11, and 12 years. Four items were selected to assess each of five dimensions: cohesion, expressiveness, conflict, achievement, and control. Children indicate whether each statement is "like my family" or "not like my family."

Child Behavior Measures

The Colorado Childhood Temperament Inventory (CCTI; Rowe & Plomin, 1977) was completed by the parents when the children were 10, 11, and 12 years of age. The CCTI was derived from the EAS (Buss & Plomin, 1975) and the nine dimensions of temperament described by Chess and Thomas (1984). Six scales constitute the CCTI: emotionality, activity, sociability, attention span, reactions to foods, and soothability. The last two scales were not included in the present report as they are more appropriate for use with younger children.

The internalizing and externalizing second-order factors of the parent-rating version of the Child Behavior Checklist (CBCL; Achenbach & Edelbrock, 1983) were examined. The CBCL was administered at years 10, 11, and 12 and has been found to be highly reliable in the CAP sample (see DeFries et al., 1994).

Two measures of competence were included: Harter's Self-Perception Scale for Children (Harter, 1982) and the CAP Social Competence Scale (CSCS). The Harter consists of six subscales: general self-worth, behavior conduct, athletic competence, scholastic competence, perceived physical appearance, and social competence. The version of the Harter used in CAP is administered as part of a battery of questions about "feelings" during a telephone interview at 10 and 11 years, and in the standard format during the laboratory assessment at 12 years. The CSCS is based on the Walker-McConnell Scale of Social Competence and School Adjustment (Walker & McConnell, 1988) and was administered at ages 10 and 11. Teacher reports on the dimensions of leadership, confidence, popularity, and behavior problems were examined.

Analyses

The full adoption design employed in CAP is a quasi-experimental design that allows genetic and environmental influences on the similarity of parents and children to be assessed. Birth mothers are genetically related to their adopted-away child, but do not provide any potential environmental influence. Adoptive parents, who are genetically unrelated to the adopted child, provide the environment in which the child grows up. Finally, the nonadoptive control mothers and children share both genetic and environmental influences.

The focus of the present report is the assessment of passive and nonpassive GE correlation. The most general approach to finding GE correlations is a multivariate genetic analysis of the association between measures of the environment and outcome (for example, Pike et al., 1996). There are two reasons that this approach will not be used in the current analyses. Most important, this method cannot differentiate between passive and nonpassive GE correlation. Second, this approach utilizes sibling-sibling correlations, not the parent-child associations that will be used to find nonpassive GE correlations. There are two methods commonly employed to search for passive GE correlation: (1) parent-offspring model-fitting designs, and (2) comparison of the variance of adopted children with nonadoptive control children. The first method is the most stringent test of passive GE correlation. The correlations between adoptive parents and adopted children are compared to the correlations between nonadoptive control parents and their biological children to obtain a path estimate of GE correlation using a model fitting approach (see Plomin et al., 1977). Although there has been evidence of some significant passive GE correlation using this method, in most cases the estimates are very small (see Hershberger, 1994, for a review of these findings).

The second method tests for variance differences between adopted children and nonadopted control children. The reasoning behind this method is as follows. Adopted children and their adoptive parents can be similar only for environmental reasons, since they share no segregating genes. Nonadopted children and their biological parents can be similar for environmental *and* genetic reasons. Therefore, if genetic factors are at least partially responsible for an association between parent and child characteristics, we would expect more variation in the nonadoptive association since genetic influences are possible there in addition to environmental influences. When *positive* passive GE correlation is present the variance of the nonadopted control children will be significantly larger than the variance of the adopted children. If *negative* passive GE correlation is present, the variance of the adopted children will be significantly larger than the variance of the nonadopted biological children. In the present report, the comparison of variances method will be employed to test for passive GE correlation at 10, 11, and 12 years. Preliminary analyses indicated that associations between parent and child measures were not large enough to warrant model-fitting analyses.

The CAP offers a unique opportunity for examining the effects of nonpassive GE correlations because it is an adoption design that has collected data on the biological mothers and the family environment. Using these two types of information, one can identify nonpassive GE correlations by correlating the biological mother's characteristics with the adopted child's family environment. Because the biological mothers do not contribute to their adopted-away child's potential environment, the only explanation for such a correlation, in the absence of selective placement, is that something about the adopted child's genetically influenced behavior is evoking an environment that is correlated with his or her genes (as represented by the biological mother). In other words, the measured characteristics of the biological mother serve as an index of the genotype of the child, while the environmental measures of the adoptive home index the environment. Because this is an exploratory analysis, and because there are many measures of both the child's environment and of the biological mother's characteristics, canonical correlations were computed. A canonical correlation represents the maximization of the relationship between two sets of variables—in this case, the multiple measures of birth mother characteristics and the measures of the adopted child's home environment. Biological mothers' characteristics were correlated with the adopted children's home environment at each assessment from age 1 to age 12 so as to identify any change over time in the presence of nonpassive GE correlations.

Results

Passive GE Correlations   The results of the comparisons of the variances of adopted and nonadopted control children's behavioral measures are reported in table 18.2. The majority of variances *were not* significantly different for adopted and nonadopted control children. There were, however, two exceptions: CBCL externalizing and CCTI sociability. Evidence of negative passive GE correlation for CBCL externalizing was indicated at ages 10, 11, and 12 and at ages 10 and 11 for the CCTI sociability. At age 12, two additional significant variance differences emerged. There was negative passive GE correlation for attention span as assessed by the CCTI at age 12, and a positive passive GE correla-

TABLE 18.2. Passive Genotype-Environment Correlation: Comparison of Standard Deviations for Adopted and Nonadopted Children

Child Age Family Type	10		11		12	
	Adoptive	Control	Adoptive	Control	Adoptive	Control
CBCL						
Internalizing	5.49	4.90	5.45	5.63	5.40	5.18
Externalizing	6.69*	5.69	6.93*	5.60	7.33*	6.08
CCTI						
Sociability	4.38*	3.76	4.49*	3.71	4.55	4.11
Emotionality	4.13	4.04	4.20	4.16	4.05	4.07
Activity	3.84	3.71	4.11	3.93	4.40	3.95
Attention span	4.17	3.69	4.24	3.79	4.32*	3.52
CSCS						
Leadership	7.22	7.11	7.00	7.16	—[a]	—
Problem behavior	6.29	6.55	5.88	5.92	—	—
Popularity	5.56	6.15	5.32	6.15	—	—
Confidence	7.50	7.91	7.20	7.91	—	—
Harter						
Scholastic competence	3.40	3.34	3.55	3.25	3.40	3.17
Social competence	3.46	3.78	3.46	3.69	3.43	3.78
Athletic competence	3.46	3.50	3.62	3.76	3.63	3.80
Physical appearance	3.88	3.52	3.40	3.12	3.80	3.58
Behavioral conduct	3.18	3.32	3.02	3.01	3.72	3.75
General self-worth	2.69	2.56	2.37	2.17	2.31*	2.77

Note: *$p < .05$. [a]The CSCS was not administered at age 12

tion for general self-worth at age 12. Findings of negative passive GE correlations are consistent with previous reports (Hershberger, 1994), which found negative passive GE correlations in similar domains. In comparison to earlier ages, there was little evidence for passive GE correlations at ages 1 to 4 outside of the cognitive realm (Plomin et al., 1988). During middle childhood, evidence for passive GE correlations emerged for the CCTI and the CBCL, the same scales for which there were significant differences in early adolescence. Overall, these data do not support Scarr's theory of the decreasing importance of passive GE correlation. Although there are a few significant differences, a pattern of decreasing magnitude in differences does not emerge.

Nonpassive GE Correlations Table 18.3 presents the unrotated first canonical correlations for biological mother characteristics and adoptive mother report FES obtained at ages 1, 3, 5, 7, 9, 10, 11, and 12. Table 18.4 presents the unrotated first canonical correlations for biological mother characteristics and child report FES obtained at ages 9, 10, 11, and 12 years. Correlations for the nonadopted control families are also provided for the purpose of comparison. In contrast to the findings reported by Hershberger (1994), these correlations do not indicate

TABLE 18.3. Canonical Nonpassive Genotype-Environment Correlations for Mother Report of the FES

	Biological Mother—Adopted Child FES		Control Mother—Control Child FES	
	Canonical Correlation	F	Canonical Correlation	F
Year 1				
EASI	.21	1.10	.48*	7.79
IQ[a]	.12	1.48	.11	1.33
16PF	.20	1.44	.51*	16.34
Talents	.28	1.64	.33*	2.59
Psychopathology	.18	.69	.40*	5.08
All measures	.41	1.41	.57*	5.10
Year 3				
EASI	.35	1.20	.48*	2.52
IQ	.15	.44	.31*	2.08
16PF	.29	.72	.59*	4.26
Talents	.37	.98	.44*	1.57
Psychopathology	.41	.90	.43	1.48
All measures	.47	.94	.64*	2.16
Year 5				
EASI	.35	.90	.45*	1.85
IQ	.17	.50	.23	1.01
16PF	.30	.05	.55*	3.00
Talents	.41	1.18	.50*	1.94
Psychopathology	.42	.82	.45*	1.74
All measures	.50	1.01	.66*	1.80
Year 7				
EASI	.37	1.03	.52*	2.44
IQ	.26	1.32	.28	1.61
16PF	.27	.95	.50*	2.84
Talents	.38	1.17	.40*	.41
Psychopathology	.49	.86	.41	1.19
All measures	.48	1.00	.64*	1.84
Year 9				
EASI	.31	.81	.54*	2.76
IQ	.17	.67	.24	1.40
16PF	.32	1.14	.45*	3.17
Talents	.34	.94	.39*	1.67
Psychopathology	.53	1.21	.30	1.10
All measures	.45	.93	.62*	1.79
Year 10				
EASI	.20	.79	.43*	1.83
IQ	.31*	4.12	.16	.63
16PF	.25	1.12	.43*	2.79
Talents	.35	1.42	.34	1.39
Psychopathology	.37	.99	.39	1.48
All measures	.46	1.12	.53*	1.44

TABLE 18.3. (*Continued*)

	Biological Mother—Adopted Child FES		Control Mother—Control Child FES	
	Canonical Correlation	F	Canonical Correlation	F
Year 11				
EASI	.16	.42	.49*	2.54
IQ	.23	2.02	.14	.46
16PF	.18	.61	.45*	2.67
Talents	.39*	1.68	.45*	1.80
Psychopathology	.46*	.98	.44	1.56
All measures	.43	1.00	.60*	1.84
Year 12				
EASI	.29	1.43	.47*	2.21
IQ	.21	1.60	.29*	2.31
16PF	.22	.64	.47*	3.12
Talents	.32	1.10	.37*	1.61
Psychopathology	.29	.61	.45*	1.89
All measures	.67	.85	.59*	1.63

Notes: [a]Parent IQ is assessed using the first principal component score from the cognitive scales.
*$p < .05$.

an increasing role of nonpassive GE correlation from 1 to 12 years. Few correlations between the biological mother's characteristics and mother ratings of the FES are significant: year 10 IQ and year 11 talents and psychopathology are the only significant correlations. Using child report FES, only year 10 psychopathology and year 11 talents are significant. While the correlations using mother report at year 12 are greater than those at year 1, they are not substantially greater and the magnitude of correlations does not systematically increase from age 1 to age 12. A great change in nonpassive GE correlations from year 9 to 12 would not be expected; the results using child report FES meet these expectations. While year 9 correlations are lower than year 12 correlations for some, but not all, measures, there is no systematic increase from year 9 to year 12.

Discussion

The purpose of this chapter was to use the longitudinal design of the CAP to test Scarr's theory of development regarding the role of GE correlations (Scarr, 1992; Scarr & McCartney, 1983). According to this theory, passive GE correlations would decrease as the child developed; in contrast, the role of reactive and evocative GE correlations would increase with age. Previous examinations of this topic have found some support for this theory, in that reactive and evocative correlations

TABLE 18.4. Canonical Nonpassive Genotype-Environment Correlations for Child Report of the FES

	Biological Mother—Adopted Child FES		Control Mother—Control Child FES	
	Canonical Correlation	F	Canonical Correlation	F
Year 9				
EASI	.24	.91	.21	.74
IQ[a]	.19	1.35	.19	1.50
16PF	.28	1.21	.23	1.13
Talents	.35	1.45	.25	.95
Psychopathology	.31	.63	.23	.95
All measures	.50	1.09	.39	.65
Year 10				
EASI	.20	.79	.23	.62
IQ	.31	4.12	.13	.73
16PF	.26	1.12	.24	1.23
Talents	.35	1.42	.24	1.09
Psychopathology	.48*	2.01	.20	.77
All measures	.46	1.12	.37	.90
Year 11				
EASI	.16	.42	.28	1.39
IQ	.23	2.02	.13	.68
16PF	.18	.61	.24	1.41
Talents	.39*	1.68	.21	.74
Psychopathology	.25	.57	.20	.57
All measures	.43	1.00	.37	.94
Year 12				
EASI	.33	1.42	.20	.72
IQ	.17	1.17	.19	1.56
16PF	.28	1.13	.28	1.46
Talents	.31	1.11	.25	.92
Psychopathology	.29	.61	.23	.91
All measures	.72	1.01	.37	1.04

Notes: [a]Parent IQ is assessed using the first principal component score from the cognitive scales.
*$p < .05$.

appeared only in later childhood and not in infancy; only passive effects were present in infancy (Plomin & DeFries, 1985; Plomin et al., 1988). Evocative GE correlations in CAP have also been found for externalizing behaviors at age 7 to 12 (O'Connor et al., 1998). In contrast to Scarr's theory, passive effects do not seem to decrease over the course of childhood, at least in the CAP sample (Hershberger, 1994). It must be noted, however, that the number and magnitude of passive GE correlations is not great, so it is difficult to say if there has been a decrease or not.

The current analysis examined the roles of passive and nonpassive GE correlations from infancy to later childhood, from ages 1 to 12 years. Previous reports showed some evidence for passive GE correlations in the cognitive domain in infancy (Plomin & DeFries, 1985; Plomin et al., 1988), and for behavior problems in middle childhood (Hershberger, 1994). In the current analyses, there was some evidence for negative passive GE correlations for externalizing behaviors (years 10, 11, and 12), sociability (years 10 and 11), and attention span (year 12), and positive passive GE correlations for general self-worth (year 12). These results do not fit the pattern of decreasing passive GE correlations suggested by Scarr, though they are consistent with previous reports of CAP data that have found evidence for negative passive GE correlations in the same domains of behavior (Hershberger, 1994). It is unclear why these GE correlations are of the negative sort. It is possible that the parents recognize these particular characteristics in themselves and are not tolerant of them in their children. Future research should address the processes underlying the prevalence of negative GE correlations.

According to Scarr's theory, an increase in the presence of nonpassive GE correlations would be expected from infancy to late childhood. While there is some change for all measures from year 1 to year 12, the increase is not systematic across time, nor is the magnitude of the increase very large.

What do these findings indicate? Taken at face value, they do not offer much support for Scarr's theory of changing roles of GE correlations over time. While nonpassive GE correlations have a slightly stronger presence in later childhood than in infancy, there is neither an increasing presence of nonpassive GE correlations over time nor a decrease in the presence of passive GE correlations, as Scarr has proposed.

When examining the findings of this and other studies using the CAP sample to search for evidence of GE correlations, one pattern does emerge: the correlations appear to be somewhat domain specific. There has been little evidence of GE correlations in the cognitive domain. However, through early childhood, there exist GE correlations for traits such as activity level and impulsivity, with nonpassive GE correlations for externalizing behaviors being present. It could be that GE correlational effects are more likely to be found in some domains rather than others.

Assessments of the biological mother are the best index of the adopted child's genotype that is available to us in this study. However, they are not a particularly strong indication of the genotype. The biological mother is reflective of only *half* of the child's genotype; the biological father constitutes the remaining half. Unfortunately, assessments of biological fathers were not used in these analyses owing to their low response rate. Another drawback to not having assessments on biological fathers is that externalizing behaviors assessed by these measures would most likely be manifested by fathers, not mothers, thereby reducing our chances of finding nonpassive effects for externalizing behaviors. Another caveat to the use of the biological mothers' data is their young age at the time of assessment (mean age = 20 years). It is possible that these assessments are not reflective of all manifestations of psychopathology because the mothers were not beyond the riskiest ages for the development of psychopathology.

The CAP does provide some evidence for the presence of GE correlations. It must be noted that other reports of evidence of GE correlations come from samples at high risk for psychopathology (Ge et al., 1996). Perhaps more exploration using extreme samples rather than population-based samples will shed further light on the domains in which GE correlations are present.

REFERENCES

Achenbach, T. M., & Edelbrock, C. (1983). *Manual for the Child Behavior Checklist and Revised Child Behavior Profile.* Burlington: University of Vermont, Department of Psychology.

Buss, A. H., & Plomin, R. (1975). *A temperament theory of personality development.* New York, Wiley-Interscience.

Caldwell, B. M., & Bradley, R. H. (1978). *Home observation for measurement of the environment.* Little Rock: University of Arkansas.

Cattell, R. B., Eber, H. W., & Tatsuoka, M. M. (1970). *Handbook for the Sixteen Personality Factor Questionnaire (16 PF).* Champaign, IL: Institute for Personality and Ability Testing.

Chess, S., & Thomas, A. (1984). *Origins and evolution of behavior disorders: Infancy to early adult life.* New York: Brunner/Mazel.

DeFries, J. C., Plomin, R., & Fulker, D. W. (1994). *Nature and nurture during middle childhood.* Cambridge, MA: Blackwell.

DeFries, J. C., Vandenberg, S. G., McClearn, G. E., Kuse, A. R., Wilson, J. R., Ashton, G. C., & Johnson, R. C. (1974). Near identity of cognitive structure in two ethnic groups. *Science, 183,* 338–339.

Ge, X., Conger, R. D., Cadoret, R. J., Neiderhiser, J. M., Yates, W., Troughton, E., & Stewart, M. A. (1996). The developmental interface between nature and nurture: A mutual influence model of child antisocial behavior and parent behaviors. *Developmental Psychology, 32,* 574–589.

Harter, S. (1982). The perceived competence scale for children. *Child Development, 53,* 87–97.

Hershberger, S. L. (1994). Genotype-environment interaction and correlation. In J. C. DeFries, R. Plomin, & D. W. Fulker (Eds.), *Nature and nurture during middle childhood* (pp. 281–294). Cambridge, MA: Blackwell.

Moos, R. H., & Moos, B. S. (1981). *Family Environment Scale manual.* Palo Alto, CA: Consulting Psychologists Press.

O'Connor, T. G., Deater-Deckard, K., Fulker, D., Rutter, M., & Plomin, R. (1998). Genotype-environment correlations in late childhood and early adolescence: Antisocial behavior problems and coercive parenting. *Developmental Psychology, 34,* 970–981.

Pike, A., McGuire, S., Reiss, D., Hetherington, E. M., & Plomin, R. (1996). Family environment and adolescent depressive symptoms and antisocial behavior: A multivariate genetic analysis. *Developmental Psychology, 32,* 590–603.

Plomin, R., & DeFries, J. C. (1985). *Origins of individual differences in infancy: The Colorado Adoption Project.* Orlando, FL: Academic Press.

Plomin, R., DeFries, J. C., & Fulker, D. W. (1988). *Nature and nurture during infancy and early childhood.* New York: Cambridge University Press.

Plomin, R., DeFries, J. C., & Loehlin, J. C. (1977). Genotype-environment interaction and correlation in the analysis of human behavior. *Psychological Bulletin, 84,* 309–322.

Rowe, D. C., & Plomin, R. (1977). Temperament in early childhood. *Journal of Personality Assessment, 41,* 150–156.

Scarr, S. (1992). Developmental theories for the 1990s: Development and individual differences. *Child Development, 63,* 1–19.

Scarr, S., & McCartney, K. (1993). How people make their own environments: A theory of genotype → environment effects. *Child Development, 54,* 424–435.

Tsuang, M. T., Crowe, R. R., Winokur, G., & Clancy, J. (1977). Relatives of schizophrenics, manics, depressives and controls. *Proceedings of the Second International Conference on Schizophrenia.* New York: John Wiley.

Walker, H. M., & McConnell, S. W. (1988). *Walker-McConnell Scale of Social Competence and School Adjustment.* Austin, TX: Pro-Ed.

STEPHEN A. PETRILL
ROBERT PLOMIN
JOHN C. DEFRIES
JOHN K. HEWITT

19

Conclusions

In this book, data from the Colorado Adoption Project (CAP) have been analyzed to examine the genetic and environment influences on several key aspects of adolescent development. This concluding chapter highlights some of the major findings and their implications for our understanding of development during the transition to early adolescence.

Summary of Findings

In the past, behavioral genetic results typically have been interpreted in terms of "how much" genes and environments impact outcomes. More recently, the focus has shifted to applying behavioral genetic methods to address theoretically meaningful questions in developmental psychology—to better understand the genetic architecture of development as well as the central role of the environment and gene-environment processes. Early adolescence is a time of immense average growth, but is a developmental period where individual differences are also important. Behavioral genetic methods examine how genes and environments shape the developmental trajectory of these individual differences. We presented quantitative genetic analyses in four substantive domains: cognitive ability and achievement, adjustment and behavior problems, mood and temperament, and the environment.

Cognitive Ability and Achievement

Some of the most consistent findings in the behavioral-genetic literature involve cognitive development. These studies have shown that genetic influences account for a significant and increasing proportion of the variance in cognitive ability and achievement throughout the life span. In contrast, shared environmental influences are important in childhood but reduce to zero by adolescence, leaving the nonshared environment as the remaining source of environmental variance throughout the lifespan (see Petrill & Wilkerson, 2000, for a review). Theoretically, these findings are important because they suggest not only that genes are important but also that the most important sources of environmental variation across the life span do not contribute to familial resemblance.

In the last decade, behavioral genetic studies have examined how genes and environments impact the relationship among cognitive skills across development. Unlike most studies, CAP has assessed cognitive skills continuously from birth to early adolescence. Chapter 2 presented a dynamic view of cognitive development from ages 1 to 12. These data suggest that the genetic influences on general cognitive ability change substantially from ages 1, 2, 3, 4, and 7. However, from ages 9 to 12, genetic influences contributing to general cognitive ability remained stable. In other words, the shift to early adolescence signals a stabilization in genetic effects on general cognitive ability. The nonshared environment contributes to instability throughout this developmental period.

A related question is the extent to which the verbal, spatial, speed, and memory skills that compose general cognitive ability are influenced by the same or different genetic or environmental influences. Chapter 3 suggested that genes account for the overlap while the nonshared environment accounts for the difference among cognitive skills. Additionally, chapter 3 also suggested that this degree of genetic overlap may increase in early adolescence, a finding consistent with chapter 2.

Chapters 4 and 5 combined longitudinal analyses with multivariate analyses within age. Chapter 4 examined the links between general cognitive ability and reading at age 7 with reading at age 12. These results suggested substantial stability in the genetic influences on reading from ages 7 to 12. General cognitive ability accounted for some, but not all of this stability. Similarly, chapter 5 suggested that genes account for the stability between recall and recognition memory both within and across ages. In general, nonshared influences tend to make cognitive skills different from one another.

Adjustment and Behavior Problems

In addition to the development of cognitive skills, early adolescence is an important transition point for the development of adjustment and behavior problems. Again, the strength of CAP is its continual assessment of adjustment and behavior problems from childhood into early adolescence. Chapter 6 examined the continuity of somatic symptoms from childhood to early adolescence. In general, the literature on somatic complaints is quite mixed, with some studies suggesting substantial genetic influences on somatic symptoms while others suggesting little genetic influence (e.g., van den Oord, Boomsma, & Verhulst, 1994). There was little evidence for genetic influences in this study.

Inattention and hyperactivity are two of the most pervasive psychiatric diagnoses in childhood and early adolescence (Barkley, 1997). While these constructs are sometimes examined as diagnostic categories, the emergent literature suggests that inattention and hyperactivity may be quantitatively distributed characteristics. Chapter 7 examined the genetic and environmental influences on the range of attentional problems and hyperactivity as rated by parents and teachers. The genetic influences on hyperactivity and attentional problems appear attenuated in this adoption sample relative to twin data, perhaps due to nonadditive genetic variance. Furthermore, the links between attentional problems and hyperactivity are more tightly correlated in the CAP than what has been found in previous research.

Although the CAP has been used primarily as a behavioral genetic study, it is also one of the best studies of adoption. An important issue in adoption is the extent to which adopted children are at increased risk of behavioral and adjustment relative to nonadopted children. Chapter 8 examined differences between adopted and nonadopted children's behavioral and academic adjustment to early adolescence. In general, adopted children experienced more problems in adjustment on average than nonadopted children. However, the observed differences between these two groups were negligible. Thus, the CAP data suggest that variability in adjustment within adoptive and nonadopted groups is far greater than the average differences between these groups.

Mood and Temperament

Mood and temperament are important moderators of adolescent adjustment and behavior. As is the case for adolescent adjustment, the CAP is unique in that it measures mood and temperament in childhood and adolescence across several raters. Having multiple sources of information on mood and temperament not only allows for triangulation of common findings across respondents but also enables the comparison of divergent findings between respondents.

Chapter 9 examined the etiology of intra-individual change in internalizing problems in self- and parent-rated measures of internalizing behavior. Interestingly, stability in self-reports was due mainly to additive genetic influences while stability in parent reports was mainly a function of the shared environment. Many studies have suggested self-report data lead to more consistent results than parent reports. More conservatively, the divergence between self- and parent reports points to the importance of paying close attention to characteristics and context of measurement when interpreting developmental data.

Chapter 10 examined the relationship between loneliness and internalizing behavior problems in the transition to adolescence. It has been suggested that internalizing behavior problems result in feelings of loneliness and isolation. This chapter suggested that, although loneliness was influenced by genes and the nonshared environment, there was very little overlap between loneliness internalizing behavior problems as measured by teachers and parents. These results suggest that the key to understanding loneliness may lie not in adult caregivers' reports of internalizing behavior problems but a better understanding of the adolescent's peer environment.

Chapter 11 examined continuity in temperament as measured by teacher, parent, and self-reports. Similar to the findings of chapters 9 and 10, there was divergence in results depending on the respondent. Teacher ratings demonstrated genetic influences on the stability of temperament from childhood to adolescence while the parent and self-report data did not.

Chapter 12 examined the links between temperament and behavior problems. Twin studies have suggested a common genetic etiology between the temperamental characteristics of sociability and shyness and problem behavior (e.g., Schmitz et al., 1999). The CAP did not replicate this finding: There was a relationship between temperament and behavior problems, but genetic influences were not responsible for this covariation.

Chapter 13 examined the relationship between adolescent self-report, sibling report, and parental report of adolescent humor. This chapter was based on the Social Relations Model (Kenny & La Voie, 1984), a model designed specifically to test the convergence and variability of different respondents on the same outcome. The chapter suggested that well-being, but not temperament, is associated with interpersonal humor use. Additionally, correlates of affiliative and aggressive humor appeared to differ, providing support for the notion that these two types of humor have distinct social functions.

The Environment

Some of the most important CAP findings have involved the examination of environmental effects upon development. Despite an emphasis on "genetics," behavioral studies using genetically sensitive designs, such as CAP, have taught us much about the environment. For example, it has long been assumed that the most important environment is shared by family members (e.g., having the same parents, growing up in the same home). The behavioral genetic literature has consistently run counter to this assumption. While the shared environment is important in childhood, by early adolescence influence of the shared environment is essentially zero. In contrast, the nonshared environment remains significant throughout the life span. This does not mean that the environment is unimportant. Instead, these results suggest that it is essential to examine the environment at the level of the individual child. For example, chapter 14 examined the relationship between prenatal exposure to nicotine and later cognitive and behavioral development. The children of the CAP mothers who smoked during pregnancy had lower birth weights and were shorter at birth. Twelve years later, these children scored higher on measures of behavioral problems and lower on measures of IQ and cognitive performance.

Chapters 15 to 18 all examined gene-environment processes in adolescence. In contrast to chapter 14, where the long-term effects of a teratogen were studied, the goal of these chapters was to examine the extent to which genes influence the probability of coming into contact with particular environments. This issue is particularly important when one considers the potential for adverse, or negative, environment experiences during adolescence.

Chapter 15 examined those children who were experiencing extreme externalizing behavior problems as well as negative family environments. Extreme groups were selected with negative family environment scores one SD above the mean. These groups suggested significant and substantial group shared environmental variance for parents' reports of parental negativity, inconsistent discipline, and warmth. There were no significant group genetic effects. Chapter 16 went further and attempted to directly assess the individually experienced environment within families. This chapter examined differential parental treatment over time. In general, the CAP suggested that maternal differential treatment in early adolescence is moderately stable. Maternal differential treatment is also influenced by genetic factors and associated with children's well-being. In other words, children who were more similar genetically were less likely to be parented differently. Similarly, chapter 16 suggested that life events in early adolescence are influenced by genetic factors. Finally, chapter 17 suggested evidence for passive GE correlation for externalizing behaviors, sociability, attention span (year 12), and general self-worth.

Implications

The results of the CAP have several important implications for the study of development in children and early adolescence. First, the CAP suggests that the genetic and environmental architecture of development is not static. There is some evidence for genetic stability, but new genetic influences are manifest as children make the transition into early adolescence. The influence of the shared environment declines while the nonshared environment remains significant throughout this developmental period.

Equally important, studies like the CAP have important implications for our understanding of how children and adolescents experience environments. The finding of genetic influences on "environmental" measures such as maternal differential treatment or life events, and the suggestion that passive GE processes are operating, does not imply that environments have DNA. Instead, these data suggest that the probability of experiencing environments appears to be influenced, in part, by endogenous genetic factors. Children who are genetically more similar present more similar patterns of behavior and, in turn, are more likely to experience similar environments. This dynamic view is in sharp contrast to the traditional view of the environment as an externally imposed main effect acting upon a passive recipient. These CAP findings are important because they provide a foundation for the long processes of identifying the gene-environment processes by which the effects of particular genes and particular environment influence behavioral outcomes.

Like other longitudinal studies, CAP becomes more valuable as it continues to follow its participants and capitalizes on the increasing store of previously collected data. Most of the CAP children have passed adolescence, and the oldest CAP participants are currently being studied in the transition to adulthood as they enter the world of work, marriage, and family life. Applying the power of the adoption design to these relatively unexplored aspects of development is interesting in itself, but especially in relation to the extensive data collected in infancy, childhood, and early adolescence.

Epilogue

The transition to early adolescence is one of the most intense and dynamic periods in human development. For too long, studies of adolescence either have focused solely on the average progression of development in adolescence or have turned to a comparison of mean differences across groups. The power of the behavioral genetic method lies not only in its focus on individual differences but also in its ability to disentangle genetic and environmental influences on developmental change and continuity. The CAP and other behavioral genetic studies have begun to move beyond merely quantifying "how much" genes and environments contribute to a particular outcome or the stability or change in an outcome as a function of development. Newer studies are now attempting to identify DNA markers or a particular set of environmental measures that contribute to the variance estimated in traditional behavioral genetic models. This identification of genes and environments within the context of genetically informative designs not only offers the hope of better measurement of these particular genes and environments but also holds the promise of quantifying environmental contexts in which genes operate and the genetic background upon which environments are experienced.

REFERENCES

Barkley, R. A. (1997). *ADHD and the nature of self-control*. New York: Guilford.
Kenny, D. A., & La Voie, L. (1984). The social relations model. In L. Berkowitz (Ed.), *Advances in experimental social psychology* (pp. 141–182). Orlando, FL: Academic Press.
Petrill, S. A., & Wilkerson, B. (2000). Intelligence and achievement: A behavioral genetic perspective. *Educational Psychology Review, 12*, 195–199.
Schmitz, S., Fulker, D. W., Plomin, R., Zahn-Waxler, C., Emde, R. N., & DeFries, J. C. (1999). Temperament and problem behavior during early childhood. *International Journal of Behavioral Development, 23*(2), 333–355.
Van den Oord, E. J. C. G., Boomsma, D. I., & Verhulst, F. C. (1994). A study of problem behaviors in 10- to 15-year-old biologically related and unrelated international adoptees. *Behavior Genetics, 24*, 193–205.

Names Index

317

Subject Index

Achenbach Youths Self-Report, 113
adjustment, 311–312
  of adopted and nonadopted
    adolescents, 109–130
  emotional, 126–127
  maladjustment, 110–116
adolescent development
  and adoption, 109–130
  genetically informative samples,
    4–5
  individual differences in normal,
    5–6
  internalization of problems,
    133–149
  memory ability, 62–74
  nature, nurture, and, 3–10
  See also problem behavior
adoption
  and adolescent adjustment,
    109–130
  and attention problems, 93–104
  and internalization of problems,
    133–149
  and loneliness, 157–158
  longitudinal studies of, 13,
    114–116
  multivariate parent-offspring
    analyses of specific cognitive
    abilities, 28–46
  population-based cross-sectional
    studies, 111–114
  and reading performance, 49–60

  as risk factor for psychiatric
    morbidity and maladjustment,
    110–116
  and scholastic adjustment,
    120–121, 126–127
  studies of clinical populations,
    110–111
  studies of problem behavior, 118,
    120, 127–128, 241
  as study design, 4–5
  and temperament, 166–182, 185–197
  and transition to adolescence,
    128–129
  See also Colorado Adoption Project
aggression. See problem behavior
anxiety, 133–149
attention problems, 93–104, 223

behavioral genetics, 4, 135–137, 240
behavior problems. See problem
    behavior
body mass, 136

CAP. See Colorado Adoption Project
CAP Social Competence Scale
    (CSCS), 300–301
CBCL. See Child Behavior Checklist
CCTI. See Colorado Childhood
    Temperament Inventory
change
  and developmental process, 14, 15,
    18